NONLINEAR TIME SERIES ANALYSIS WITH R

Nonlinear Time Series Analysis with R

Ray Huffaker

Department of Agricultural and Biological Engineering,
University of Florida, USA

Marco Bittelli

Department of Agricultural Sciences, University of Bologna, Italy

Rodolfo Rosa

National Research Council (CNR), Institute for Microelectronics and Microsystems,
Section of Bologna, Italy

OXFORD
UNIVERSITY PRESS

OXFORD
UNIVERSITY PRESS

Great Clarendon Street, Oxford, OX2 6DP,
United Kingdom

Oxford University Press is a department of the University of Oxford.
It furthers the University's objective of excellence in research, scholarship,
and education by publishing worldwide. Oxford is a registered trade mark of
Oxford University Press in the UK and in certain other countries

First Edition published in 2017

Impression: 1

Published in the United States of America by Oxford University Press
198 Madison Avenue, New York, NY 10016, United States of America

British Library Cataloguing in Publication Data

Data available

Library of Congress Control Number: 2017953063

ISBN 978–0–19–878293–3 (hbk.)
ISBN 978–0–19–880825–1 (pbk.)

DOI 10.1093/oso/9780198782933.001

Printed and bound by
CPI Group (UK) Ltd, Croydon, CR0 4YY

Preface

Nonlinear Time Series Analysis with R joins the chorus of voices recommending 'getting to know your data' as an essential preliminary evidentiary step in scientific inquiry. Time series are often highly fluctuating, with a random appearance. Observed volatility is commonly attributed to exogenous random shocks to stable real-world systems. Consequently, investigators are driven to reproduce volatility with a variety of linear-stochastic and probabilistic methods. However, breakthroughs in nonlinear dynamics raise another possibility: highly complex dynamics can emerge endogenously from astoundingly parsimonious deterministic models.

Nonlinear time series analysis (NLTS) is a collection of empirical tools that allow practitioners to diagnose whether observed data are most likely generated by stochastic or deterministic dynamics. In particular, practitioners can use NLTS in an attempt to reconstruct, characterize and model real-world dynamics from a single time series or multiple causally interactive time series. This information can be used, along with scientific principles and other expert information, to guide the specification of mechanistic models used to build theory or to support high-stakes public policy. Models used for public policy are increasingly subjected to formal government audit to ascertain how well they correspond to reality. The compatibility of audited models with NLTS-detected dynamics offers evidence of proper model specification.

This book targets students and professionals in physics, engineering, biology, agriculture, and economics and other social sciences. Our major objectives are to put key concepts of NLTS – developed in the mathematical physics literature – within the operational reach of non-mathematicians with limited knowledge of nonlinear dynamics, and in this way to pave the way for NLTS to be adopted in the conventional empirical toolbox and core coursework of other disciplines. Consistent with modern trends in university instruction, the book makes readers active learners with hands-on computer experiments in R code directing them through NLTS methods. The computer code is explained in detail so that readers can adjust it for use in their own work. The book also provides readers with an explicit framework – condensed from sound empirical practices recommended in the literature – that proposes a strategy for implementing NLTS in real-world data diagnostics. Practitioners become 'data detectives', accumulating hard empirical evidence directing scientific inquiry.

We used R 3.3.1 and the following packages to construct the code in this book:

animation 2.5; boot 1.3-18; crqa 1.0.6; deSolve 1.12; extRemes 2.0-7; fields 8.4-1; fractal 2.0-1; glmnet 2.0-5; gplots 3.0.1; graphics 3.3.1; igraph 1.0.1; MESS 0.4-3; mpoly 1.0.3; multispatialCCM 1.0; nonlinearTseries 0.2.3; pdc 1.0.3; pdist 1.2; phaseR 1.3; plotrix 3.6-3; ppls 1.6-1; psych 1.7.3.21; rgl; Rssa 0.13-1; scatterplot3d 0.3-37; stats 3.5.0; tseriesChaos 0.1-13; tseriesEntropy 0.6-0

Acknowledgements

We would like to thank the following institutions for institutional and financial support: the University of Florida (USA), the University of Bologna (Italy) and the National Research Council (Italy). MB thanks Roberto Olmi and RR thanks Simone Giannerini for collaborative and fruitful research over many years of friendship and collaboration. We also wish to thank our students, whose feedback helped us to improve the material presented in this book. RGH gratefully acknowledges Dr Gerhard Schiefer for providing numerous opportunities to present this material and develop productive international collaborations at the annual IGLS-FORUM 'System Dynamics and Innovation in Food Networks', and Maurizio Canavari, Ernst Berg, Rafael Muñoz-Carpena, Klaus Frohberg and Miles Medina for valuable input and collaborations. We thank Sonke Adlung, Ania Wronski and especially Mac Clarke for expert editorial support. And thanks for their endearing support to Ann Huffaker, Andrea Vogt and AnnaMaria Bononcini.

Contents

1
Why Study Nonlinear Time Series Analysis?

'"Data!data!data!" he cried impatiently. "I can't make bricks without clay."'
Arthur Conan Doyle, The Adventure of the Copper Beeches

1.1 Introduction

Nonlinear time series analysis (NLTS) requires time series data on only a single variable to diagnose, reconstruct, characterize and model the dynamics of the real-world system generating the data. This is possible because any single variable in an interdependent dynamic system encodes the history of its interactions with other system variables. The famous naturalist John Muir (1911) intuited this result in the early twentieth century, observing that: 'When we try to pick something up by itself, we find it hitched to everything else in the universe.'

Perhaps the most compelling reason for studying NLTS is it that facilitates scientific inquiry. However, demonstrating how NLTS data diagnostics fit into the scientific method is complicated by the lack of consensus regarding what that method is in the first place. The *Stanford Encyclopedia of Philosophy* concludes that 'there is not any unique, easily described scientific method' because 'scientific activity varies so much across disciplines, times, places, and scientists that any account which manages to unify it all will either consist of overwhelming descriptive detail, or trivial generalizations' (Andersen and Hepburn, 2016).

A conventional view is that scientific inquiry cycles between inductive reasoning that converts detected regularities in data into testable hypotheses, and deductive reasoning that elevates those hypotheses to theory whose predicted consequences are corroborated by further observation or experimentation (Andersen and Hepburn, 2016). This view is roundly criticized as an overly simplistic representation of how science converts experience into knowledge. For one thing, there has been unending disagreement over epistemic issues, including the proper balance between empirical observation and deductive reasoning on the one hand and the requirements of corroboration on the other. Moreover, the view excludes important human, social and political aspects of science, including the talent, imagination and objectivity of scientists; the benefits of transdisciplinary collaboration; and disincentives created by science and political communities that discourage researchers from undertaking truly novel research projects

Nonlinear Time Series Analysis with R. Ray Huffaker, Marco Bittelli and Rodolfo Rosa, Oxford University Press (2017).
© Ray Huffaker, Marco Bittelli and Rodolfo Rosa. DOI: 10.1093/oso/9780198782933.001.0001

(Haack, 1999; Geman and Geman, 2016). Also, it does not account for how science progresses cataclysmically as new paradigms overthrow old ones in scientific revolutions (Kuhn, 1962). The lack of consensus leaves us between two extremes: the narrow epistemic framework of conventional views and the nihilistic positions that science is merely politics with no epistemic authority.

Haack (1999) provides a middle ground:

'An adequate account of scientific knowledge and scientific inquiry must acknowledge a subtle interplay of logical, personal, and social aspects. The interplay begins at the beginning, of course, with talented individuals coming up with imaginative conjectures on which others build and which are subject to the scrutiny of the whole relevant community, and it is present at every stage. The warrant of any empirical proposition depends in part on experimental evidence, i.e., on what some individual observe(s) see(s) or hear(s), etc, and so, on how justified others are in thinking the observer(s) reliable.'

The chief epistemic value of this middle ground is a 'respect for evidence' that meets general standards of good inquiry, namely 'good, strong, supportive evidence and . . . well-conducted, honest, thorough, imaginative inquiry' (Haack, 1999).

Data provide an evidentiary portal to the real world to which there is only limited access. For example, Charles Darwin consolidated several months of studying numerous marine and terrestrial samples collected in the Galapagos into his famous sketch of the evolutionary tree (Berra, 2009). In another example, Leonardo da Vinci compiled a lifetime of observing nature into notebooks explaining diverse behaviours, including water and sediment movement in river systems, waves in ponds, sound waves in air, and even whether spirits can speak (da Vinci, 1519). The contributions of both scientists are especially renowned because they detected patterns that were not obvious in the data, and processed that information into astounding knowledge of systematic natural behaviour that has withstood the test of time.

Haack (1999) sets 'realism' as the goal of scientific inquiry. The results of scientific inquiry must agree with reality if they are to be put to reliable use. For example, policymakers reasonably expect that theory supporting high-stakes public policy adequately represents the real world that they are charged with regulating. Otherwise, policies 'leave the real problem unaddressed, waste resources, and impeded learning' (Saltelli and Funtowitz, 2014). Recently, a 2010 US Congressional special hearing, *Building a Science of Economics for the Real World* (US Congress Subcommittee on Science and Technology, 2010), criticized the performance of macroeconomic models for failing to reproduce temporal patterns of booms and busts observed in the 2008 financial crisis. Congress concluded that 'if major crises are a recurrent feature of the economy then our models should incorporate this possibility', and expressed frustration that 'because our experts' way of looking at the economy left them blind to the crisis that was building, we were unprepared to deal with the crisis'. Most ominously, Congress broadly questioned why 'we continue to rely upon [theoretical models] for so many critical decisions, so much practical policy advice'.

An economist empanelled at the 2010 hearing recommended that policy models be formally audited by the National Science Foundation. An earlier recommendation by Oreskes *et al.* (1994) would place the burden on the modeller 'to demonstrate the degree of correspondence between the model and the material world it seeks to represent' when 'public policy and public safety are at stake'. The European Commission's Joint Research Centre conducts a formal audit of models used to assess the impacts of EU initiatives, legislation and policy.

NLTS facilitates well-conducted evidentiary scientific inquiry by providing a collection of mathematically rigorous procedures that help practitioners to extract information on real-world dynamics from observed data that often have a complex, highly variable and random appearance. Applied science disciplines conventionally presume that apparent randomness of volatile data must result from exogenous shocks to inherently stable dynamic systems (Feder, 1979; Uusitalo *et al.*, 2015), and turn to stochastic methods without further justification. Alternatively, the theory of randomness teaches us that mathematically random output can be generated by both physically random (indeterministic) and physically nonrandom (deterministic) processes (Horan, 1994), and breakthroughs in nonlinear dynamics demonstrate that parsimonious deterministic models can produce surprisingly irregular and complex behaviour (Glendinning, 1994).

The possibility of deterministic volatility should not be surprising. Many essential biophysical processes exhibit strong patterns of dynamic behaviour to align with environmental regularities. Eating and sleeping in animals (including humans), as well as leaf movements and photosynthetic reactions in plants, exhibit built-in circadian (roughly 24-hour) rhythms that sunlight adjusts to the local environment. Tidal transitions exhibit a tidal (12.4-hour) rhythm, and tidal amplitudes a lunar (29.5-day) rhythm. Climate exhibits regular (diurnal and seasonal) cycles, quasi-periodic cycles (e.g. El Nino) and highly irregular cycles (e.g. volcanic winters).

NLTS allows us to replace presumption of stochasticity with rigorous empirical evidence. The data themselves can serve as our initial guide for ascertaining whether observed volatility is driven by stochastic or deterministic nonlinear real-world dynamics. This distinction matters critically both for theory and its practical application. For example, the *efficient-markets hypothesis* in economics is based on the presumption that market instability results from exogenous shocks to otherwise stable markets. The hypothesis holds that markets tend to equilibrate in response to these shocks as economic agents process all available information in adjusting supply to demand conditions (Fama, 1970). Observed volatility reflects corrective supply and demand adjustments. The hypothesis supports *laissez faire* market policies that do not interfere with corrective adjustments.

The failure of the efficient-markets hypothesis to predict the 2008 financial crisis awoke the profession to another possibility: real-world markets do not naturally equilibrate but may be inherently unstable. Consequently, *The Economist* recommended that 'like physicists, [economists] should study instability instead of assuming that economies naturally self-correct' (Economist, 2016*a*). In striving to understand the economics of market instability, economists returned to the earlier work of Minsky (1992), who developed the *financial instability hypothesis* explaining how financial booms systematically breed their own busts (see also (Economist, 2016*b*)). According to this hypothesis, markets do not provide a natural corrective mechanism, and public intervention should be geared to smoothing systematic boom and bust cycles, and buffering their negative impacts on consumers and producers (Huffaker *et al.*, 2016*a*).

The 2008 crisis led *The Economist* to endorse a more evidentiary approach for the profession in an article entitled *If Economists Reformed Themselves* (Economist, 2016*c*):

'Economists are good at reducing a complicated world to a few assumptions, then adding bells and whistles to make their models more realistic. But problems arise when they mistake the map for the territory ... In future, big data and new "machine learning" techniques could help test the relative power of competing theories.'

1.2 Nonlinear Dynamics and a Strategy for Applying NLTS

Sophisticated application of NLTS requires a firm theoretical foundation in basic nonlinear dynamics that we provide in Part 1 of this book (Chapters 2–5). We investigate how complex behaviour can arise from a simple deterministic nonlinear specification (Chapter 2), how system dynamics can be reconstructed from a single solution variable with *phase space reconstruction* techniques (Chapter 3) and how tools from nonlinear dynamics can be used to measure characteristics of reconstructed dynamics (Chapters 4 and 5).

Phase space reconstruction is the centrepiece of NLTS (Kantz and Schreiber, 1997). Phase space records the level of system (*state*) variables at each point in time. For example, assume that a dynamic system is fully expressed with two state variables x_t and y_t that change with time t. Phase space plots y_t against x_t, so that one point could be their values in the year 2000: (x_{2000}, y_{2000}). Running a line through this point and past and future points results in a phase space *trajectory* depicting the evolution of the system through time. In *dissipative* dynamic systems, trajectories converge to a subset of phase space where they oscillate aperiodically along an *attractor* – a geometric structure with noticeable regularity – forever after (Brown, 1996) . Consequently, an attractor encapsulates the long-term dynamic behaviour of the system. Prior to the 1980s, researchers assumed that data on all system variables were needed to construct phase space representations of system dynamics. This is problematic in practice because we cannot reasonably identify all of the variables at work in real-world dynamic systems, and we might not be able to adequately measure some variables that we can identify. Researchers then discovered that phase space dynamics could be reconstructed from time series data on a single variable using delay coordinates (Breeden and Hubler, 1990). As a result, we can potentially reconstruct real-world system dynamics from time series data on a single observed variable.

Consider a simple preliminary example of phase space reconstruction using data on snowshoe hares (prey) and lynx (predator) collected by the Hudson Bay company in Canada from 1845 to 1935 (Odum, 1953). The time series for each population cycles through time (Figure 1.1a). System dynamics are portrayed in phase space by plotting lynx against hares at each point in time (Figure 1.1b). The populations co-evolve repeatedly along a predator–prey cycle constituting the attractor for the system. A large predator population over-consumes available prey and crashes for want of food. This allows the prey to recover until pressed again by recovering predators, and cycling continues. A shadow version of the predator–prey attractor is reconstructed in phase space from a single variable by plotting either the prey or predator population against its level a period later (Figure 1.1c). This is the *time delay embedding* method of phase

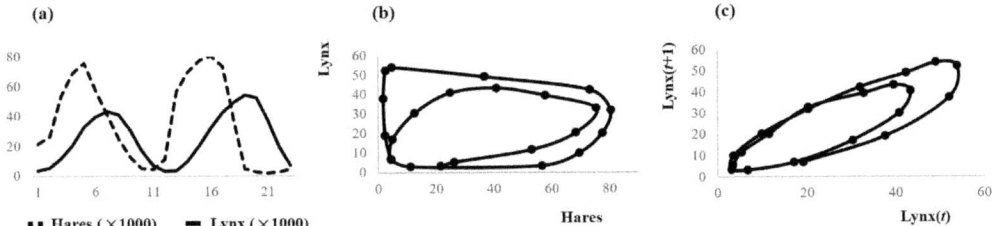

Fig. 1.1 Example of phase space reconstruction: (a) snowshoe hare (prey) and lynx (predator) population time series data; (b) phase space solution (from model to dynamics); (c) reconstructed 'shadow' phase space (from data to dynamics). The data were collected by the Hudson Bay Company in Canada from 1845 to 1935. The figure uses data from 1908 to 1930.

space reconstruction. Takens (1981) derived sufficient conditions guaranteeing that a *shadow* phase space preserves essential dynamic properties of the original phase space.

In this clear-cut example, the shadow dynamics reconstructed from the lynx population are already obvious in the cycling time series plots. The full potential of phase space reconstruction is appreciated when reconstructing deterministic dynamics concealed in a volatile and random-appearing time series. Consider an example from Kaplan and Glass (1995). There are two time series: x_t (plotted in Figure 1.2a) and y_t (plotted in Figure 1.2d). One of these time series is randomly generated, while the other is the solution to a deterministic difference equation. We apply three methods to distinguish between the two: casual observation, autocorrelation functions testing for linear correlations in the data and NLTS time-delay embedding plots. First, neither plot exhibits obvious behavioural patterns on casual observation – any dynamic structure is well concealed. Second, the autocorrelation functions do not exhibit patterns indicative of corresponding patterns in either time series (Figure 1.2b, e). Time-delay embedding plots succeed in distinguishing between random and deterministic dynamics. The time-delay plot using x_t is randomly distributed on the plane, correctly detecting that this series was randomly generated (Figure 1.2c). In contrast, the time-delay plot using y_t shows structure: a parabolic serial correlation missed by the autocorrelation functions (Figure 1.2f). Indeed, y_t was generated by the deterministic parabolic logistic map whose dynamics we investigate in Chapter 2:

$$x_{t+1} = 4x_t(1 - x_t), \qquad x_t \in [0, 1]$$

In Part 2 (Chapters 6–11), we focus on the application of NLTS to reconstruct real-world dynamic structure from observed time series data. We propose a strategy for implementing NLTS that is modified from previous versions in Huffaker (2015), Huffaker *et al.* (2016*b*) and (Huffaker *et al.*, 2016*a*) (Figure 1.3). We emphasize from the outset that NLTS is capable of reconstructing linear as well as nonlinear deterministic system dynamics, and of diagnosing the presence of linear stochastic dynamics. Our objective is not limited to finding evidence pointing to nonlinear deterministic structure, but extends to diagnosing the structure most closely corresponding to reality whether that be linear, nonlinear, deterministic or stochastic.

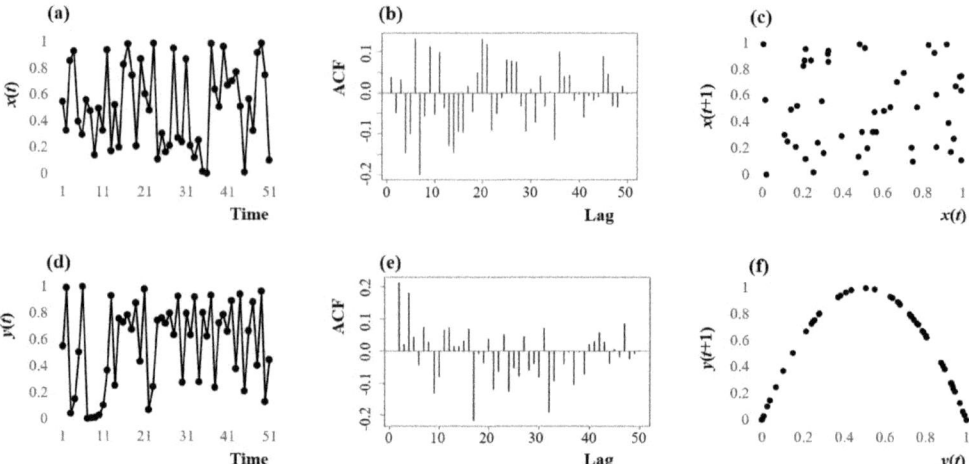

Fig. 1.2 Detection of nonlinear dynamics in data: (a) plot of fifty uniform random variates; (b) autocorrelation plot of random variates; (c) time-delay embedding plot of random variates; (d) plot of nonlinear logistic map; (e) autocorrelation plot for logistic map; (f) time-delay embedding plot for logistic map. Conventional linear methods fail to detect nonlinear deterministic structure in data.

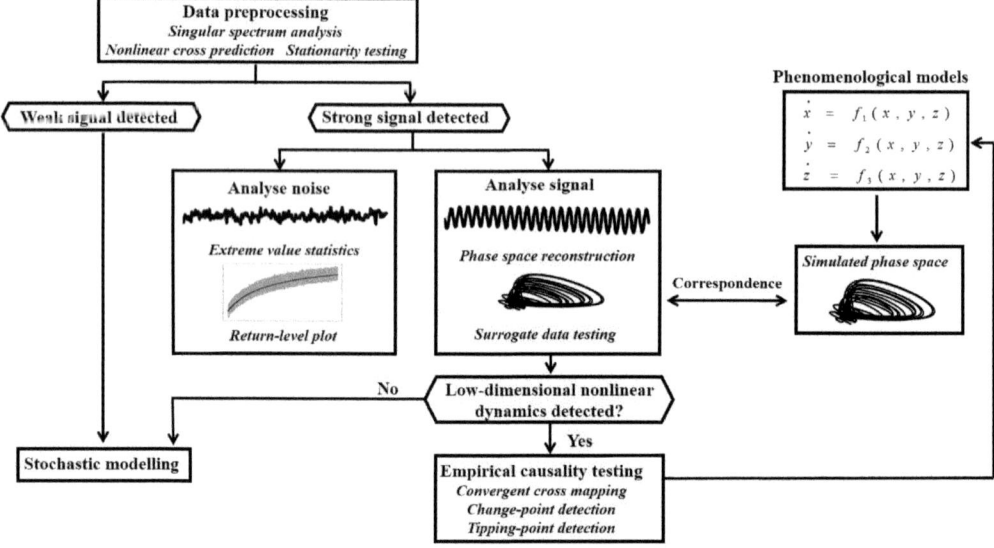

Fig. 1.3 NLTS diagnostics strategy.

In Chapter 6, we investigate how data preprocessing contributes to the successful application of NLTS. The success with which NLTS reconstructs real-world system dynamics depends critically on whether data are informative. This is a concern in practice because data are often noisy and of short duration. Similar to an out-of-focus photograph, noisy time series data obscure the resolution of reconstructed phase space attractors (Kot *et al.*, 1988). Short time series data may not adequately sample dominant cycles occurring at lower frequencies (Schreiber, 1999). If so, these cycles are lumped together as linear or nonlinear trends, and the time series may be nonstationary within the sampled time frame. Similar to other time series methods, NLTS demands stationary data. Problems with noise and nonstationarity can be mitigated by initially applying signal processing techniques that separate data into structured variation (signal) and unstructured variation (noise), remove trends, and measure signal strength as the percentage of variation explained in the observed data. Strong signals found to be stationary are used to reconstruct real-world system dynamics. Weak signals indicate that conventional linear stochastic methods may be more effective. We model the noise separated from a time series probabilistically with *extreme value statistics* (Katz, 2010). This generates an informative diagnostic (return-level plots) estimating the time periods expected before a particular noise level is realized (Chapter 11).

In Chapter 7, we present *surrogate data testing*. Successful phase space reconstruction of real-world system dynamics provides only circumstantial evidence of deterministic real-world dynamic structure. We must further test the null hypothesis that noticeable geometric regularity observed in a reconstructed attractor is more likely mimicked by linear stochastic dynamics. In surrogate data tests, the observed time series is used to randomly generate a set of surrogate data vectors formulated to destroy intertemporal patterns in the data while preserving various statistical properties. A phase space attractor is reconstructed for each surrogate vector and the measures of nonlinear structure developed in Chapters 4 and 5 are estimated. The null hypothesis of linear stochastic dynamics is rejected if there is statistically significant difference between the discriminating measures taken from the surrogate attractors and corresponding measures taken from the attractor reconstructed from the observed time series (Schreiber and Schmitz, 2000; Theiler *et al.*, 1992).

If surrogate data tests do not eliminate the possibility of nonlinear deterministic dynamics in observed data, we proceed to further nonlinear techniques that analyse reconstructed system dynamics in more depth. In Chapter 8, we present *convergent cross mapping* methods (Sugihara *et al.*, 2012), which test whether multiple observed time series belong to the same dynamic system and thus causally interact. We also investigate *change-point* and *tipping-point* methods testing whether variables with continuous time records undergo abrupt changes in dynamics corresponding to the irregular timing of variables exhibiting only occasional activity. In Chapter 9, we investigate phenomenological (data-driven) modelling (Baker *et al.*, 1996; Dong *et al.*, 2015; Brunton *et al.*, 2016), which extracts a set of ordinary differential equations governing empirically reconstructed system dynamics from a single time series or from multiple time series on causally interacting variables. We apply *regularized* regression techniques to extract a mathematical model of nonlinear system dynamics from the data, which can be used to provide valuable data-driven insight into real-world biophysical processes

(Huffaker *et al.*, 2016*b*; Huffaker *et al.*, 2016*a*).

In Chapter 10, we provide a capstone to Part 2 by taking the reader step by step through the application of NLTS to diagnose and model system dynamics generating real-world infectious disease epidemic data.

1.3 The Contribution of NLTS Diagnostics to Theoretical Modelling

We view NLTS diagnostics as a precursor to (and not a replacement for) theoretical modelling attempting to explain observed natural behaviour. NLTS provides a mathematically rigorous means of extracting as much information as we can from data. NLTS data diagnostics provide a rigorous empirical benchmark to guide specification and testing of theoretical models corresponding to reality. Signal processing identifies patterns in data that should be reproduced by theoretical models. Phase space reconstruction provides a geometric picture of real-world dynamics that should be reproduced by attractors simulated by theoretical models. For example, simulated dynamics approaching a stable equilibrium would not correspond to aperiodic cycling characterizing an empirically reconstructed attractor. Moreover, the dimensionality of an empirically reconstructed attractor indicates the minimum dimensionality required by a theoretical model to reproduce diagnosed long-term dynamics. Convergent cross mapping methods provide empirical evidence of causal interactions among observed variables that belong to the same real-world dynamic system. Phenomenological modelling sheds light on the type of interactions among system variables capable of reproducing an empirically reconstructed attractor. This offers valuable guidance in selecting among the wide variety of functional responses available to model theoretically posited interactions among system variables.

NLTS diagnostics open the door to true interdisciplinary theoretical modelling. Real-world dynamic systems integrate climatic, environmental, economic, geopolitical and sociological processes whose interactions determine natural behaviour over time and space. Yet, conventional practice in applied science disciplines tends to relegate processes within the purview of outside disciplines to random exogenous forces. For example, the risk literature in agricultural economics relegates natural processes to the realm of real uncertainty, generally interpreted to mean that agricultural decision-making environments are inherently random owing to uncontrollable and random exogenous events, including weather, diseases, insect infestations, technological innovations, government policies and so on (Feder, 1979). These events are certainly not viewed as inherently random by climatologists, entomologists, political scientists and so on. If we diagnose deterministic nonlinear dynamics in observed data, we must integrate disciplinary knowledge to formulate theoretical explanations corresponding to reality.

1.4 Caveats in Application

There are important caveats to applying NLTS diagnostics to short, noisy, and non-stationary time series data frequently encountered in empirical work. Faced with these data limitations, one gives up the ambition of reconstructing the very fine structure of the real-world attractor including complex folding and fractal patterns (Vautard, 1999). One is limited to reconstructing a sampling or a skeleton of a real-world attractor (Ghil *et al.*, 2001), and cannot hope to recover full dynamics from a recording of a single variable (Schreiber, 1999). In the worst case, NLTS may be unable to provide an adequate sampling of a real-world attractor if available data are insufficiently informative.

Consequently, the literature strongly warns against using NLTS diagnostics as positive evidence that natural systems are chaotic. Chaotic attractors exhibit distinct topological features, including self-similar geometry, and sensitive dependence on initial conditions. Topological measures of these features are based on asymptotic properties best met with vast amounts of high-quality data provided by laboratory experiments designed specifically for investigating chaos. The measures are unreliable when estimated from short and noisy real-world data (Schreiber, 1999; McSharry, 2011). In contrast, the topological measures perform reliably when the focus of inquiry is relaxed from attempting to prove the presence of deterministic chaos in observed data to gauging whether deterministic nonlinear dynamics might possibly explain observed volatility (Schreiber, 1999). Our NLTS strategy follows this path.

Other potential caveats are that available data may not lie on the real-world attractor or that a low-dimensional nonlinear attractor may not exist in the first place (Williams, 1997). When NLTS diagnostics fail, linear stochastic approaches remain a viable alternative. However, we propose that NLTS diagnostics be applied before presuming linear stochastic structures that potentially misrepresent real-world dynamics.

1.5 Summary

Nonlinear time series analysis (NLTS) provides a mathematically rigorous collection of techniques designed to reconstruct real-world system dynamics from time series data on a single variable or multiple causally related variables. NLTS facilitates scientific inquiry that emphasizes strong supportive evidence, well-conducted and thorough inquiry, and realism. Data provide an essential evidentiary portal to a reality to which we have only limited access. Random-appearing data do not prove that underlying dynamic process are subject to exogenous inherently-random forces. The possibility exists that observed volatility is generated by inherently unstable, deterministic and nonlinear real-world dynamic systems. NLTS allows the data to speak regarding which type of system dynamics generated them. It is capable of detecting linear as well as nonlinear deterministic system dynamics and of diagnosing the presence of linear stochastic dynamics. Our objective is to use NLTS to uncover the structure most closely corresponding to reality, whether that be linear, nonlinear, deterministic or stochastic. Accurate diagnosis of real-world dynamics from observed data is crucial to develop valid theory and to formulate effective public policy based on theory. Because of the short and noisy observed time series data that are frequently encountered in empirical work, NLTS diagnostics may fail to provide an adequate sampling of real-world dynamics. However, we propose that NLTS diagnostics be applied before presuming linear stochastic structures that may not correspond to real-world dynamics.

2

Linear and Nonlinear Dynamic Behaviour

'Quelli che s'innamoran di pratica sanza Scienza son come il nocchieri che entra
ai navilo senza timone o bussola, cha mai ha certezza dove si vada.'

'Those who fall in love with application without science are like a sailor who enters
a ship without a helm or a compass, and who never can be certain whither he is going.'

Leonardo da Vinci

2.1 Introduction

The objective of applied dynamics is to formulate mathematical models that describe
data collected in the real-world systems of interest. Such models generate a range of
dynamic behaviours that may be able to be observed again in corresponding real-world
systems. The utility of nonlinear time series analysis (NLTS) is to provide empirical
evidence of the range of dynamic behaviour that a model should be capable of reproducing.

In this chapter, we investigate the dynamic capabilities of well-known linear and
nonlinear deterministic models. We see how linearity restricts behaviour to stable dynamics that equilibrate system variables in the long run, to explosive unstable dynamics or to periodic dynamics. Linear models must be exogenously affected to generate
aperiodic behaviour often observed in real-world time series data. While exogenous effects may succeed in simulating irregular behaviour, they are a dissatisfying approach
if it matters that irregular behaviour might occur systematically.

We proceed by investigating the dynamic behaviour of discrete linear difference
equations. We investigate how a simple nonlinear transformation results in the *logistic
map*, which expands endogenously generated dynamics to include aperiodic (nonrepeated) cycles that, while appearing highly erratic, are governed by surprising order.
Importantly, the numerical and analytical tools developed to characterize the order in
these irregular dynamics form the basis of NLTS methods used to detect deterministic
order in observed data.

2.2 Discrete Linear Dynamics

Consider a first-order linear difference equation measuring, for example, the transition
of the current population x_t to the next period's population x_{t+1} through time t:

Nonlinear Time Series Analysis with R. Ray Huffaker, Marco Bittelli and Rodolfo Rosa, Oxford University Press (2017).
© Ray Huffaker, Marco Bittelli and Rodolfo Rosa. DOI: 10.1093/oso/9780198782933.001.0001

$$x_{t+1} = rx_t + b \tag{2.1}$$

where r and b are constant parameters, and the initial population is x_0. The solution gives the sequence of population levels over time, $x_0, x_1, x_2, \ldots$, satisfying eqn (2.1):

$$x_t = \begin{cases} (x_0 - x^*)r^t + x^* & \text{if } r \neq 1 \\ x_0 + bt & \text{if } r = 1 \end{cases} \tag{2.2}$$

The population x^* is the steady-state equilibrium defined by $x(t) = x(t+1)$. Once the population reaches x^*, it remains there throughout time. Generally speaking, a steady state is a *fixed point* of a function $f(x_t)$; that is, the value such that

$$x^* = f(x^*) \tag{2.3}$$

We can solve for the steady state by inserting x^* into eqn (2.1):

$$x^* = rx^* + b \quad \Rightarrow \quad x^* = \frac{b}{1 - r}$$

An equilibrium solution x^* is characterized by its stability. An equilibrium is globally stable if

$$\lim_{t \to \infty} x_t = x^* \quad \forall \; x_0$$

That is, as time approaches infinity, the solution of eqn (2.1) asymptotically converges to x^* from all initial values x_0.

Code 2.1 numerically solves eqn (2.1) to produce time series plots of x_t for selected values of r. It defines the model equation, sets model parameters, and formulates an iteration loop to solve the model through time.

```
#Code 2.1  Numerical solution of linear difference equations (lin_soln.R)
rm(list=ls(all=TRUE)) #clear values between runs

f.x<- function(x,r,b) {r*x+b}   #Define model equation

r<-0.8          #stability parameter
b<-0.5          #model parameter
x_initial<-10   #initial condition
nstep<- 25      #number of iterations

#Formulate iteration loop
xt<- numeric()
x<- x_initial   #initialize loop
xt[1]<- x
for(i in 2:nstep)  {
  y<- f.x(x,r,b)
  x<- y
  xt[i]<- x
  }

plot(xt,type="b",xlab="time",ylab="x(t)",ylim=c(0,10),cex.lab=1.5,cex.axis=1.2,lwd=2)
```

Figure 2.1 displays plots for values of the stability parameter r generating stable solutions (a, b), unstable solutions (c, d) and two-period cycling (e).

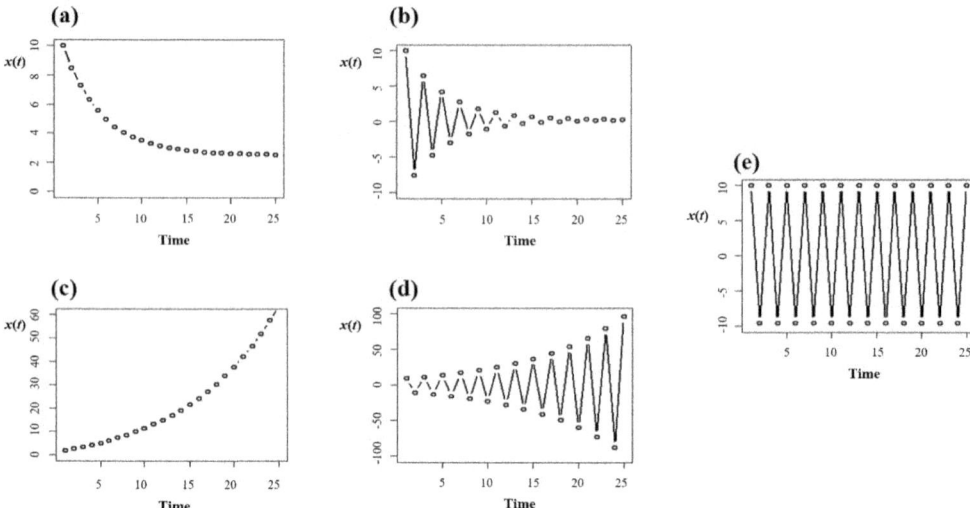

Fig. 2.1 Time series plots of eqn (2.1): (a) $r = 0.8$ (monotonic convergence to equilibrium); (b) $r = -0.8$ (oscillatory convergence to equilibrium); (c) $r = 1.1$ (monotonic divergence); (d) $r = -1.1$ (oscillatory divergence); (e) $r = -1$ (two-period cycling)

Solutions to the difference equation (2.1) also can be portrayed graphically in cobweb plots of x_{t+1} against x_t. A cobweb plot is a step diagram that is graphically equivalent to spreadsheet iteration. We start at x_0 on the horizontal axis and find the subsequent value, x_1, on the vertical axis with eqn (2.1). Iteration occurs as x_1 is reflected by a $45°$ line from the vertical axis to the same value on the horizontal axis. The process is repeated to obtain x_2, and so on. Code 2.2 generates cobweb plots for a given r. It defines the model equation, formulates the cobweb iteration loop and sets parameter values. Finally, it sets plotting limits on the axes most fully depicting the cobweb plot for the selected r.

```
#Code 2.2  Cobweb plots for linear difference equations (lin_cob.R)
rm(list=ls(all=TRUE)) #clear values between runs

##Define model equation
f.x<- function(x,r,b) {r*x+b}

##Formulate cobweb iteration loop
cobweb<- function(x_initial,nstep,r){
  iterations<-nstep  #number of iterations
  x<- x_initial #initial condition for x
  y<- f.x(x,r,b)  #subequent value of x
  segments(x,0,x,y,lty=1,lwd=2)
  for(i in 1:iterations){
    points(x,y,pch=19,cex=1.5)  #iterated points along model equation
    segments(x,y,y,y,lty=1,lwd=2) #segments from model to reflection line
    x<- y
    y<- f.x(x,r,b)
    segments(x,x,x,y,lty=1,lwd=2) #segments from reflection line to model
  }
```

```
}     # end cobweb function

## Set parameters and initial condition for cobweb function
r<- -0.5  #stability parameter
b<-0.5
x_initial<-10  #initial condition
nstep<-25  #number of iterations

## Set axis limits for cobweb plots
x_upper<-x_initial+400  #set upper limit if solution explodes
x_upper_cycle<-x_initial+4  #set upper limit for two-cycle

if(r>0&&r<1){ #solution decays monotonically
  plot(0,0,type="n",xlim=c(0,x_initial),ylim=c(0,x_initial),xlab="x(t)",
      ylab="x(t+1)",cex.lab=1.5,cex.axis=1.2)
  #model equation
  curve(f.x(x,r,b),from = 0, to =x_initial,lty=1,lwd=2,col="blue",add=T)
  #reflection line
  segments(0,0,x_initial,x_initial,lty=1,lwd=2,col="red")}

if(r>-1&&r<0){ #solution decays cyclically
  plot(0,0,type="n",xlim=c((-x_initial),x_initial),
      ylim=c((-x_initial),x_initial),xlab="x(t)",
      ylab="x(t+1)",cex.lab=1.5,cex.axis=1.2)
  curve(f.x(x,r,b),from = (-x_initial), to =x_initial,lty=1,lwd=2,col="blue",add=T)
  segments((-x_initial),(-x_initial),x_initial,x_initial,lty=1,lwd=2,col="red")}

if(r>1){ #solution explodes monotonically
  plot(0,0,type="n",xlim=c(x_initial,x_upper),ylim=c(x_initial,x_upper),
      xlab="x(t)",ylab="x(t+1)",cex.lab=1.5,cex.axis=1.2)
  curve(f.x(x,r,b),from = x_initial, to =x_upper,lty=1,lwd=2,col="blue",add=T)
  segments(x_initial,x_initial,x_upper,x_upper,lty=1,lwd=2,col="red")}

if(r<=(-1)){ #solution explodes cyclically
  plot(0,0,type="n",xlim=c((-x_upper_cycle),x_upper_cycle),ylim=c((-x_upper_cycle),
      x_upper_cycle),xlab="x(t)",ylab="x(t+1)",cex.lab=1.5,cex.axis=1.2)
  curve(f.x(x,r,b),from =(-x_upper_cycle),to =x_upper_cycle,lty=1,lwd=2,col="blue",add=T)
  segments((-x_upper_cycle),(-x_upper_cycle),x_upper_cycle,x_upper_cycle,lty=1,
      lwd=2,col="red")}

cobweb(x_initial,nstep,r)  #call for cobweb function
```

Figures 2.2a–e show the cases corresponding to the plotted time series solutions in Figures 2.1a–e. Consider first Figure 2.2a. We start at $x_0 = 10$ on the horizontal axis and find the subsequent value, x_1, on the vertical axis with the plot of eqn (2.1) (the line with slope less than one). Iteration occurs as x_1 is reflected by the 45° line (with slope equal to one) from the vertical axis to the same value on the horizontal axis. The process is repeated to obtain x_2, and so on. The equilibrium x^* is located where the difference equation intersects with the reflection line. Stability depends on the slope of the difference equation r in comparison with the 45° line. In Figure 2.2a, the slope of the difference equation ($r = 0.5$) is lower than that of the reflection line. Iterations decay as the cobweb approaches equilibrium (stability). In Figure 2.2b, the opposite cobweb adjustments occur, since the slope of the difference equation ($r = 1.5$) is greater than that of the reflection line. Iterations explode away from the initial condition. When $r < 0$, the logistic map equation has a negative slope that gives rise to oscillatory adjustments. When r is a negative fraction (Figure 2.2c),

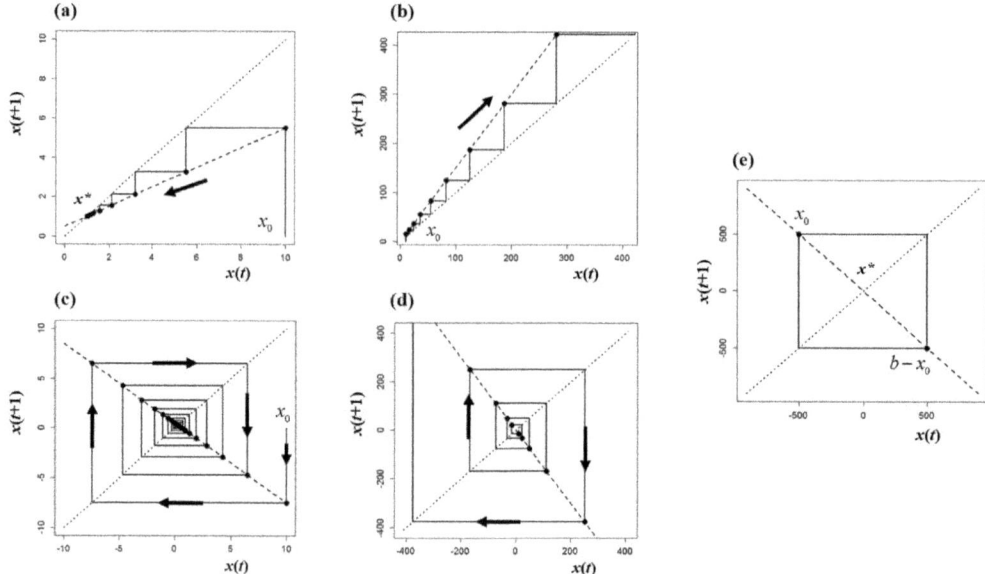

Fig. 2.2 Cobweb plots of eqn (2.1): (a) $r = 0.5$ (monotonic convergence to equilibrium); (b) $r = 1.5$ (monotonic divergence); (c) $r = -0.8$ (oscillatory convergence to equilibrium); (d) $r = -1.5$ (oscillatory divergence); (e) $r = -1$ (two-period cycling). The dotted lines in the figures are the $45°$ reflection lines and the solid lines are the model equations.

the cobweb cycle dampens towards equilibrium (stability). In contrast, when $r < -1$ (Figure 2.2d), the cobweb oscillates with ever-increasing amplitude away from the initial value. Finally, when $r = -1$ (Figure 2.2e), the cobweb cycles forever between the initial value x_0 and $b - x_0$.

2.3   The Nonlinear Logistic Map

We now consider a nonlinear version of eqn (2.1) that allows for a broader range of dynamic behaviours:

$$N_{t+1} = r(N_t)N_t \tag{2.4}$$

We build in a negative feedback loop that accounts for the impact of crowding on slowing population growth. In particular, the proportional growth rate $r(N_t)$ is assumed to decrease at a linear rate with population:

$$r(N_t) = r(1 - N_t/K) \tag{2.5}$$

where r is a constant maximum (*intrinsic*) growth rate existing at very small population levels, $N_t \to 0$ and K is a constant environmental carrying capacity. If the population exceeds the carrying capacity, proportional growth becomes negative owing to lack of vital resources such as food or water. Substituting eqn (2.5) into eqn (2.4) gives

$$N_{t+1} = r(1 - N_t/K)N_t \tag{2.6}$$

which is parabolic in N_t. This specification can be simplified by transforming it to a scaled variable $x_t = N_t/K$, which measures the population as a fraction of carrying capacity. The new variable x_t is scaled between zero and one. It is zero when no population is present ($N_t = 0$) and one when the population reaches carrying capacity ($N_t = K$). Rewriting the scaled variable as $N_t = Kx_t$ and substituting it into eqn (2.6) gives the Verhulst (1838) logistic growth population model:

$$x_{t+1} = rx_t(1 - x_t) \tag{2.7}$$

Mathematically speaking, this is the nonlinear logistic map. It is a unimodal map in $[0, 1]$. A map is unimodal in the interval $[a, b]$ if it presents a unique maximum value in $[a, b]$ and is a strictly increasing function before the maximum and a strictly decreasing one after the maximum. It was in a study of a nonlinear logistic map very similar to the above that the term 'chaotic' was introduced for the first time in a mathematical context by Li and Yorke (1975, p. 986): 'In this paper we analyze a situation in which the sequence $F^n(x)$ is non-periodic and might be called "chaotic".' (The sequence of iterates $F^n(x)$ in our notation will be $f^{[n]}(x)$; see below.) The map has fixed points x^* satisfying

$$x_t = rx_t(1 - x_t) \quad \Rightarrow \quad x_t(r - rx_t - 1) = 0$$

so that

$$x^* = 0 \quad \text{and} \quad x^* = \frac{r - 1}{r} \tag{2.8}$$

When $x^* = 0$, there are no individuals in the population. If $x_t = (r - 1)/r$, the population has reached an equilibrium state and it does not increase or decrease. Clearly, the population cannot be in both states. The value x^* in the equilibrium state does not depend on the initial conditions, but only on the value of r. In general, if there is more than one fixed point (and they are not cyclical), the system will reach only one of those fixed points, depending on the initial conditions. Following the logic of Codes 2.1 and 2.2, Code 2.3 portrays the time series and cobweb solutions to logistic map (2.7) for selected values of r:

```
#Code 2.3 Nonlinear logistic map: time series and cobweb plots (logistic.R)
f.x<- function(x,r){
  r*x*(1-x)
}
##### function to draw the time plot #####
f.temp<-function(xinit,nstep,r){   # starting function f.temp
  xt<- numeric()
  x<- xinit
  xt[1]<- x
  for(i in 2:nstep){
    y<- f.x(x,r)
    x<- y
    xt[i]<- x
  }
  plot(xt,type="b",xlab="time",ylab="x(t)",
       cex.lab=1.7,cex.axis=1.3,lwd=2)
  #xt              # comment to skip iterates
}    # ending function f.temp
```

```
##### function to draw the cobweb plot ####
iter<- function(xinit,nstep,r){    # starting function iter
  x<- xinit
  y<- f.x(x,r)
  segments(x,0,x,y,lty=1,lwd=2)
  for(i in 1:nstep){
    points(x,y,pch=19,cex=1.5)
    segments(x,y,y,y,lty=1,lwd=2)
    x<- y
    y<- f.x(x,r)
    segments(x,x,x,y,lty=1,lwd=2)
  }
}    # ending function iter
### parameters and initial conditions
r<- 2.8
#vary r: r<- 2.8,  r<- 3.2,  r<- 3.5,  r<- 4
xinit<- 0.002
nstep<- 25
# nstep<- 25: r=2.8, 3.2; nstep<- 50:  r=3.5, 4
f.temp(xinit,nstep,r)                   # call up the time plot
### preparation of the cobweb plot
windows()
plot(1,1,type="n",xlim=c(0,1),ylim=c(0,1),xlab="x(t)",ylab="x(t+1)",
     cex.lab=1.5,cex.axis=1.2)
# plot of the function f.x
curve(f.x(x,r),from = 0, to = 1,lty=5,lwd=2,col="black",add=T)
# more aesthetically: col="blue"
segments(0,0,1,1,lty=3,lwd=2,col="black") # bisector
# more aesthetically: col="magenta"
iter(xinit,nstep,r)                     # call up the cobweb plot
```

Code 2.3 initially sets $r = 2.8$. From eqn (2.8), the nonzero fixed point is $x^* = (2.8-1)/2.8 = 0.6429$. After about twenty iterations, the time series plot (Figure 2.3a) shows that x_t converges to $x^* = 0.6429$. The cobweb plot (Figure 2.3b) displays the same convergent oscillatory behaviour.

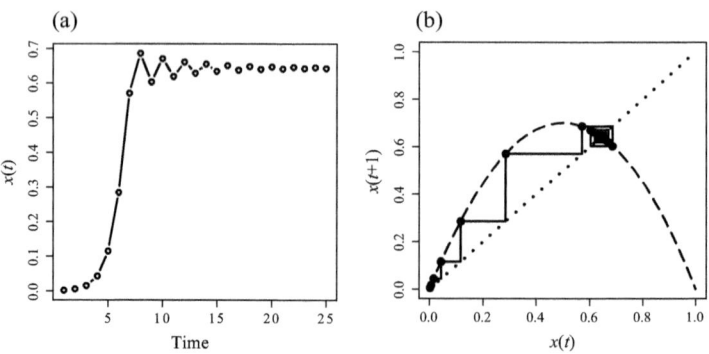

Fig. 2.3 Time series plot (a) and cobweb plot (b) of the logistic map with $r = 2.8$.

From this example, we can see the importance of the cobweb plots. They provide a visual description of the general behaviour of the equation. The cobweb plots also show that the fixed points ($x^* = 0$ and $x^* = 0.6429$) occur where the function $f(x_t)$ intersects the 45° reflection line. Consequently, the fixed points have coordinates (x_{t+1}, x_t), with $x_{t+1} = x_t \; \forall t$.

If we experiment with different initial starting points x_0, we see that the solution always converges to the nonzero fixed point $x^* = 0.6429$, which indicates that it is an *attractor* or *sink* of period one. The iterates reach values larger than $x^* = 0.6429$, but eventually converge on this final value. The growth process does not preserve the memory of the previous states and, in particular, of the initial starting point. In contrast, the fixed point $x^* = 0$ repels initial conditions. More precisely, for x_0 close to 0 (i.e. for $x_0 = 0 + \epsilon$, where ϵ is an arbitrarily small precise number), the iterates diverge from the origin towards the nonzero fixed point. The fixed point $x_0 = 0$ is referred to as a *repeller* or *source*.

The iterates converging towards the nonzero fixed point are in a *transient state*. Figure 2.4 illustrates this concept by displaying two trajectories starting from different starting points. The time series plot (a) and cobweb plot (b) show that, while the iterates follow different oscillating trajectories, each trajectory converges to $x^* = 0.6429$. If we used Code 2.3 to print the iterated solutions, we would see that that approximately 20–30 iterations are required to reach the fixed point at a precision of three decimal places: $x^* = 0.643$. We would also see that far more iterations are required to achieve greater precision: for example, approximately 70 iterations to reach $x^* = 0.6428571$. This illustrates that $f(x_t)$ *asymptotically* approaches a stable fixed point.

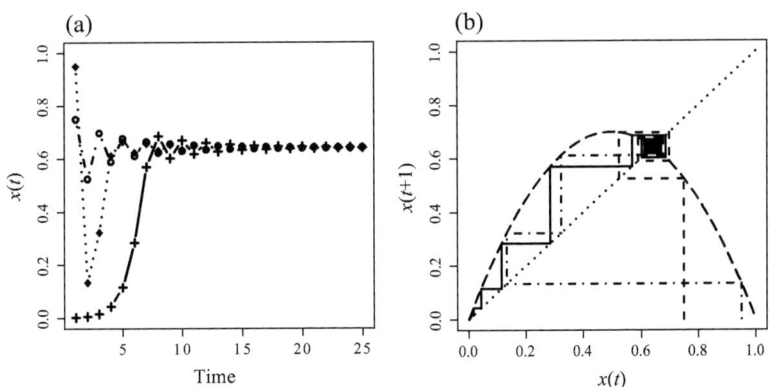

Fig. 2.4 Time series plot (a) and cobweb plot (b) of three trajectories with different initial values for the logistic map with $r = 2.8$.

2.4 Stability of Fixed Points

We have seen graphically how solutions to the logistic map (2.7) diverge from an unstable fixed point and converge towards a stable one. We now consider how to analytically determine the stability of a fixed point. This will help us to analyse the complex dynamics arising from the logistic map for increasing values of r.

We proceed by linearizing the map around the fixed point. Let x_t be the value of an iterate very close to the fixed point x^* and let δx_t be the distance $|x^* - x_t|$, with $\delta x_t \ll 1$. If $x_t > x^*$, for example, then $x_t = x^* + \delta x_t$. The iterate at the subsequent time step is

$$x_{t+1} = x^* + \delta x_{t+1} = f(x^* + \delta x_t) \tag{2.9}$$

Expanding $f(x_t)$ as a Taylor series about x^* gives

$$x_{t+1} = f(x_t) \approx f(x^*) + \left. \frac{df(x_t)}{dx_t} \right|_{x_t = x^*} \delta x_t$$

$$\approx x^* + \left. \frac{df(x_t)}{dx_t} \right|_{x_t = x^*} \delta x_t$$

where we have substituted the definition of a fixed point from eqn (2.3) into the second equation. Equating the second equation with eqn (2.9) results in the linearized map

$$\delta x_{t+1} = \left. \frac{df(x_t)}{dx_t} \right|_{x_t = x^*} \delta x_t = f'(x^*)\, \delta x_t \tag{2.10}$$

The term $f'(x^*)$ is the *fixed point multiplier*:

$$m = \left. \frac{df(x_t)}{dx_t} \right|_{x_t = x^*} = f'(x^*) \tag{2.11}$$

Equation (2.10) has the same features as the linear equation (2.1), $N_{t+1} = rN_t$, with $r = f'(x^*)$. Therefore, the same reasoning as applied to eqn (2.1) can be repeated. Now x_t converges towards, or diverges from, the fixed point x^* if $|f'(x^*)| < 1$, or $|f'(x^*)| > 1$, respectively. The quantity $f'(x^*)$ measures the contraction or expansion rate of the map.

We will use the following procedure to assess the stability of fixed points:

1. Search for fixed points x^* by solving the equation $x_t = f(x_t)$. Linear equations have a single fixed point, while nonlinear equations may have multiple fixed points.
2. Compute the fixed point multiplier m, eqn (2.11).
3. The value of m characterizes the fixed point:
   - $m > 1$: unstable fixed point, exponential divergence from x^*.
   - $0 < m < 1$: stable fixed point, monotonic convergence towards x^*.
   - $-1 < m < 0$: stable fixed point, oscillatory convergence towards x^*.
   - $m < -1$: unstable fixed point, oscillatory divergence from x^*.

In sum, a fixed point is attracting (repelling) if its multiplier has modulus less (greater) than 1.

We now assess stability in the logistic map with the value $r = 2.8$ used in Figures 2.4 and 2.5. The fixed point multiplier is

$$m = \frac{df(x_t)}{dx_t} = r(1 - 2x_t) \tag{2.12}$$

The fixed point $x^* = 0$ is unstable, since $m = r = 2.8 > 1$. The nonzero fixed point $x^* = 0.6429$ is stable with oscillatory convergence, since $m = -0.800$ satisfies $(-1 < m < 0)$. Figure 2.5 zooms in on the iterates around the fixed points $x^* = 0$ (a) and $x^* = 0.6429$ (b) from an initial point of `xinit <- 0.002`. In Figure 2.5a, the slope of the map is positive and very steep at $x^* = 0$, so that iterates x_t are pushed further away from `xinit`. If instead the slope of the map were lower than the 45°, then $x^* = 0$ would convert to an attractor. In Figure 2.5b, the slope of the map is negative and less than 45° in absolute value at $x^* = 0.6429$, so that the steady state attracts initial points.

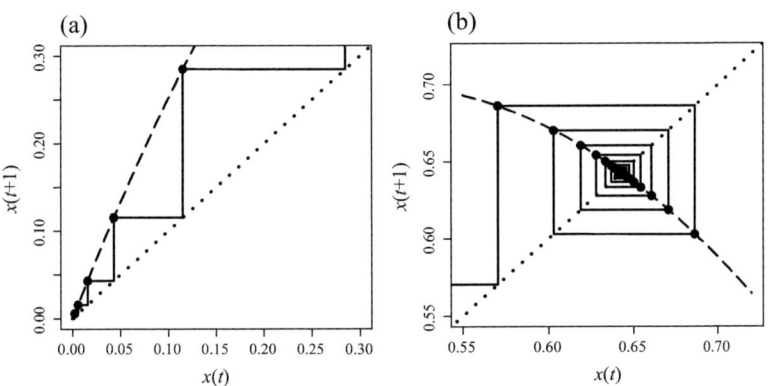

Fig. 2.5 Zoom of the cobweb plot around the fixed points $x_1^* = 0$ (a) and $x_2^* = 0.6429$ (b) of the logistic map with $r = 2.8$.

If we cannot analytically solve for the fixed point multiplier (2.11), we can find attractors by iterating a map from different initial conditions. Initial conditions attracted towards a fixed point are said to be in its *basin of attraction*.

2.5 Dynamics of the Logistic Map

The logistic map (2.7) generates a wide variety of dynamic behaviours for increasing values of r. We can analyse the stability implications by substituting the fixed point $x^* = (r-1)/r$ into the fixed point multiplier (2.11):

$$m = r(1 - 2x^*) = r\left(1 - 2\frac{r-1}{r}\right) = 2 - r \qquad (2.13)$$

The fixed point is stable for $1 < r < 3$, since $|2 - r| < 1$, and unstable for $3 < r < 4$, since $|2 - r| > 1$.

We now consider logistic map dynamics when r is increased into the beginning of the unstable range with value $r = 3.2$. This introduces the behaviour of *periodic fixed point attractors*. Code 2.3 produces the resulting time series and cobweb plot from an initial value $x_0 = 0.002$ in Figure 2.6.

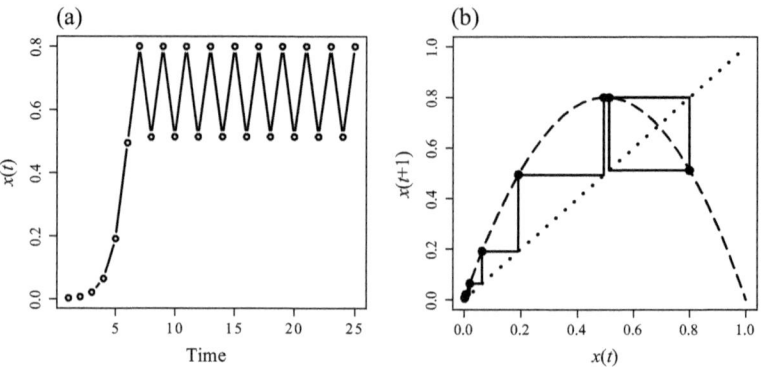

Fig. 2.6 Time series plot (a) and cobweb plot (b) of the logistic map with $r = 3.2$.

In comparison with Figure 2.3, there is no longer a unique stable fixed point. Instead, the solution of eqn (2.7) is *cyclic* with period 2. Starting from x_0, the iterates are

$$0.00200000, \quad 0.00638720, \quad 0.02030849, \quad 0.06366738,$$
$$0.19076431, \quad 0.49399452, \quad 0.79988459, \quad 0.51222155, \quad \quad (2.14)$$
$$0.79952203, \quad 0.51291697, \quad 0.79946609, \quad 0.51302420, \quad \dots$$

After transient iterations, x_t oscillates between two values: 0.7995 and 0.5130. Iterates bounce between one point and the other, and a point is repeated after two periods (not one):

$$x_{t+2} = x_t, \quad \text{while} \quad x_{t+1} \neq x_t, \quad \forall t$$

By increasing r to 3.2, the logistic map shifts from a configuration with a single stable fixed point to one with *periodic fixed points* characterized by a period-2 cycle. Considered together, the couple (0.7995,0.5130) constitute a unique period-2 attractor. The fixed points associated with $r = 3.2$ are $x^* = 0$ and

$$x^* = \frac{r-1}{r} = \frac{3.2-1}{3.2} = 0.6875$$

(eqn 2.8). The fixed point $x^* = 0$ continues to be unstable, because $m = r = 3.2$ (eqn 2.13). The fixed point $x_t = 0.6875$ becomes unstable, because $m = -1.2$, so that initial conditions oscillate away from it. Figure 2.7 shows how increasing r to 3.2 generates instability because the slope of the map at the fixed point is greater in absolute value than the 45° slope of the reflection line.

We can better understand why the logistic map converts to a 2-cycle configuration when $r = 3.2$ by rewriting the equation dictating that x_t repeats every two periods, $x_{t+2} = x_t$, in iterative form:

$$x_{t+2} = f(x_{t+1}) = f\big(f(x_{t+1})\big) = x_t$$

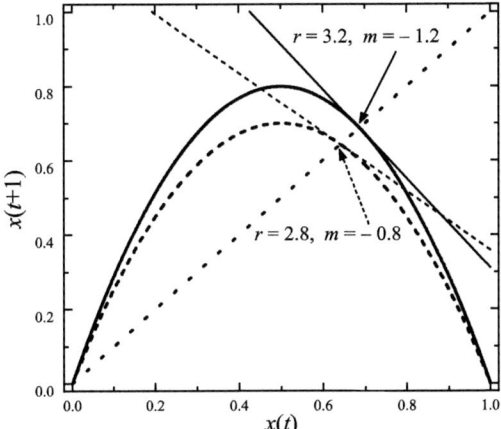

Fig. 2.7 Plots of $f(x_t)$ (logistic map) with $r = 2.8$ (dashed line) and $r = 3.2$ (solid line). The dash–dot lines are the tangent lines to $f(x_t)$ at the intersection with the bisector, i.e. at the fixed point, and m is the slope of $f(x_t)$ at this point.

From now on, we will write $f\big(f(x_{t+1})\big)$ as $f^{[2]}(x)$. In terms of the logistic map (2.7) iterates, we solve for $f^{[2]}(x)$ as follows:

$$f^{[2]}(x_t) = f(x_{t+1}) = rx_{t+1} - rx_{t+1}^2$$
$$= r(rx_t - rx_t^2) - r(rx_t - rx_t^2)^2$$
$$= r^2 x_t - (r^2 + r^3)x_t^2 + 2r^3 x_t^3 - r^3 x_t^4$$

Solution of this equation gives four fixed points:

$$x_1^* = 0$$
$$x_2^* = \frac{r-1}{r}$$
$$x_3^* = \frac{r+1+\sqrt{r^2-2r-3}}{2r} \tag{2.15}$$
$$x_4^* = \frac{r+1-\sqrt{r^2-2r-3}}{2r}$$

Fixed points x_3^* and x_4^* arise when the discriminant under the square root, $r^2 - 2r - 3 = (r+1)(r-3)$, is positive, which holds for $r \geqslant 3$. Otherwise, the fixed points are complex, and only x_1^* and x_2^* exist (as in the case $r = 2.8$).

The stability of the fixed point $x_{t+2} = f^{[2]}(x_t)$ depends on the derivative:

$$f^{[2]\prime}(x^*) = f'(x^*) \cdot f'\big(f(x^*)\big) \tag{2.16}$$

For the fixed points of $f(x_t)$, we have $f(x^*) = x^*$, so that eqn (2.16) becomes

$$f^{[2]\prime}(x^*) = [f'(x^*)]^2$$

or, in terms of the logistic map (2.7),

$$[f'(x^*)]^2 = [r(1 - 2x^*)]^2$$

$$= (r - 2rx^*)^2 \tag{2.17}$$

The fixed point $x_1^* = 0$ is unstable for both $f(x_t)$ and $f^{[2]}(x_t)$, since the multiplier (2.17) is $3.2^2 = 10.24$. The same holds for the fixed point $x_2^* = 0.6875$, since the multiplier is

$$[f'(x_2^*)]^2 = \left[r - 2r\left(\frac{r-1}{r}\right)\right]^2$$

$$= (-r + 2)^2 = r^2 - 4r + 4$$

which equals 1.44 for $r = 3.2$. Consequently, the period-2 cycle $(0, 0.6875)$ is unstable (a repeller). Analogously to the unstable fixed point, a point near the repelling orbit (i.e. near to 0 or to 0.6875) tends to move away.

Although the points x_3^* and x_4^* are not fixed points of $f(x_t)$, they are *stable* fixed points of $f^{[2]}(x_t)$ and correspond to the period-2 stable cycle $(0.7995, 0.5130)$ of $f(x_t)$. Since $f(x_t)$ alternates between values x_3^* and x_4^*, we have

$$f^{[2]\prime}(x_3^*) = f'(x_4^*) \cdot f'(x_3^*) = f^{[2]\prime}(x_4^*) \tag{2.18}$$

indicating that the slope of $f^{[2]}(x_t)$ in the two fixed points is the same: $f^{[2]\prime}(x_3^*) = f^{[2]\prime}(x_4^*)$. Thus, we can compute either slope by computing $f'(x_3^*)$, for example, and squaring the result. Using the fixed point multiplier (2.13), substituting for x_3^* from eqn (2.15) and denoting $\sqrt{\cdot} = \sqrt{r^2 - 2r - 3}$ gives

$$f^{[2]\prime}(x_3^*) = f^{[2]\prime}(x_4^*) = \left(r - 2r\frac{r+1+\sqrt{\cdot}}{2r}\right) \times \left(r - 2r\frac{r+1-\sqrt{\cdot}}{2r}\right)$$

$$= (-1 + \sqrt{\cdot}) \times (-1 - \sqrt{\cdot})$$

$$= 1 + \sqrt{\cdot} - \sqrt{\cdot} - r2 + 2r + 3$$

$$= 4 + 2r - r^2 \tag{2.19}$$

Computing eqn (2.19) for $r = 3.2$ results in $f^{[2]\prime}(x_3^*) = f^{[2]\prime}(x_4^*) = 0.16$, indicating that x_3^* and x_4^* are stable fixed points of $f^{[2]}(x_t)$.

The correspondence between fixed points and cycles is shown in Figure 2.8, which plots both $f^{[2]}(x_t)$ (solid line), and $f(x_t)$ (dashed line). We see that $f^{[2]}(x_t)$ intersects the 45° straight line at the four fixed points identified in eqn (2.15), and $f(x_t)$ intersects the 45° straight line at the unstable fixed points 0 and 0.6875. Moreover, the slope of $f^{[2]}(x_t)$ is the same at the points x_3^* and x_4^*.

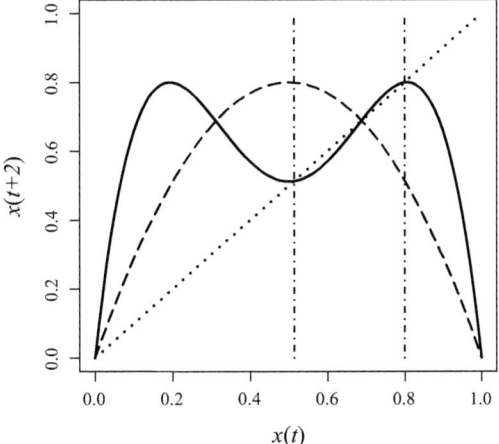

Fig. 2.8 Plots of $f^{[2]}(x_t)$ (solid line) and $f(x_t)$ (dashed line), the logistic map with $r = 3.2$. The vertical lines locate the stable fixed points of $f^{[2]}(x_t)$, which correspond to the period-2 cycle of $f(x_t)$.

To summarize, both fixed points and cycles can be stable (attractors) or unstable (repellers). The following rule holds: the stable (unstable) *fixed points* of $f^{[2]}(x_t)$ correspond to the stable (unstable) *cycles* of $f(x_t)$.

A further increase of r in eqn (2.7) from 3.2 to 3.5 introduces 'period-doubling' behaviour. Code 2.3 generates the time series and cobweb plots displayed in Figure 2.9. The plots show that the iterated solution doubles from a period-2 to a period-4 attractor with points 0.3828, 0.5009, 0.8269 and 0.8750.

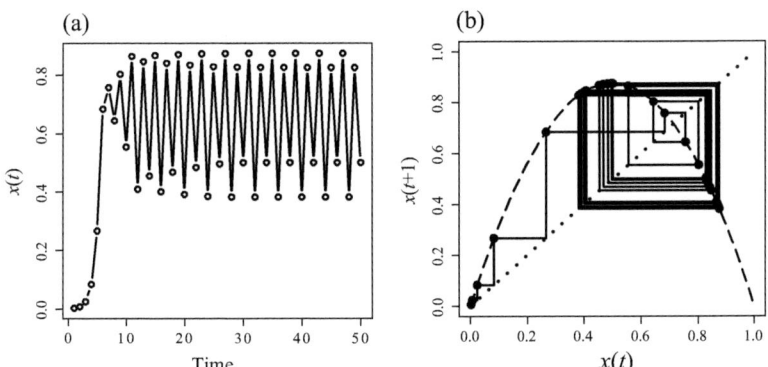

Fig. 2.9 Time series plot (a) and cobweb plot (b) of the logistic map with $r = 3.5$.

Since x_t repeats every four periods ($x_{t+4} = x_t$), we have

$$x_{t+4} = f^{[4]}(x_t) = x_t$$

The curve $f^{[4]}(x_t)$ (solid line) intersects the 45° reflection line eight times (Figure 2.10). The intersections along the reflection line that $f^{[4]}(x_t)$ shares with $f^{[2]}(x_t)$ (dashed line) are unstable fixed points at $(0, 0.4286, 0.7143, 0.8571)$ that define an unstable period-4 attractor repelling initial conditions. The remaining four intersections that $f^{[4]}(x_t)$ and $f^{[2]}(x_t)$) do not share are stable fixed points at $(0.3828, 0.5009, 0.8269, 0.8750)$ (identified by vertical lines in Figure 2.10). These points define a stable period-4 attractor for $f^{[2]}(x_t)$ that attracts initial conditions.

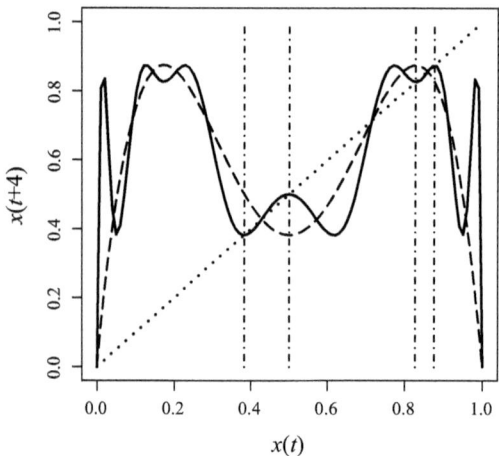

Fig. 2.10 Plots of $f^{[4]}(x_t)$ (solid line) and f $f^{[2]}(x_t)$ (dashed line) for $r = 3.5$. The vertical lines locate the stable fixed points of $f^{[2]}(x_t)$.

The period-doubling behaviour has the following pattern. As r increases from 2.8 to 3.2, a stable fixed point becomes unstable and two new stable fixed points appear. A further increase to $r = 3.5$ causes two stable fixed points to become unstable, with the appearance of two new couples of stable 'twins'. We see from Figure 2.10 that the twins are born at points where $f^{[2]}(x_t)$ has the same slope as $f^{[4]}(x_t)$, so that 'parents' become unstable at the same time. When the slope at a certain fixed point reaches 45° (or −45°), so that $m = \pm 1$ in eqn (2.12), a stable fixed point becomes unstable and gives birth to twins.

In general, to locate the n-period cycle, we numerically solve

$$f^{[n]}(x_t) = \underbrace{f\big(f(\ldots f(x_t))\big)}_{n \text{ times}} = x_t$$

where $f^{[n]}(x)$ is the nth iterate. A point on a period$-n$ cycle of the map $f(x)$ is a fixed point for the nth iterate $f^{[n]}(x)$. To assess the stability of the detected cycle, we compute the product M of n multipliers evaluated at the n-period points of the cycle $(x_1^*, x_2^*, x_3^*, \ldots, x_n^*)$:

$$M = f'(x_1^*) \cdot f'(x_2^*) \cdot f'(x_3^*), \ldots, f'(x_n^*) \qquad (2.20)$$

If $|M| < 1$, the cycle is a period$-n$ attractor; otherwise, it is a period$-n$ repeller.

The main conclusion of this section is the correspondence between fixed points and cycles: a point on a period$-n$ cycle of the map $f(x)$ is a fixed point for the nth iterate $f^{[n]}(x)$.

2.6   Analyzing Period Doubling with Bifurcation Diagrams

With the R Code 2.4 `logistic_bif.R`, we can study what happens at increasing values of r.

```
#Code 2.4 Nonlinear logistic map: bifurcation diagram

f.x<- function(x,r){
    r*x*(1-x)          }
ntrans<- 1000                # transient
rin<- 2.8
rfin<- 4
n<- 400                      # number of iterations after the transient
nt<- ntrans+n                # total number of iterations
nr<- 241                     # number of r step
xinit<- 0.2
r<- seq(rin,rfin,length=nr)
plot(0,0,type="n",xlim=c(rin,rfin),ylim=c(0,1),xlab="r",ylab="x(t)",
    cex.lab=1.5,cex.axis=1.2)
for(i in r) {                # starting loop on r values
x<- xinit
    for(j in 1:nt) {         # starting loop on the iterations
       y<- f.x(x,i)
       if(j > ntrans) points(i,y,pch='.',cex=3)
       x<- y
                }            # ending loop on the iterations
            }                # ending loop on r values
segments(3.569944,0,3.569944,1,lty=4,lwd=2,col="black")
```

The code is based on two loops. In the first, r varies, for instance from 2.8 (`rin`) to 4 (`rfin`). The instruction `r<- seq(rin,rfin,length=nr)` produces a vector with 241 (`nr`) values of r that are utilized by the loop `for(i in r)`. Then, we start from $r = 2.8$ and go up to 4, with steps equal to 0.005. Inside the loop over r is 'nested' a second loop over the iterations, already used to obtain the cobweb plots.

Now, however, the aim is not a cobweb plot for each r, but rather to compute, for each r, the stable fixed points. For this reason, the transient part of the trajectory (`ntrans`) is deleted from the plot. The value 1000 chosen for `ntrans` is certainly (better: almost always) redundant, but this is unimportant. Note the instruction `if(j > ntrans) points(i,y,pch='.',cex=2)`, which is used to plot the points only after `ntrans` iterations.

Let us follow the procedure step by step. We start with `rin` = 2.8 and with x_0 (`xinit`) equal to 0.2. Obviously, any initial value may be chosen, since all the iterations before arriving on the attractor have no interest. Let us begin to iterate within the loop `for(j in 1:nt)`. The number of total iterations `nt` is given by the number of transient iterations (1000) plus a further 400 ones (`n<- 400`). As already mentioned, the *transient* is the part of the trajectory preceding the attractor. After 1000 iterations, the attractor is reached (it is unimportant if it is reached with fewer than 1000 iterations), so that the 400 values from x_{1001} to a x_{1400} are all equal to $x^* = 0.6429$ (Figure 2.3). Therefore, the instruction `points(...)` plots 400 dots (`pch='.'`) upon each other, all at the point on the graph with coordinates (i, y), that is, $(2.8, 0.6429)$. The same thing occurs for the following value $r = 2.805$, so that the point on the graph, very near to the first one, has coordinates $(2.8, 0.6435)$. Everything else proceeds in a similar manner, until r assumes the value 3 at the first bifurcation. After that, half of the 400 values of the iterates are on x_1^* and half on x_2^*. Figure 2.11, called a *bifurcation diagram* shows the mechanism of the *period doubling* up to the vertical dash–dot line at $r \approx 3.56994$.

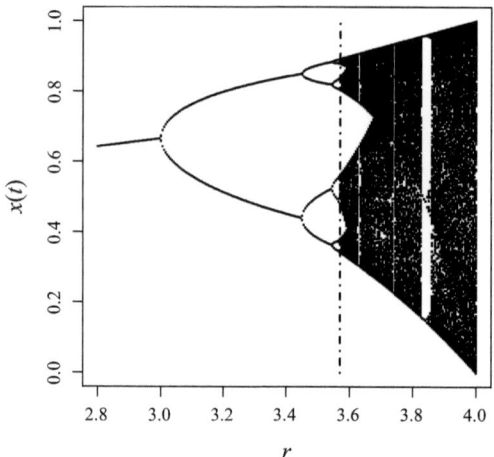

Fig. 2.11 Bifurcation diagram of the logistic map; the vertical dash–dot line at $r \approx 3.56994$ denotes the beginning of the chaotic regime.

As the period doubles, the 400 values of the iterates split at the fixed points. For instance, when the period is 4, there are 100 iterates on each of the four attractors. In Figure 2.11, one can clearly see the transitions to period 2 and to period 4, and less clearly to period 8. Many similar figures, with greater definition, can be found on websites and in books about chaos.

Two technical notes are in order here. We discarded the first 1000 iterations since, near the bifurcation points, the approach to fixed points is very slow, as we have already noted. Hence, it is true that, far from the fixed points `ntrans<- 100` is more than enough, but near the attractors with `ntrans<- 1000`, the plot is 'cleaner'. Also, `n<- 400` might seem redundant – indeed, after `ntrans`, all the iterates are on the attractors and, at most, we can 'see' period 8. But to have at least a summary depiction of the structure beyond the line $r \approx 3.56994$, some hundreds of iterates must be used, as we will see later. The study of the diagram shown in Figure 2.11 comes within the field of *bifurcation theory*.

What is a bifurcation? A dynamical system has a bifurcation when a tiny variation of a parameter causes a sudden *qualitative* change in the system's dynamics. The parameter value for which the bifurcation occurs is called the **bifurcation point**. For instance, in the case of the logistic map, the bifurcation can be shown schematically as in Figure 2.12. We can see how, for a certain value r^* of the parameter, the fixed point x^* goes from stable (solid line) to unstable (dashed line), and at the same time gives birth to a period-2 cycle (solid line). This kind of bifurcation is called a *flip bifurcation* or a *period-doubling bifurcation*. The same scheme also represents the **pitchfork bifurcation**. Also, in this latter type of bifurcation, a 'stable arm', that is, a sequence of stable fixed points, becomes unstable and, at the same time, two new stable arms are born. The difference between the two bifurcations lies in the following: after the change in the stability of the fixed point, for the flip bifurcation the system visits alternately the states of both arms, while for the pitchfork bifurcation the system stays only on one arm, and does not visit the states of the other one. However, in the literature, sometimes both bifurcations are called 'pitchfork bifurcations'.

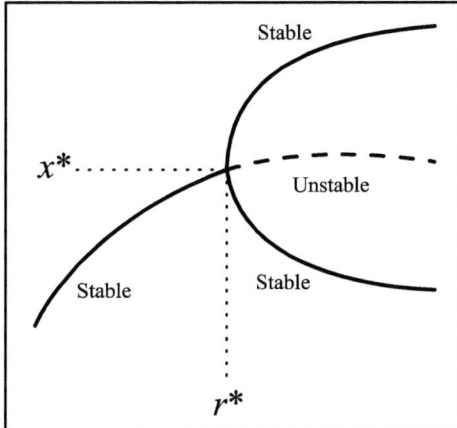

Fig. 2.12 Scheme of a flip bifurcation: for a parameter value equal to r^*, the fixed point x^* goes from stable (solid line) to unstable (dashed line), also giving birth to a stable 2-period cycle (solid lines).

Table 2.1 lists the period of the attractor and the value of r_n^* (with six decimal digits) at the point at which the cycle with period 2^n is born. This cycle will live up to r_{n+1}^*, when the cycle with double the period (2^{n+1}) will be born. For instance, the stable cycle with period $2^2 = 4$ is born from the cycle with period $2^1 = 2$ when $r_2^* = 3.449490$ and lives up to $r_3^* = 3.544090$, when the stable cycle with period $2^3 = 8$ is born. As r increases, there are cycles with greater and greater period; that is, the number of visited states becomes larger and larger. In the interval (r_n^*, r_{n+1}^*), the period is equal to 2^n; thus, the period increases exponentially with r. One might think that if $n \to \infty$, then r also approaches infinity. But this is not true.

Table 2.1 Period-doubling cascade for the logistic map. The attractor period and the value of r_n^* at the moment of birth of the cycle of period 2^n are listed.

Attractor period	Bifurcation point
$1 \times 2^0 = 1$	$r < 3$
$1 \times 2^1 = 2$	$r_1^* = 3$
$1 \times 2^2 = 4$	$r_2^* = 3.449490$
$1 \times 2^3 = 8$	$r_3^* = 3.544090$
$1 \times 2^4 = 16$	$r_4^* = 3.564407$
$1 \times 2^5 = 32$	$r_5^* = 3.568759$
$1 \times 2^6 = 64$	$r_6^* = 3.569692$
$1 \times 2^\infty = \infty$	$r_\infty^* = 3.569945718\ldots$

For $n \to \infty$, the sequence of bifurcation points converges rapidly to an accumulation point given by

$$r_\infty^* = \lim_{n \to \infty} r_n^* = 3.569945718\ldots$$

At this value (the dash–dot vertical line in Figure 2.11), attractors converge more and more rapidly and with ever-increasing period tending to infinity. But, however great the periods of the attractors, their configurations are determined only by the value of r, not by the initial state. As we will see, beyond r_∞^*, that is, in the *chaotic regime*, the opposite occurs, owing to the *sensitivity to initial conditions*, an essential property of chaotic systems.

The convergence to the critical value r_∞^* obeys specific underlying rules. Figure 2.13 (modified from Hofstadter (1981)) shows a qualitative aspect. We selected r values very close to the critical values reported in Table 2.1. Notice that in the period-16 cycle ($r = 3.5686$), some fixed points have values very close to each other. For instance, the two values on the far right-hand side are almost overlapping, with values 0.890509 and 0.892106.

A further interesting aspect of Figure 2.13 is its repeated geometric scheme. The basic scheme consists of four points, as in the line corresponding to $r = 3.544$: two points far apart (left) and two closer together (right) The same scheme appears below on the line $r = 3.5642$, but now the single point is replaced by a couple of points. On the line $r = 3.5686$, the same scheme is realized by four small groups of four points each. Every small group is a midget 'copy' of the basic scheme, as highlighted by the dotted rectangle around the second small group. Notice also the alternating distances inside each small group.

As r increases, the bifurcations become more frequent and on a smaller scale: this is a property of fractal geometry and is named *self-similarity*. These concepts will be discussed further in later chapters. The bifurcation diagram of Figure 2.11 is thus a fractal. It contains infinite copies of itself and on zooming a portion of it an infinite number of times, no detail is lost. Of course, this is true only from an analytical mathematical point of view – in numerical computations, the resolving power depends on various technical factors.

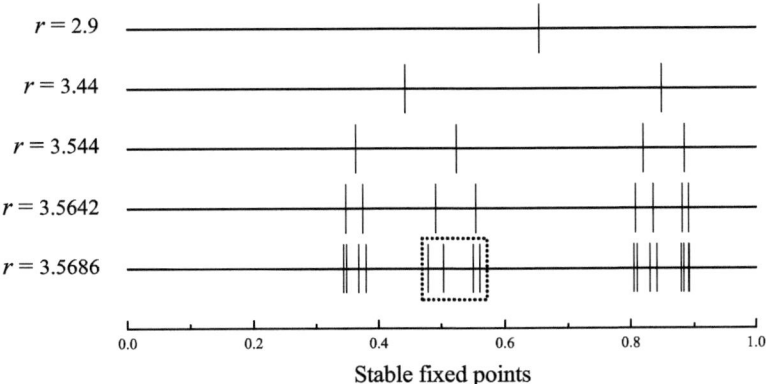

Fig. 2.13 Scheme of the period-doubling cascade for the logistic map from the period-1 cycle up to the period-16 cycle. The small vertical bars show the values of the stable fixed points corresponding to the r values on the left.

Figure 2.14 shows two parts of Figure 2.11. Figure 2.14a (obtained with Code 2.5 `logistic_bif_dett1.R`) is limited by $3.4 \leqslant r \leqslant 3.5699$ (very close to r_{∞}^{*}) and $0.80 \leqslant x_t \leqslant 0.90$, in which the 2-period cycle has already been born.

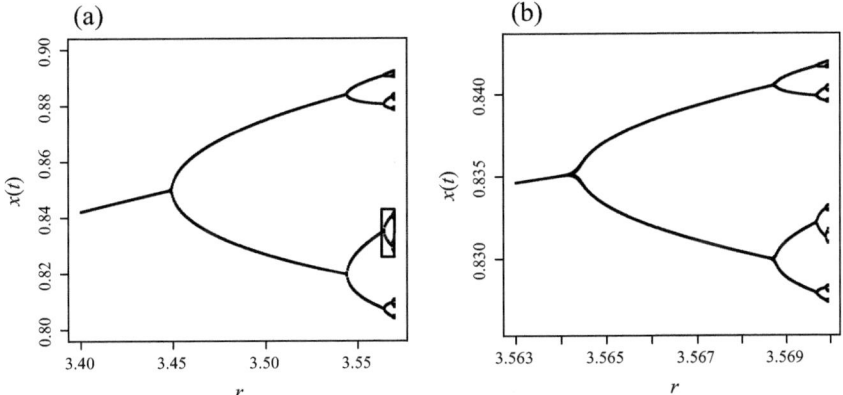

Fig. 2.14 Details of Figure 2.11 highlighting the self-similarity. The first bifurcation involves a period-2 arm that splits in two to give rise to the period-4 cycle (a) and a period-8 arm that splits in two to give rise to the period-16 cycle (b)

```
#Code 2.5 Nonlinear logistic map: detail of the bifurcation diagram (logistic_bif_dett1.R)

f.x<- function(x,r){
      r*x*(1-x)            }
ntrans<- 1000               # transient
rin<- 3.4
rfin<- 3.56994
# faster, but a bit worse:    n<- 50   and  nr<- 200
```

```
# a bit slower, but more detailed: n<- 400   and   nr<- 400
n<- 400            # number of iterations after the transient
nt<- ntrans+n                  # total iterations
nr<- 400        # number of r step
xinit<- 0.2
r<- seq(rin,rfin,length=nr)
plot(0,0,type="n",xlim=c(rin,rfin),ylim=c(0.80,0.90),xlab="r",ylab="x(t)",
    cex.lab=1.7,cex.axis=1.3)
for(i in r) {                    # starting loop on r values
x<- xinit
    for(j in 1:nt) {         # starting loop on iterations
        y<- f.x(x,i)
        if(j > ntrans) points(i,y,pch='.',cex=2)
        x<- y
                }           # ending loop on iterations
            }               # ending loop on r values

segments(3.563,0.826,3.56997,0.826,lty=1,lwd=2,col="black")
segments(3.563,0.843,3.56997,0.843,lty=1,lwd=2,col="black")
segments(3.563,0.826,3.563,0.843,lty=2,lwd=2,col="black")
segments(3.569944,0.826,3.56997,0.843,lty=1,lwd=2,col="black")
# more highlighted: col="black" --> col="red"
```

Figure 2.14b zooms in further on the region displaying an 8-period cycle: $3.563 \leqslant r \leqslant 3.5699$ and $0.826 \leqslant x_t \leqslant 0.843$. This region is identified in Figure 2.14a by the boxed area. The additional zooming is done by adding additional instructions to Code 2.5:

```
. . . . . . . . . . . . . . . . . . . . . . . . . . . . . . . . . . . . . . . . . . . . .
rin<- 3.563
rfin<- 3.56994
. . . . . . . . . . . . . . . . . . . . . . . . . . . . . . . . . . . . . . . . . . . . .
plot(0,0,type="n",xlim=c(rin,rfin),ylim=c(0.826,0.843),xlab="r",ylab="x(t)",
    cex.lab=1.7,cex.axis=1.3)
. . . . . . . . . . . . . . . . . . . . . . . . . . . . . . . . . . . . . . . . . . . . .
```

It should be noted how these particular sectors of Figure 2.11 have the same form as the whole figure (up to r^*_{∞}).

Cycles double with increasing frequency in accordance with a mathematical scheme detailed in Table 2.2. Let us determine the transition time from r^*_n to r^*_{n+1}, with respect to the transition time from r^*_n and r^*_{n+1}. Let Δ_n be the stability interval for the cycle of period 2^n, that is, the distance between the two values r^*_n and r^*_{n+1}. For instance, the period-2^2 cycle begins at $r^*_2 = 3.449490$ and ends at $r^*_3 = 3.544090$; hence, $\Delta_2 = 3.544090 - 3.449490 = 0.094600$. Similarly, $\Delta_3 = 3.564407 - 3.544090 = 0.020316$. Let us compute the ratios Δ_n / Δ_{n+1}, for instance $\Delta_2 / \Delta_3 = 4.6564$. The first six values of r^*_n are listed in Table 2.2.

It is apparent that the ratio Δ_n / Δ_{n+1} remains almost constant as n increases. For instance, the distance between the first and second period doublings is $3.449490 - 3 \approx 0.449$, while the distance between the second and third period doublings is $3.564407 - 3.449490 \approx 0.095$. The latter distance is smaller by a factor of 4.7 with respect to the former. Similarly, the distance between the third and fourth period doublings (≈ 0.00203) is still a factor of ≈ 4.7 shorter than the previous distance (≈ 0.095). And so on, with period doublings following one another with shorter and shorter distances by a factor of ≈ 4.7. In fact, there exists a limit

Table 2.2 Logistic map. Regularity in period doubling: attractor period, bifurcation point r_n^*, length of stability interval Δ_n and ratio between lengths.

n (period $= 2^n$)	r_n^*	Δ_n	Δ_n/Δ_{n+1}
1	3		
2	3.449490	4.49489×10^{-1}	
3	3.544090	9.46000×10^{-2}	4.7514
4	3.564407	2.03170×10^{-2}	4.6564
5	3.568759	4.35200×10^{-3}	4.6684
6	3.569692	9.33000×10^{-4}	4.6645

$$\lim_{n \to \infty} \frac{\Delta_n}{\Delta_{n+1}} = \lim_{n \to \infty} \frac{r_n^* - r_{n-1}^*}{r_{n+1}^* - r_n^*}$$

$$\approx 4.6692016091029909 \equiv \delta \tag{2.21}$$

This result was obtained at the end of the 1970s by Mitchell Jay Feigenbaum (Feigenbaum, 1979), and the above limit is known as the *Feigenbaum delta*. The bifurcation diagram in Figure 2.11 is also called the *Feigenbaum tree*. Ironically, in German, *Baum* means 'tree' and the noun *Feigenbaum* literally means 'fig tree'. Feigenbaum not only found the value of δ, but also demonstrated that this value is not exclusive to the logistic map, but rather is a *universal* constant, like π or the Napier number e.

A wide class of one-dimensional maps (the unimodal maps) follow the same *route to chaos* as the logistic map, namely through a *bifurcation cascade* with the distances between doubling periods becoming shorter and shorter, and in the limit approaching the Feigenbaum δ. The same dynamics is followed also by real systems as they approach the chaotic regime.

There is a further constant characteristic of the Feigenbaum tree – it concerns the width of the bifurcations, as shown in Figure 2.15. From the bifurcation diagram, one can see that, as n increases, not only do the period doublings follow each other at shorter and shorter distances, but also the widths of bifurcations becomes smaller and smaller. Let a_n be the width of the bifurcation in the period-2^n cycle, measured with respect to the line $x_t = 0.5$. Then, as $n \to \infty$, the ratio between the widths of consecutive bifurcations tends to the following limit:

$$\lim_{n \to \infty} \frac{a_n}{a_{n+1}} \approx 2.5029078750 \equiv \alpha$$

Both of the constants δ and α are known as *Feigenbaum numbers*.

In conclusion, we can say that it is true that the values of the bifurcation points are different for different maps, since the values of r_∞^* differ, but for unimodal maps the Feigenbaum numbers are the same.

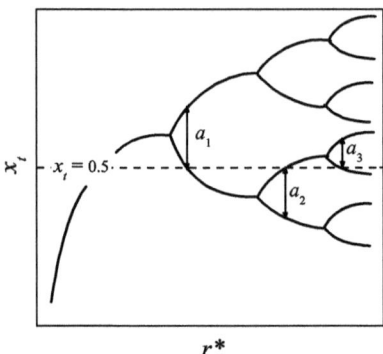

Fig. 2.15 Scheme, not to scale, of the Feigenbaum tree for the first bifurcations; the distances between the branches are increased for better visualization.

2.7 Chaotic Behaviour

In Figure 2.11, after the dot–dash line at $r \approx 3.56994$, it appears that there are no arms showing any periodicity, and the 400 points of the iterates never overlap, but nevertheless such points are not uniformly distributed and only for $r = 4$ are they distributed on the whole interval $[0, 1]$. Moreover, one can clearly see that at values of r just after 3.8 there appears to be a 'window' in which there is a return to periodicity. We will discuss this feature later, but for now let us see what happens at $r = 4$. In Code 2.3, we change the parameters to `r<- 4`, `xinit<- 0.002`, `nstep<- 50`. Then, with these parameters, we obtain Figure 2.16. There is still the same parabola, but the points now are not fixed points or on stable cycles, and the time plot also shows no periodicity. Such 'chaotic' behaviour – we will return to this – is due to the fact that the trajectories 'bounce' continuously within the interval $[0, 1]$, from which they cannot escape (the iterates have values in this interval) and where they are repelled by infinite repeller period points.

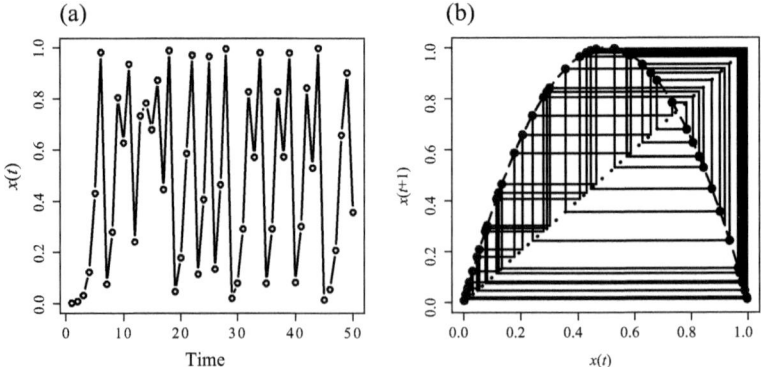

Fig. 2.16 Time series plot (a) and cobweb plot (b) of the logistic map with $r = 4$.

A first characterization of chaos is in terms of its dynamics, which describes how a system in a chaotic regime changes over time. Chaotic dynamics is *aperiodic*; that is, the system never returns to the same state. A population from generation to generation never has the same number of individuals. It might happen in computer simulations that one finds iterates with the same values, but this is due to the rounding errors of computers. We have seen that with respect to fixed points or cycles, the initial conditions are irrelevant, since trajectories starting from different points converge on the same attractor. On the other hand, in a chaotic regime, differences in initial conditions are exponentially amplified.

A further characterization of chaotic dynamics is that it is *limited*. *Limited dynamics* means that the variables defining the system states vary in finite intervals – in the case of the logistic map, they vary in the interval $[0, 1]$. This is in contrast to linear equations, where, for example, we have seen that in eqn (2.1) the variable N_t can go to infinity if $r > 1$. Thus, chaotic dynamics is aperiodic, limited and also *deterministic*. In fact, there are no random variables, no stochastic components, in the description of chaotic systems. Nevertheless, chaotic system are unpredictable, because of a third characterization of their dynamics, namely the so-called *sensitive dependence on initial conditions*.

Unlike fixed points and cycles (where initial conditions are irrelevant), differences in initial conditions in a chaotic regime are amplified exponentially. This sensitivity is demonstrated in Figure 2.17. We plot two solution trajectories beginning from slightly different initial points. One trajectory (plotted with circles) starts from $x_0 = 0.002$. The second (plotted with crosses) starts from $x_0' = 0.00202$. The two trajectories remain close for about six or seven steps, but then begin to diverge along completely different paths. For example after 19 steps $x_0 = 0.0468$, while $x_0' = 0.9633$. The sensi-

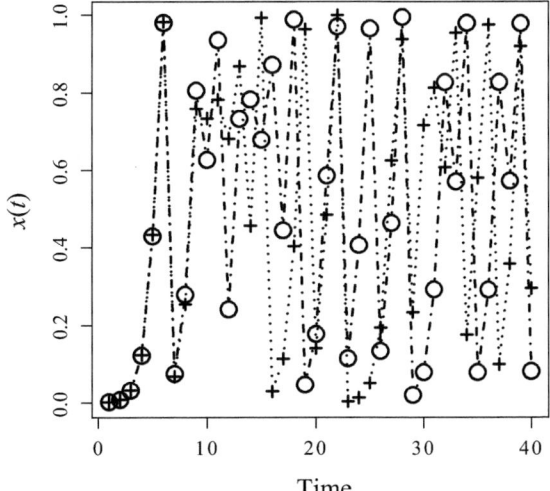

Fig. 2.17 Time series plot of two trajectories of the logistic map with $r = 4$ and initial conditions $x_0 = 0.002$ (circles) and $x_0' = 0.00202$ (crosses)

tive dependence on initial conditions renders chaotic dynamics unpredictable. A tiny imprecision in measuring initial conditions results in exponentially large imprecision regarding future states. There is a well-known metaphor for this sensitivity to to initial conditions, the so-called *butterfly effect*. In 1972, Edward Norton Lorenz, at the 139th Annual Meeting of the American Association for the Advancement of Science, delivered a lecture with the title: *Predictability: Does the Flap of a Butterfly's Wings in Brazil set off a Tornado in Texas?* (printed in Lorenz (1995)). The phrase refers to the idea that a tiny change in the initial conditions of the system, namely the flapping wing, can cause catastrophic events through a chain of amplifications.

Finally, these erratic and unpredictable dynamics surprisingly emerge from a relatively simple deterministic model. Chaotic dynamics is not generated exogenously by random variables or stochastic components.

The above characteristics derive from the stretching and folding of chaotic trajectories in phase space. This is often compared with how a baker stretches dough and folds it over repeatedly. During kneading, points in the dough that are initially close may diverge substantially. This offers an intuitive picture of why chaotic dynamics exhibits sensitivity to initial conditions that precludes prediction and of how trajectories that never cross remain bounded in phase space.

Figure 2.18 illustrates stretching and folding for the logistic map in the chaotic regime ($r = 4$). The iterate of $x_t = 0$ is $f(x_t) = 0$, and the iterate of $x_t = \frac{1}{2}$ is $f(x_t) = 1$. Consequently, the half-segment $[0, \frac{1}{2}]$ is stretched on the entire interval $[0, 1]$ after one iteration. In another example, the iterate of $x_t = 1$ is $f(x_t) = 0$, and we again see where a half-segment $[\frac{1}{2}, 1]$ is stretched on the whole interval $[0, 1]$, but in the opposite direction. In sum, the segment $[0, 1]$ doubles its length *on average* by bending back on itself. We emphasize *on average* because subsegments close to the endpoints $[0, 1]$ (as in Figure 2.18) are markedly stretched while those in the central part of $[0, 1]$ may be contracted.

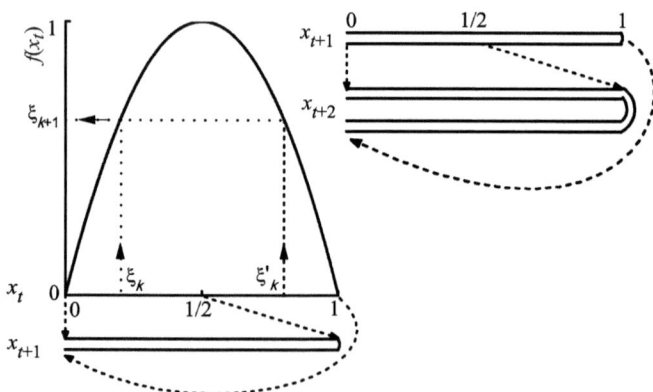

Fig. 2.18 (a) Graphical representation of the mechanism of stretching and folding for the logistic map. (b) The same process at the following iterate.

Figure 2.18 also demonstrates how the *noninvertibility* of the logistic map precludes predicting backwards through time (i.e. *retrodiction*). A map $x_{t+1} = f(x_t)$ is invertible if it is possible to define an inverse map f^{-1} such that $x_t = f^{-1}(x_{t+1})$. The logistic map in not invertible, because every point ξ_{k+1} has two precursors, one to the right, ξ_k, and one to the left, ξ'^k, of $x^{[t]} = 0.5$. Moreover, every precursor has two other precursors. For example, ξ_{k+1} has $2^{100} \approx 1.3 \times 10^{30}$ precursors going back 100 iterations. In general, a one-dimensional map must be noninvertible to be chaotic. However, even invertible multidimensional maps may exhibit chaotic behaviour.

2.7.1 Quantifying Sensitive Dependence with Lyapunov Exponents

To estimate the rate at which the trajectories diverge, we account for how the distance between them changes at each time (Figure 2.19). Recalling eqn (2.10), let $|x_0 - x'_0| = \delta x_0$ be the very small distance (strictly, $\delta x_0 \to 0$) between the points x_0 and x'_0. The initial distance is

$$\delta x_1 = |x_1 - x'_1| = |f(x_0) - f(x'_0)| \approx |f'(x_0)|\, \delta x_0$$

At the second iteration, the distance is

$$\delta x_2 = |x_2 - x'_2| \approx |f'(x_1)|\, \delta x_1 \approx |f'(x_1) \cdot f'(x_0)|\, \delta x_0$$

The distance at the nth iteration is

$$\delta x_n = |x_n - x'_n|$$
$$\approx |f'(x_{n-1}) \cdot f'(x_{n-2}) \cdots f'(x_1) \cdot f'(x_0)|\, \delta x_0 \tag{2.22}$$

We want to estimate the extent to which two trajectories diverge (or converge) for progressively increasing times. The behaviour at short time intervals is of no interest.

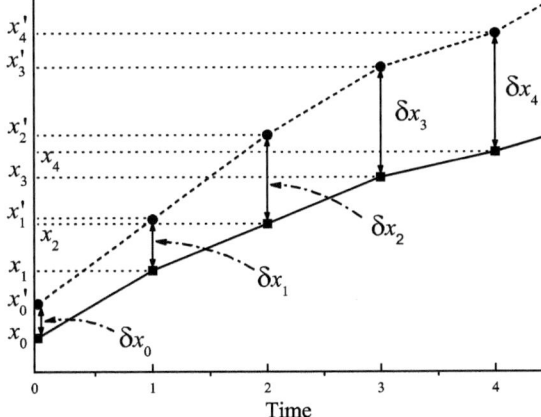

Fig. 2.19 Increase with time of the distance between two trajectories (solid and dashed lines) starting from very close points (initial distance δx_0).

We have to take into account the behaviour at each step; that is, we have to compute the average of the product of the derivatives. Taking the logarithm of both sides of eqn (2.22) gives

$$\log\left[|f'(x_{n-1}) \cdot f'(x_{n-2}) \cdots f'(x_1) \cdot f'(x_0)|\right]$$

$$= \log|f'(x_0)| + \log|f'(x_1)| + \ldots + \log|f'(x_{n-2})| + \log|f'(x_{n-1})|$$

Considering the whole trajectory, that is, $n \to \infty$, and remembering that $\delta x_0 \to 0$, the *Lyapunov exponent* is defined as

$$\lambda = \lim_{n \to \infty} \lim_{\delta x_0 \to 0} \frac{1}{n} \sum_{t=0}^{n-1} \log|f'(x_t)| \tag{2.23}$$

We take the limit $n \to \infty$ to exclude the transient part of the trajectory, and we assume that the initial discrepancy in initial conditions is very small: $\delta x_0 \to 0$. For one-dimensional maps, λ is independent of all initial conditions (strictly speaking, excluding a set of measure zero), so we denote δx_0 simply by δ_0. In summary, the *Lyapunov exponent* represents the *average rate* of separation (expansion or contraction) *per iteration* of two trajectories on an attractor with very close initial conditions.

From eqn (2.22), by taking the logarithms of both sides, we obtain

$$\delta_n \approx e^{\lambda n} \delta_0 \tag{2.24}$$

It is clear that if $\lambda < 0$, the trajectories converge, while if $\lambda > 0$, they diverge. If $\lambda = 0$, there is neither expansion nor contraction. In Chapter 4, we will see that multidimensional dynamical systems have multiple Lyapunov exponents, in which case one positive Lyapunov exponent is a signature of a chaotic regime.

There is no unambiguous definition of 'chaos' that is universally accepted. Here, we shall simply summarize the features that we have discussed. System dynamics is chaotic if it is

(1) deterministic
(2) limited
(3) aperiodic
(4) sensitive to initial conditions

2.7.2 Lyapunov Exponent of the Logistic Map

For the logistic map, the derivative $|f'(x_t)|$ can be analytically computed as (see eqn 2.12)

$$f'(x_t) = r(1 - 2x_t)$$

For $1 < r < 3$, the stable fixed point is $x^* = (r-1)/r$. So eqn (2.23) gives (see also eqn 2.13)

$$\lambda = \lim_{n \to \infty} \lim_{\delta x_0 \to 0} \frac{1}{n} \sum_{t=0}^{n-1} \log |f'(x_t)| = \log |f'(x^*)|$$

$$= \log \left| r \left(1 - 2 \frac{r-1}{r} \right) \right| = \log |2 - r|$$

If $r = 1$ or $r = 3$, then $\lambda = 0$, and the system is exactly on the attractor, where there is no expansion or contraction. If $r = 2$, then $|f'(x^*)| = 0$, and hence $\lambda \to -\infty$. If $|f'(x^*)| = 0$, then x^* is called **superstable**.

Within the interval $(3, 1 + \sqrt{6})$, where the period-2 cycle lives, we have

$$\lambda = \tfrac{1}{2} \log |4 + 2r - r^2|$$

from eqn (2.19). Then $\lambda = 0$ if $r = 3$ or $r = 1 + \sqrt{6}$. A further superstable point results for $r = 1 + \sqrt{5} \approx 3.236$, where $\lambda \to -\infty$.

Now we implement the R Code 2.6 `logistic_Lyap.R`, which is similar to Code 2.4 `logistic_bif.R` for computing the bifurcation diagram, but at each r, the Lyapunov exponent, rather than the value of x_t, is computed.

```
#Code 2.6 Nonlinear logistic map: Lyapunov exponent
lambdai<-numeric()
f.x<- function(x,r){
        r*x*(1-x)              }
ntrans<- 1000                # transient
rin<- 2.8
rfin<- 4
n<- 400                      # number if iterations after the transient
nt<- ntrans+n                # total number of iterations
xinit<- 0.2
r<- seq(rin,rfin,by=0.002)      #0.002: step of r
ymin<- -4.2
plot(0,0,type="n",xlim=c(rin,rfin),ylim=c(ymin,1),xlab="r",ylab="Lyapunov exp.",
     cex.lab=1.5,cex.axis=1.2)
ii<-0
for(i in r) {                # starting loop on r values
x<-xinit
lambda<-0
ii<-ii+1
    for(j in 1:nt) {         # starting loop on the iterations
        y<-f.x(x,i)
        # compute Lyapunov exponent only if j > ntrans
        if(j > ntrans)lambda<-lambda+log(abs(i*(1-2*y)))
        x<-y
                }            # ending loop on the iterations
        lambdai[ii]<-lambda/n  # averaging lambda over n steps and recording it
                }            # ending loop on r values
#r              # comment to not write r values
#lambdai        # comment to not write lambdai vs r
lines(r,lambdai,type='l',lwd=2)
segments(3.569944,ymin,3.569944,1,lty=4,lwd=2,col="black")    # or: col="red"
segments(rin,0,rfin,0,lwd=2,lty=8,col="black")                # or: col="blue"
```

Obviously, eqn (2.23) is used with finite δx_0 and n. So a *sample* of $\log |f'(x_t)|$ is computed and the final result is an *estimate* $\hat{\lambda}$ of λ. Within the loop on r, the loop on the iterations `for(j in 1:nt)` is nested, in which, after discarding the transient part of the trajectory `ntrans`, the instruction `lambda<-lambda+log(abs(i*(1-2*y)))` is executed, where `i` enumerates the values assumed by r, and `y` is the value of $f(x_t)$ computed `nt` times for that value of r. Therefore, for each r, the code computes `nt` values of `lambda`, which are added at each step. After the end of the loop on r, the sum of these `nt` values of `lambda` is divided by `nt` to obtain their average, and the result is $\hat{\lambda}$.

Figure 2.20 shows results obtained using the code `logistic_Lyap.R`. Notice that $\hat{\lambda}$ is less than 0 (or equal to 0 for certain values of r) before the chaotic regime, that is, for $r < r_\infty^*$. For $r > r_\infty^*$, $\hat{\lambda}$ is greater than 0, apart from certain values of r.

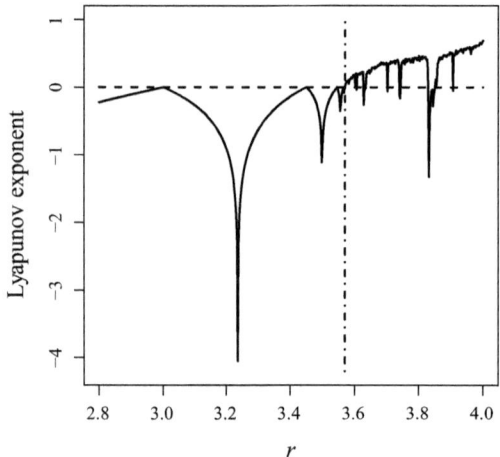

Fig. 2.20 Lyapunov exponent estimate for the logistic map; the dash–dot line at $r \approx 3.56994$ denotes the beginning of the chaotic regime.

On comparing Figures 2.20 and 2.11, one can see that the values of r for which $\hat{\lambda} = 0$ correspond to the bifurcation points; that is, $r_1^* = 3$ corresponds to the point at which the period-2 attractor begins, $r_2^* \approx 3.449$ to that at which the period-4 attractor begins and $r_3^* \approx 3.544$ to that at which the period-8 attractor begins (after these values, the figure has insufficient resolution). The deep minimum at $r \approx 3.236$ corresponds to a superstable point. Further superstable points can be seen at $r \approx 3.498$ and $r \approx 3.554$. If the step in r (at present 0.002) is slightly varied, the shapes of these minima change significantly, since at superstable points $\lambda \to -\infty$ and so in the neighbourhood of such a point the estimate $\hat{\lambda}$ is very sensitive to the value of r.

In summary, if $r < r_\infty^*$, the Lyapunov exponent is never positive, so there is no exponential divergence – indeed, on the contrary, there is asymptotic stability. In the chaotic regime, $r > r_\infty^*$, the Lyapunov exponent is positive (except in some intervals), indicating the sensitive dependence on initial conditions.

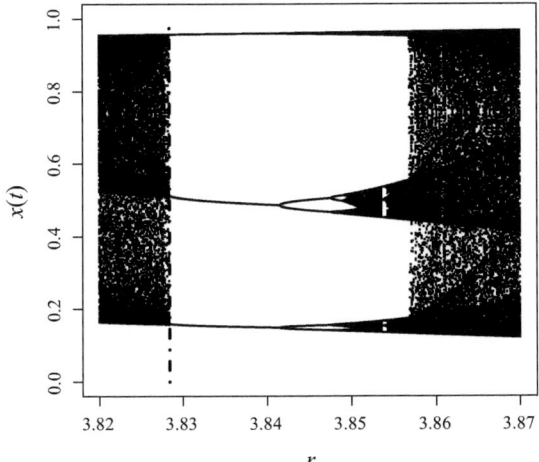

Fig. 2.21 Blow-up of the bifurcation diagram in Figure 2.11 in the vicinity of the period-3 window beginning at $r = 1 + \sqrt{8} \approx 3.828427$ (dot–dash line).

2.7.3 Periodic Windows

Figure 2.21 zooms in on the periodic window in the chaotic region of the bifurcation diagram in Figure 2.11. At the bifurcation point $r = 1 + \sqrt{8} \approx 3.828427\ldots$, cycles are born having periods equal to 3×2^n (Saha and Strogatz, 1995). After about 200 iterations, the logistic map has periodic values: 0.1597, 0.5139, 0.9564.

Figure 2.22 investigates the dynamic behaviour of the logistic map entering the window $r = 3.81 < 1 + \sqrt{8}$, at the entrance $r = 1 + \sqrt{8} \approx 3.82847$ and inside the window $r = 3.835 > 1 + \sqrt{8}$ by plotting the third iterate $f^{[3]}(x_t)$ against x_t. Entering the period-3 window (Figure 2.22a), there are two intersections between $f^{[3]}(x_t)$ and the 45° reflection line. These correspond to fixed points at which $f(x^*) = x^*$: one at the origin and the other at $x^* = 0.7368$. Both can be shown to have positive fixed point multipliers (2.13), so that both are unstable fixed points. A stable cycle has yet to emerge. At the beginning of the periodic window, Figure 2.22b shows that $f^{[3]}(x_t)$ is tangent to the 45° reflection line at three points (vertical dot–dash lines). The fixed point multiplier (2.13) at these points is equal to one, indicating that they are *nonhyperbolic* fixed points. Inside the periodic window, Figure 2.22c shows that each *nonhyperbolic* fixed point gives birth to two more fixed points, one of which is stable (filled circle) and the other unstable (open circle). These points correspond to an attractive and a repelling period-3 cycle, respectively. They solve the equation $f^{[3]}(x^*) = x^*$, which, as an eighth-degree polynomial, has eight solutions. Six of the solutions are period-3 fixed points, the other two are period-1 'imposters' (Saha and Strogatz, 1995):

$$x_1^* \approx 0.152074 \quad stable, \qquad x_2^* \approx 0.167205 \quad unstable$$
$$x_3^* \approx 0.494514 \quad stable, \qquad x_4^* \approx 0.534015 \quad unstable$$
$$x_5^* \approx 0.954313 \quad stable, \qquad x_6^* \approx 0.958635 \quad unstable$$

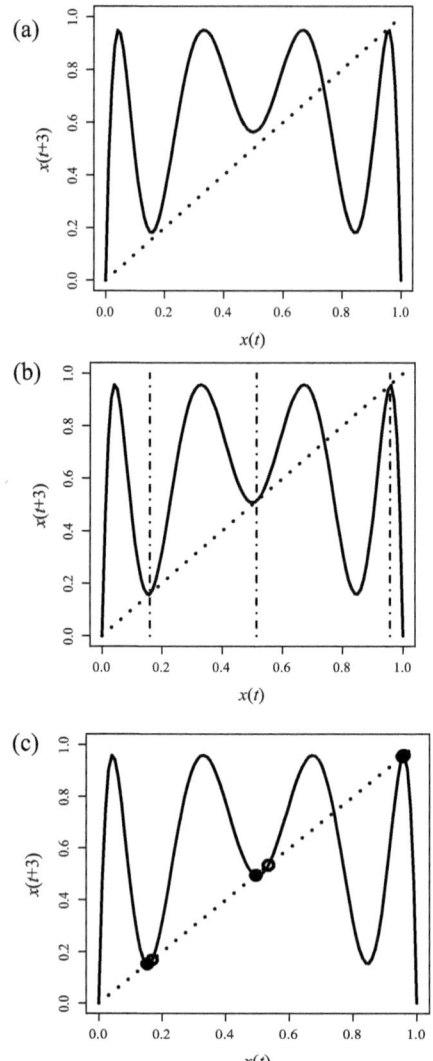

Fig. 2.22 Plot of $f^{[3]}(x_t)$ for r close to the critical value $1 + \sqrt{8}$: (a) $r = 3.81 < 1 + \sqrt{8}$ (before the period-3 window); (b) $r = 1 + \sqrt{8}$ (the beginning of the period-3 window); (c) $r = 3.835 > 1 + \sqrt{8}$ (period-3 cycles).

Figure 2.22c shows that x_5^* and x_6^* are very close each other, so that the corresponding circles are almost superimposed.

Zooming in further into the pre-window interval by setting $r = 3.8283$, we find that the logistic map follows an *intermittency route* to chaos. Figure 2.23 plots 400 iterates of the map against time. We see lengthy *quasiperiodic* intervals of period-3 cycling interrupted by short chaotic bursts. The corresponding cobweb plot (Figure 2.24) shows that the iterates are tunnelling between $f^{[3]}(x_t)$ and the 45° reflection line in

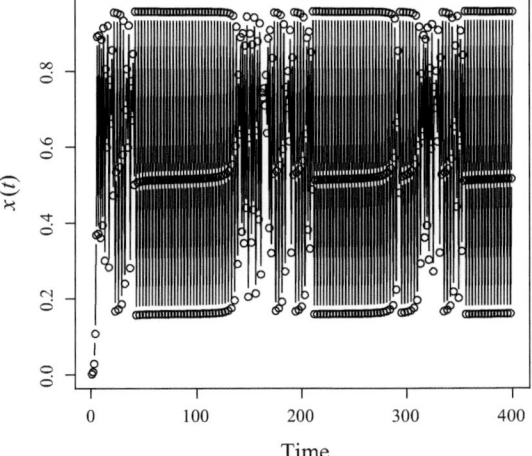

Fig. 2.23 Time series plot of $f^{[3]}(x_t)$ for $r = 3.8283$ showing *quasiperiodic* intervals separated by short chaotic bursts.

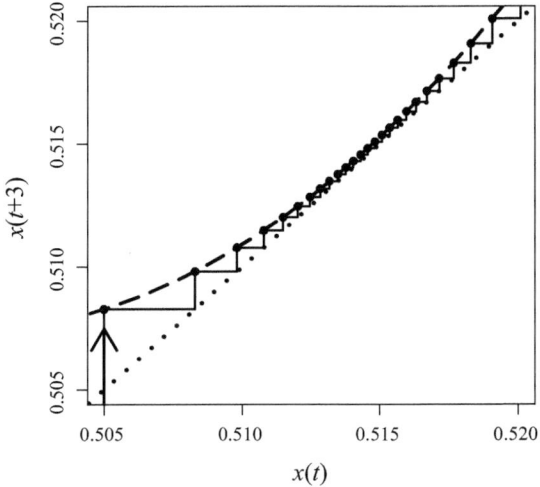

Fig. 2.24 Blow-up of the cobweb plot of $f^{[3]}(x_t)$ with $r = 3.8283$.

this interval. The map behaves as if it were in a period-3 cycle, but then bursts into chaos at the end of the tunnel. As described by Cvitanovic (1989), 'the system is being fooled into believing that it is converging toward a fixed point only to discover that the fixed point is not there after all; it then wanders away again, in the hope of finding a true fixed point'.

At the critical value $r = 1 + \sqrt{8} \approx 3.82847$, the logistic map undergoes a *tangent bifurcation*. Figure 2.25 compares the *tangent bifurcation* (a) associated with the *intermittency route* to chaos with the *flip bifurcation* (b) associated with the *period-*

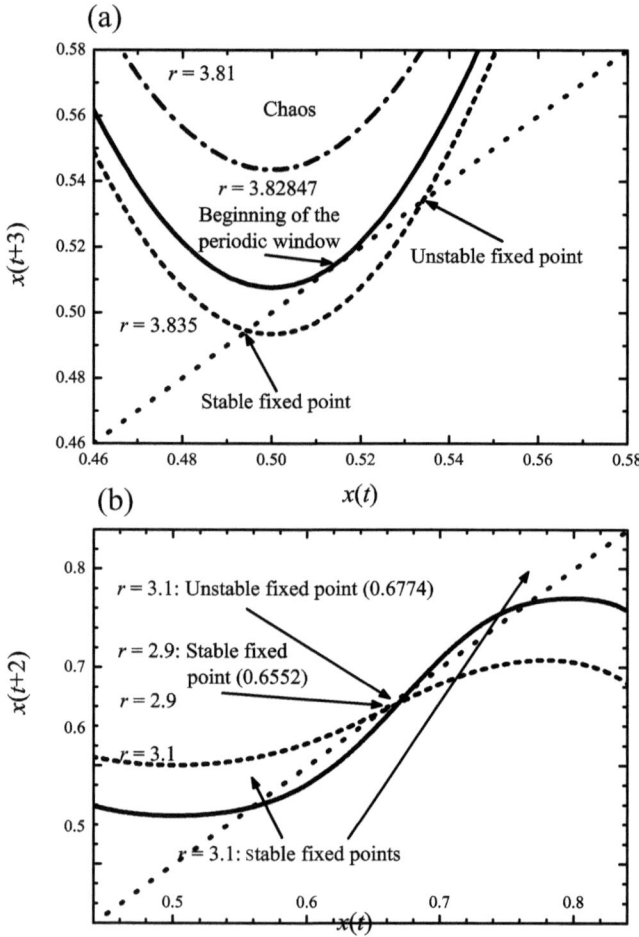

Fig. 2.25 Comparison between the tangent bifurcation (a) and the flip bifurcation (b) for different values of r in the logistic map.

doubling route discussed earlier. The plot in (a) zooms in on the graph of x_{t+3} versus x_t for values of r around the onset of the period-3 window. We can see how the *tangent bifurcation* splits a nonhyperbolic fixed point into stable and unstable fixed points corresponding to attractive and repelling period-3 cycles. The plot in (b) zooms in on the graph of x_{t+2} versus x_t for values of r around the onset of the period-doubling regime. We can see how, in the *flip bifurcation*, a stable fixed point becomes unstable and gives rise to a period-2 cycle.

2.8   Statistical Description of Chaotic Dynamics

Let us return to Figures 1.2c and 1.2f. We said that the scatter plots can reveal the deep difference between random and deterministic dynamics. Are there further ways to do this? The answer is affirmative and will bring us face to face with the fundamental notion of *ergodicity* of dynamical systems. We will see one of the most impressive properties of chaos: the dynamics of an individual trajectory is highly unstable – hence the unpredictability of chaotic systems – but, from a statistical perspective, some form of predictability can emerge, even for a single trajectory.

The interval $[0, 1]$ is divided into N subintervals I_k, with

$$I_k = \big[(k-1)/N, k/N\big], \quad k = 1, \ldots, N$$

and the number m_k of iterates in each interval I_k is computed with Code 2.7 Logistic map: Density, a slight variant of Code 2.3.

```
#Code 2.7 Density of the logistic map, 1 trajectory
f.x<- function(x,r){
     r*x*(1-x)
                 }
##### function to compute  the time trajectory #####
f.temp<-function(xinit,nstep,r){ # starting function f.temp
xt<- numeric()
x<- xinit
xt[1]<- x
for(i in 2:nstep){
y<- f.x(x,r)
x<- y
xt[i]<- x
                 }
return(xt)
                          } # ending function f.temp
### parameters and initial conditions
r<- 4
xinit<- 0.2
nstep<- 1000
xt<- f.temp(xinit,nstep,r) # iterates
lbin<- 0.05
par(mai=c(1.02,1.,0.82,0.42)+0.1)
hist(xt,freq=F,xlab="x(t)",ylab="density",main="",br=seq(0,1,by=lbin),
cex.axis=1.2,cex.lab=1.5)
curve(dbeta(x, 0.5,0.5),lty=2,col="black",add=T,lwd=2)
# random process
windows()
set.seed(1)    # seed of the sequence of (pseudo) random numbers
xt<-runif(nstep,0,1)
par(mai=c(1.02,1.,0.82,0.42)+0.1)
hist(xt,freq=F,xlab="x(t)",ylab="density",main="",br=seq(0,1,by=lbin),
ylim=c(0,3),cex.axis=1.2,cex.lab=1.5)
```

Using this code, the logistic map with $r = 4$ and a uniform random process (xt<-runif(nstep,0,1) are compared, as shown in Figure 2.26. Note that the option freq=F in hist(.) is used to plot the probability densities, not the absolute or relative frequency, given by $m_k/(N \times \text{lbin})$, where lbin = 0.05 is the width of the bins. In this manner, the histogram has a total area equal to 1.

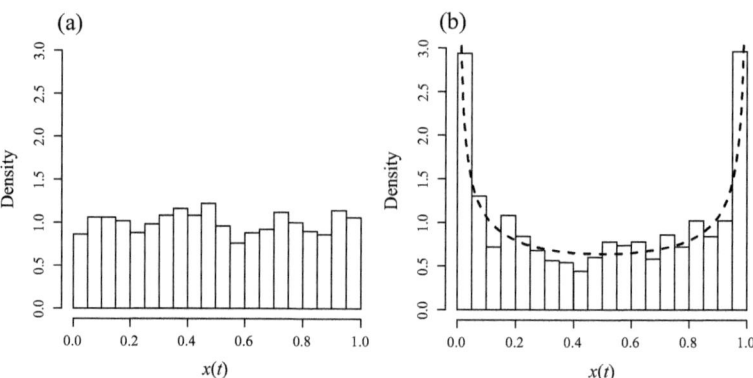

Fig. 2.26 Histogram of the positions x_t taken by 1000 iterates of two time series: (a) realizations of a uniformly distributed random variable; (b) the logistic map with $r = 4$. The dashed curve is the probability density function of the random variable $Beta(0.5, 0.5)$.

The realizations of the uniformly distributed random variable are, as expected, uniformly distributed in the whole interval $[0, 1]$, while the iterates of the logistic map do not cover the interval $[0, 1]$ uniformly, but with a variable density, namely, as we will see later, the density of the random variable $Beta(0.5, 0.5)$. This result prompts us to state that there are iterates whose values are more probable than others, a claim that, in turn, leads us to deal with the problem of the statistical description of chaotic dynamics. Let us examine some further aspects of this problem, such as what happens with different initial conditions.

Figure 2.27 shows histograms obtained with three different initial points and with only 200 iterates. It is clear that the final distribution does not depend on the starting point of the trajectories. The option `col="."` is added to `hist(.)` to fill in the bars.

We known that in the chaotic regime, trajectories diverge exponentially and the system described by the logistic map is unpredictable; nevertheless, the histograms of the coordinates of the trajectories are quite predictable. They converge to a unique distribution, whatever the initial point. Why does a deterministic chaotic system, a single trajectory of which is unpredictable, exhibits regular statistical behaviour? This is a very good question, and perhaps the final answer has yet to be given. Ergodic theory has clarified many aspects of the problem, as we will mention in Appendix A, but for now we keep the exposition at a somewhat intuitive level.

We introduce the *invariant density*, also called the *natural invariant* or *natural density* $\rho(x, x_0)$. We define $\rho(x, x_0)$ as the density of points in the interval $[x, x + dx]$, with initial point x_0, after $n \to \infty$ iterations. Letting $m(x, x_0)$ be the number of points in dx, given x_0, we have

$$\rho(x, x_0) = \lim_{n \to \infty} \frac{1}{n} \times m(x, x_0)$$

or, introducing the Dirac delta function (see Appendix A),

(a)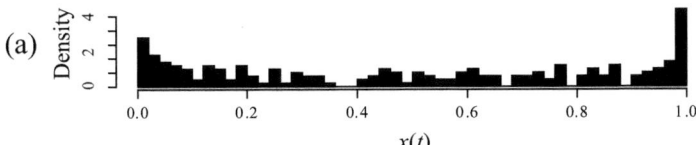

(b)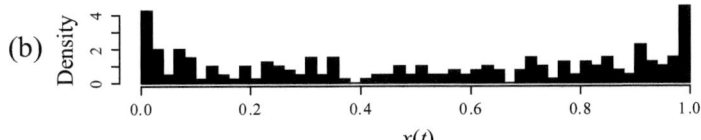

(c)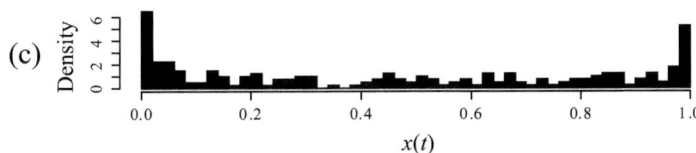

Fig. 2.27 Histograms of the positions x_t assumed by 200 iterates of the logistic map, with $r = 4$, with three different initial points: (a) 0.2; (b) 0.95; (c) 0.002.

$$\rho(x, x_0) = \lim_{n \to \infty} \frac{1}{n} \sum_{i=1}^{N} \delta\big(x - f^{[i]}\big)(x_0)$$

As we said about the Lyapunov exponent, we expect $\rho(x, x_0)$ not to depend on x_0, for 'almost all' initial conditions x_0, that is, except for those x_0 that correspond exactly to an unstable periodic point. The invariant density $\rho(x)$ of the attractor is obtained. As for stochastic processes, $\rho(x)\,dx$ represents the frequency of visits in the interval $[x, x + dx]$. Then the integral

$$\int_a^b \rho(x)\,dx$$

represents the time spent by the trajectory in the interval $[a, b]$. The probability density $Beta(0.5, 0.5)$ in Figure 2.26b is $\rho(x)$ for the logistic map, with $r = 4$; that is,

$$\rho(x) = \frac{1}{\pi\sqrt{x(1-x)}} \tag{2.25}$$

This result is derived in Appendix A.

The structure of the invariant density reflects the dynamical structure of the attractor. For instance, let us take $r = 3.8283$. We have already seen in Figure 2.23, obtained with the same value of r, the quasiperiodic intervals of period 3. In Figure 2.28a, three distinct peaks appear around the x_t values 0.163, 0.518 and 0.963, corresponding to

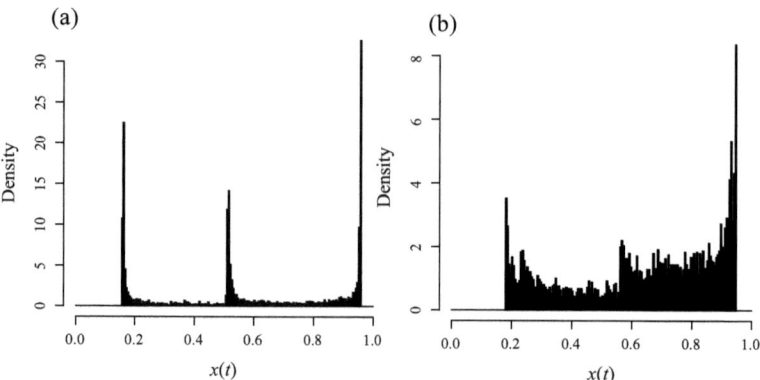

Fig. 2.28 Invariant density of the logistic map with $r = 3.8283$ (a) and $r = 3.8$ (b): histograms of the positions x_t obtained by 5000 iterations.

points that are not yet periodic, but are very near to periodicity.

With $r = 3.8$, the invariant density is more jagged, as shown in Figure 2.28b. The iterates are limited to the interval $[0.1805, 0.95] \subset [0, 1]$. The value 0.95 corresponds to the maximum of the map, $f(0.5) = 0.95$, and the value 0.1805 is given by the first iteration of 0.95, $f(0.95) = 0.1805$. With $r = 4$, the invariant density shows only two singularities, at 0 and 1, while with $r = 3.8$, a large number of singularities is visible. It can be demonstrated that as the length of the series approaches infinity, the invariant density $\rho(x)$ exhibits an enumerable set of singularities, which are dense in $[0.1805, 0.95]$.

It should be emphasized that, first, the shape of the invariant density is different for different maps. It is also different for the same map in different dynamical regimes, as we have seen for the logistic map with $r = 4$, 3.8283 and 3.8. Second, to derive the invariant density, numerically or analytically, we have to follow a proper statistical methodology, in the sense that we have to consider *samples of iterates* in finite regions of phase space, in our case the interval $[0, 1]$, and not the particular points reached by the single iterate. We can ask ourselves whether it might be more appropriate to turn our attention to the evolution of the invariant density, rather than the evolution of single trajectories. We now present some numerical experiments. We will not compute the histogram of the single iterate values; rather, we will follow an *ensemble* of M trajectories, say 1000, and memorize the values of these M iterates at successive time steps. Starting from M different initial points, we will obtain histograms varying in time; that is we will be able to visualize how the density of the iterates evolves from the initial ensemble up to the final ensemble that is reached when the density no longer changes and has thus become invariant. More intuitively, we imagine 'taking a picture' of all M trajectories at determinate instants in time and computing at each of these instants the density of the M points.

Figure 2.29 shows the evolution of an ensemble of $M = 100$ trajectories of the logistic map, with $r = 4$. The 100 trajectories for 15 iterations are shown in (a), while (b) shows the 'pictures' taken at $t = 0$, 3, 5, 10 and 15, that is, the values assumed by

(a)

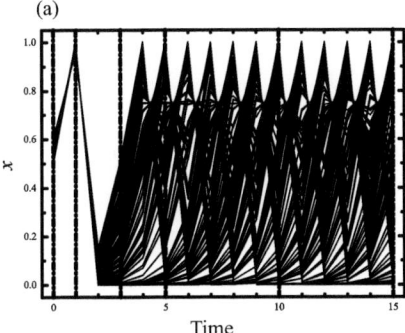

Time

(b)

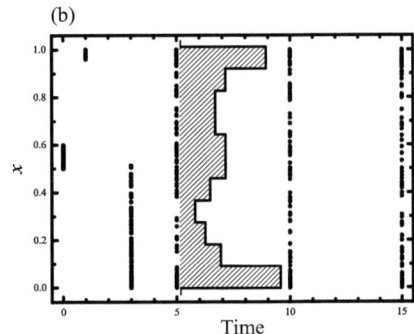

Time

Fig. 2.29 Logistic map with $r = 4$: (a) 100 trajectories with initial points uniformly distributed in the interval $[0.5, 0.6]$; (b) density of the 100 trajectories at $t = 0, 3, 5, 10$ and 15 represented as points in the interval $[0, 1]$. For $t = 5$, the histogram is also shown.

the iterations at these instants. The starting points are uniformly distributed in the interval $[0.5, 0.6]$; thus, at $t = 0$, 100 points are within this interval.

After one time step, we can see the 100 points gathered near $x(1) = 1$. After more time steps, the points move towards a nonuniform distribution in $[0, 1]$. Such distributions of points can be shown as histograms, as in Figure 2.29b for $t = 5$.

The essential result is that the estimated invariant density is the same whether it is derived from the iterates of a single trajectory followed for a finite time interval or from the iterates of an ensemble of trajectories computed at a precise instant. This expresses the property of ergodicity, as we will see more clearly in Appendix A. For now we present an example, with further examples being left as exercises for the reader.

Figure 2.30 shows the evolution of the invariant density of the logistic map with $r = 4$. The starting ensemble ($t = 0$) shows the initial iterates uniformly distributed in $[0, 1]$; at the first step ($t = 1$), the iterates in the central portion of $[0, 1]$ are mapped close to $x(1) = 1$; at the second step ($t = 2$), many points close to $x(1) = 1$ are mapped around $x(2) = 0$, while the points around $x(1) = 0.5$ are mapped again around $x(2) = 1$. The two modalities tend to balance, and we can thus see the birth of the invariant density described by the density function of the random variable $Beta(0.5, 0.5)$, whose shape is already visible at the second step.

Figure 2.30 was obtained with Code 2.8 **Logistic map: Density, M trajectories**.

```
#Code 2.8 Logistic map: density, M trajectories, initial distribution: uniform in (0,1)

f.x<- function(x,r){
     r*x*(1-x)
                 }
r<- 4
M<- 1000                 # number of trajectories
nstep<- 25               # number of iterations
xt<- numeric()
xiniz0<- numeric()
```

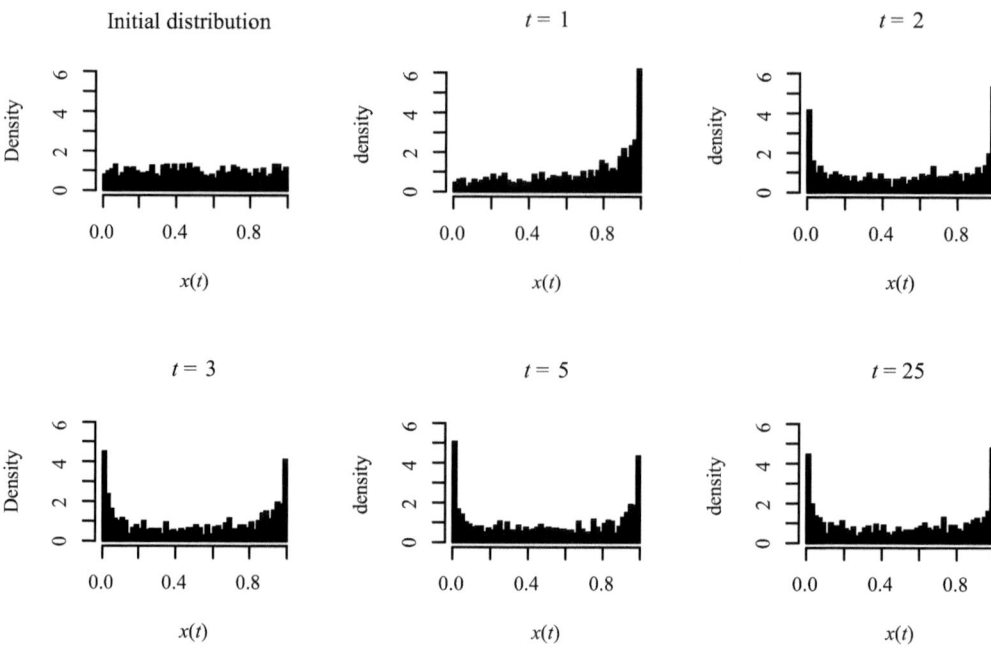

Fig. 2.30 Evolution of the density of an ensemble of 1000 trajectories of the logistic map with $r = 4$. The initial density is uniformly distributed in $[0, 1]$.

```
xens<-matrix(,M,nstep)    # to memorize the single trajectory
set.seed(1)               # seed of the sequence of (pseudo) random numbers
for(l in 1:M){            # starting loop on the trajectories
# it is possible to change the initial distribution
xiniz0[l]<- runif(1)      #  uniform distribution in (0,1)
x<- xiniz0[l]
xt[1]<- x
    for(i in 1:nstep){    # starting loop on the iterations
    y<- f.x(x,r)
    x<- y
    xt[i]<- x
    xens[l,i]<- xt[i]
               }          # ending loop on the iterations
           }              # ending loop on the trajectories
mstep1<-1
mstep2<-2
mstep3<-3
mstep4<-5
mstep5<-nstep
lbin<- 0.02
windows()
par(mfrow=c(3,3),cex.main=0.8)
hist(xiniz0,probability=T,xlab="x(t)",ylab="density",main="initial distr.",
xlim=c(0,1),ylim=c(0,6),br=seq(0,1,by=lbin),col="black",border="black")
hist(xens[,mstep1],probability=T,xlab="x(t)",ylab="density",main="t = 1",
xlim=c(0,1),ylim=c(0,6),br=seq(0,1,by=lbin),col="black",border="black")
hist(xens[,mstep2],probability=T,xlab="x(t)",ylab="density",main="t = 2",
```

```
xlim=c(0,1),ylim=c(0,6),br=seq(0,1,by=lbin),col="black",border="black")
hist(xens[,mstep3],probability=T,xlab="x(t)",ylab="density",main="t = 3",
xlim=c(0,1),ylim=c(0,6),br=seq(0,1,by=lbin),col="black",border="black")
hist(xens[,mstep4],probability=T,xlab="x(t)",ylab="density",main="t = 5",
xlim=c(0,1),ylim=c(0,6),br=seq(0,1,by=lbin),col="black",border="black")
hist(xens[,mstep5],probability=T,xlab="x(t)",ylab="density",main="t = 25",
xlim=c(0,1),ylim=c(0,6),br=seq(0,1,by=lbin),col="black",border="black")
```

The $M = 1000$ initial points are uniformly distributed, with the instruction `xiniz0[l] <- runif(1)`, where the index `l` identifies the lth trajectory. Notice the two nested loops. The inner loop on the iterates computes the `xens[l,i]` values, that is, the values of each iterate at successive time steps i, for each trajectory l. At the end of the outer loop (over l), the histograms of the estimated invariant density are plotted at $t = 0, 1, 2, 3, 5, 25$.

The reader can change the value of r and both the starting interval and the starting distribution; moreover, the invariant densities of other maps can also be studied. Here we summarize the main results on the properties of the invariant density:

- The invariant density describes the stationary behaviour of an ensemble of trajectories with different initial conditions. It can be derived both from the temporal evolution of a single trajectory and from an ensemble of trajectories considered at fixed instants in time.
- The invariant density does not depend on the initial density.
- The invariant density is different for different maps.
- The speed of convergence to the invariant density depends on the map.
- The speed of convergence to the invariant density depends on the initial conditions.
- For the same map, both the invariant density and the speed of convergence are different for different dynamical regimes.

The most important aspect of what we have discussed can be summarized as follows. In a chaotic system, we cannot make predictions about a single trajectory if we address it from an exclusively deterministic point of view. However, if we apply a statistical approach, we can then distinguish regions that are more or less probable, so we should observe an ensemble of trajectories with different initial conditions, even though we know only one trajectory.

2.9 Summary

In this chapter, we have seen how highly erratic dynamic behaviour can arise from a nonlinear logistic map and how this apparently random behaviour is governed by a surprising order. With this lesson in mind, we should not be overly surprised that highly erratic and random-appearing observed data might also be generated by parsimonious deterministic dynamic systems. As a minimum, we contend that researchers should apply NLTS to test for this possibility.

We have also introduced tools to analyse the dynamic behaviour that forms the foundation for NLTS. In particular, we have stressed the quite unexpected capability of achieving some form of predictability even with only one trajectory at hand. In subsequent chapters, we shall treat known nonlinear dynamical systems as unknown, and investigate how NLTS methods rely on a single solution (or multiple solutions) generated by them to reconstruct equivalent systems. This is a conventional approach in the literature for seeing how NLTS methods work, since we know what needs to be reconstructed.

3
Phase Space Reconstruction

'"My good Adso," my master said, "during our whole journey I have been teaching you to recognize the evidence through which the world speaks to us like a great book. Alanus the Insulis said that omnis mundi creatura, quasi liber et pictura, nobis est in speculum."'

Umberto Eco, The Name of the Rose

3.1   Introduction

We introduce some basic concepts concerning the study of both discrete and continuous dynamical systems. The first concept is that of *phase space* or 'state space'. This is an abstract mathematical construction with important applications in statistical mechanics, where it allows the time evolution of a dynamical system to be represented in geometric form. Phase space has as many dimensions as the number of variables needed to define the instantaneous state of the system. For instance, the state of a material point moving on a straight line is defined by its position and velocity at each instant, so that the phase space for this system is a plane in which one axis is the position and the other the velocity. In this case, the phase space is also called the 'phase plane'. For a planet orbiting around the Sun, or a satellite around the Earth, the phase space has six coordinates: (x, y, z) defining the instantaneous position and (v_x, v_y, v_z) defining the components of the instantaneous velocity vector $\mathbf{v}$. More generally, for a system consisting of N particles, the phase space is a $6N$-dimensional space, whose axes are $3N$ generalized coordinates and $3N$ generalized momenta, so we may have linear coordinates, or angles, angular velocities and other variables. In an ecological model, the axes of the phase space may be, for instance, the number of individuals of a certain species and their rate of growth.

In short, every single point of phase space represents the entire system. Corresponding to the motion of the system in real space, there is a trajectory in phase space traced out by the point representing the instantaneous state of the system (the *representative point*). For a dynamical system, the trajectory may be a closed curve, as we will see, but intersections are not possible. If that were the case, then the state represented by the intersection point would have more than one successive point, and hence the temporal evolution of the system would not be unique. When we speak about stable or unstable fixed points, convergent or divergent trajectories, basin of attraction, fractal dimension, attractors (strange or not), etc., we always refer to this abstract space.

Nonlinear Time Series Analysis with R. Ray Huffaker, Marco Bittelli and Rodolfo Rosa, Oxford University Press (2017).
© Ray Huffaker, Marco Bittelli and Rodolfo Rosa. DOI: 10.1093/oso/9780198782933.001.0001

3.2   Ideal Simple Pendulum

We choose, as a guide, an ideal simple pendulum with small-amplitude oscillations (for a detailed study of pendulums in different regimes, see Baker and Gollub (1996)). We use this simple system to introduce some R codes dealing with the study of various properties of dynamical systems. The simple pendulum has to be considered as a point mass (the bob) suspended from a massless wire or string. There is no friction at the suspension point and no air resistance and it is also assumed that the swings are small (a few degrees). The motion of the pendulum can be computed if the angle $\theta(t)$ that the pendulum makes with the vertical and the angular velocity $d\theta(t)/dt$ are known (see Figure 3.1).

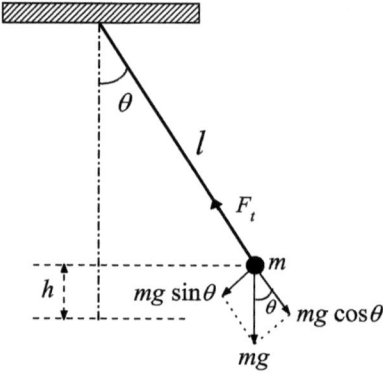

Fig. 3.1 Schematic diagram of the pendulum, where l is the length of the wire, g the acceleration due to gravity, θ the angle with respect to the vertical, m the mass of the bob and F_t the tension force pulling on the bob.

In the following and also in the R codes, we will denote $\theta(t)$ and $d\theta(t)/dt$ also by $x(t)$ and $y(t)$:

$$\theta(t) \quad \Leftrightarrow \quad x(t)$$

$$\frac{d\theta(t)}{dt} \quad \Leftrightarrow \quad y(t)$$

From Newton's second law, the system of equations describing the motion can be derived:

$$\begin{cases} \dfrac{dx}{dt} = y \\[2mm] \dfrac{dy}{dt} = -\omega^2 x \end{cases} \tag{3.1}$$

which are equivalent to

$$\frac{d^2 x}{dt^2} + \omega^2 x = 0 \tag{3.2}$$

The quantity ω is the angular frequency (or *pulsatance*), and $\omega = 2\pi/T$, where T is the period, the time for one complete oscillation. As a young man, Galileo Galilei, seventeen years old in 1581 (or thereabouts), observing the oscillations of a votive lamp hanging from the ceiling of the Cathedral of Pisa, realized that the oscillation period of a pendulum is independent of its amplitude. This property is called *isochronism* (remember that we are speaking of an ideal pendulum in the small-angle approximation). Recall that the period T increases with the square root of the length of the wire, $T = 2\pi\sqrt{l/g}$, and do not confuse the *angular frequency* $\omega^2 = g/l$, measured in radians per unit time, with the ordinary frequency ν, which is the number of complete oscillations per unit time:

$$\nu = \frac{1}{T} = \frac{\omega}{2\pi}$$

If T is measured in seconds, then ν is measured in s^{-1}, that is in *hertz* (Hz). If $\nu = 1$ Hz, this means that there is one complete oscillation per second and $\omega = 2\pi/1$ rad/s. The solution of eqn (3.1) can be obtained analytically under the assumptions that we have made; nevertheless, we present here the R Code 3.1, **pend_id_s_o.R**, to show how differential equations can be solved numerically with R.

```
#Code 3.1 Ideal pendulum: small oscillations (pend_id_s_o.R)
# parms: 1, omega2  parameters
omega2<- 3
parms<- c(1,omega2)
tinit<-0
tfin<-10    # tfin<-1 if one wishes to write xy, otherwise 10 or 20
step<-0.01 # 0.01 is OK, but try with 0.5 and 1 to see that they are bad values.
times<-seq(tinit,tfin,by=step)
funct<- function(t,integ,p){                   # the system of equations
x<-integ[1]
y<-integ[2]
dx<-  parms[1]*y    # that is: dx/dt = 1 y
dy<- -parms[2]*x    # that is: dy/dt = -omega2 x
list(c(dx,dy))
                       }    # end of funct
require(deSolve)            # ODE solver package
cinit<-c(1,1)
xy<-rk4(cinit,times,funct,parms)  # also xy<-lsoda(cinit,times,func,parms)
#print (xy)    # comment if you do not wish the xy values printed
#myData <- data.frame(xy[,2],xy[,3])
#write.csv(myData, "MyData.csv")       #to save the matrix xy as a comma-separated file

plot(xy[,2],xy[,3],type="l",xlim=c(-3,3),ylim=c(-3,3),xlab="x(t)",ylab="y(t)",
cex.lab=1.7,cex.axis=1.3,lwd=3,lty=1)   # phase space portrait

plot(xy[,1],xy[,2],type="l",xlab="t",ylab="x(t)",cex.lab=1.7,
cex.axis=1.3,lwd=4,lty=1,ylim=c(-2,2))  # x(t) vs. t
segments(0.,0,10.,0,lty=2,lwd=3)
# to add to the time plot the period T and the the amplitude oscillation A
# better: col="black" -> col="red"
segments(0.944,0.50,4.58,0.50,lty=4,lwd=3,col="black")
segments(7.57,0,7.57,1.15,lty=4,lwd=3,col="black")
text(2.76,0.66, labels ="T",cex=2,col="black" )
text(7.15,0.25, labels ="A",cex=2,col="black" )
```

The package deSolve (Soetaert *et al.* (2016)) is a family of codes that solve initial-value problems for systems of first-order ordinary differential equations (ODEs), of partial differential equations, of differential algebraic equations and of delay differential equations. If necessary deSolve can be installed with install.packages("deSolve", dep = TRUE). In pend_id_s_o.R, to solve the system of first-order ODEs, Runge–Kutta fourth-order integration is used with the command xy<-rk4(cinit,times, funct,parms) The function **rk4** has many optional parameters, but here only the essential ones are described. The vector cinit contains the initial conditions for the ODE system: in our case, the pendulum starts at $x = 1, y = 1$. The quantities tinit and tfin represent the extremes of the time interval for the system evolution; they are physical times, measured, for instance, in seconds. The vector times contains the points at which the integration of the system (3.1) is computed. The integration step, step, has to be chosen with a bit of good sense, as we will see. The parameters parms in our case are 1 and $\omega^2 = 3$. The function **funct** represents the system, t is the current time point in the integration, integ represents the current computed variables and p is the parameter vector. The result is the matrix xy. In the code, there is the possibility of saving the matrix as a comma-separated file, with extension csv. The first column xy[,1] contains the components of the vector times: $0.00, 0.01, 0.02, \ldots, 9.99, 10$. The second column xy[,2] and third column xy[,3] contain the computed values of $x(t)$ and $y(t)$, respectively. The first 11 and last 3 rows of the output matrix xy are as follows:

```
       time          1          2
1      0.00   1.0000000  1.0000000
2      0.01   1.0098495  0.9698515
3      0.02   1.0193961  0.9394121
4      0.03   1.0286368  0.9086908
5      0.04   1.0375690  0.8776969
6      0.05   1.0461899  0.8464398
7      0.06   1.0544969  0.8149287
8      0.07   1.0624876  0.7831731
9      0.08   1.0701596  0.7511826
10     0.09   1.0775105  0.7189668
11     0.10   1.0845382  0.6865352
.............................
999    9.98  -0.5702282  1.7391144
1000   9.99  -0.5527525  1.7559596
1001  10.00  -0.5351109  1.7722779
```

Since the integration step (step) is 0.01, the time xy[,1] is equal to 10 after 1001 integration steps.

To obtain the time plot, i.e. $x(t)$ versus t, xy[,1] and xy[,2] are used, while for the phase space portrait, i.e. $x(t)$ versus $y(t)$, xy[,2] and xy[,3] are used, as shown in Figure 3.2. Other numerical integrators can be used in place of **rk**. The function **lsoda** is better, since it works for both stiff and non-stiff systems of ODEs. The stiffness is related to instability of solutions of the differential system itself.

From Figure 3.2a, it can be seen that $x(t)$ is a sinusoid with period $T = 2\pi/\omega$:

$$x(t) = c\cos(\omega t + \alpha)$$

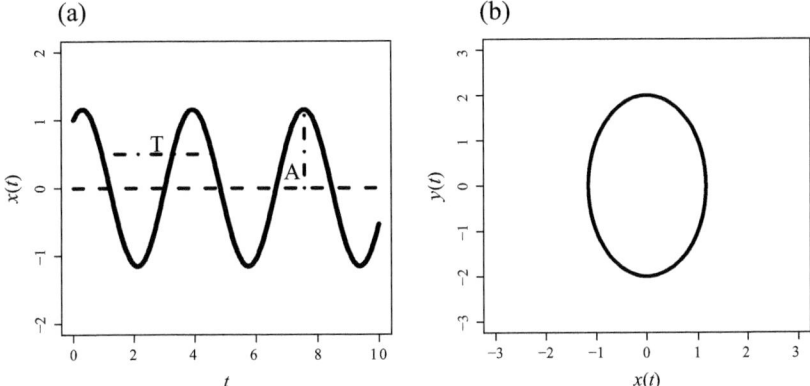

Fig. 3.2 Time plot (a) and phase space portrait (b) of the time evolution of an ideal pendulum with small oscillations, eqn (3.1). In the time plot, the period T and the amplitude of oscillation A (dash–dot lines) are indicated.

where c and α are constants that can be determined from the initial conditions. α is the value of the angle θ at time $t = 0$. It is possible to choose the origin of the axes such that $\alpha = 0$, and since $y(t) = dx(t)/dt$, the solution is then

$$\begin{cases} x(t) = & c\cos(\omega t) \\[2mm] \dfrac{y(t)}{\omega} = & -c\sin(\omega t) \end{cases} \tag{3.3}$$

The constant c represents the amplitude of oscillation, i.e. the maximum value periodically taken by the angle θ (A in Figure 3.2a).

Figure 3.2b represents the time evolution of the system in phase space. Clearly, the trajectory of the representative point in phase space is by no means the same as the trajectory described by the bob in real space.

Figure 3.3 shows just the beginning of the trajectory in phase space with integration step $\Delta t = 0.1$, where Δt is `step` in the code `pend_id_s_o.R`. (In Figure 3.2, $\Delta t = 0.01$ and the points are joined by a line.) At $t = t_0 = 0$, the starting point is $(1, 1)$; after the integration step 0.1 (i.e. at $t = t_0 + 0.1$), the representative point is at $(1.085, 0.687)$; at $t = t_0 + 0.2$, the representative point is at $(1.137, 0.353)$; and so on. In this way, the trajectory in phase space is constructed. Thus, at any step, we can know the exact state, angle and angular velocity of the system. Why does such a trajectory turn out to be an ellipse? The answer is easy: we can write eqn (3.3) as

$$x^2(t) + \frac{y^2(t)}{\omega^2} = c^2\cos^2(\omega t) + c^2\sin^2(\omega t) = \text{constant}$$

which is the equation of an ellipse.

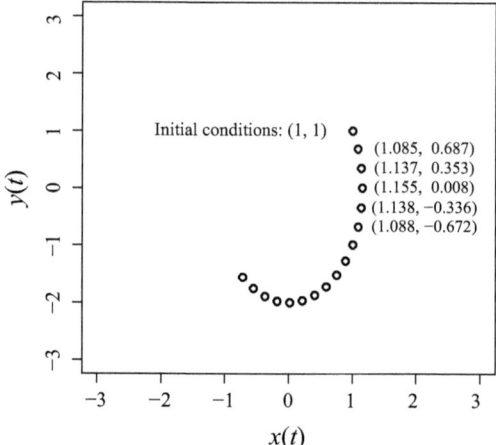

Fig. 3.3 First steps of the trajectory in phase space of the system (3.1).

To end this section, we show what can happen with a bad choice of integration step in the integration of eqn (3.1). Recall that the plots in Figure 3.2 were obtained with an integration step $\Delta t = 0.01$ (`step` in the code `pend_id_s_o.R`). Figure 3.4 shows the results obtained with integration steps of 0.5 (continuous line) and 1 (dashed line). Therefore, in practice, it is best to try different values of the integration step.

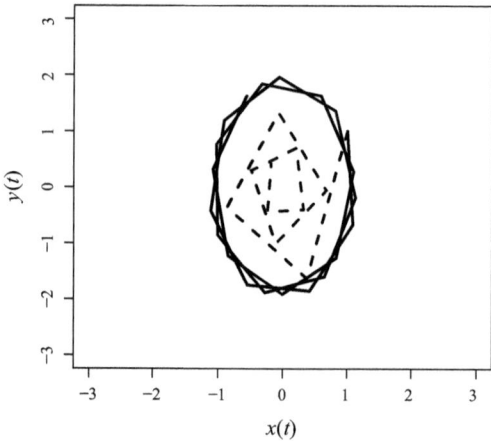

Fig. 3.4 Integration of the system (3.1) with integration steps of 0.5 (continuous line) and 1 (dashed line).

We will continue with pendulums by presenting an important tool for studying both chaotic and non-chaotic time series.

3.3 Embedding Procedure

Suppose that we have measured the angle θ every second and also that we are not acquainted with eqn (3.1). Can we obtain the information necessary to plot the phase space portrait of Figure 3.2? The answer is affirmative and the procedure is called *embedding*. Embedding a time series means reconstructing (approximately) the phase space in which the system evolves in time; in other words, showing how 'all' the variables describing the system evolve in time. By 'all' here we mean more precisely 'the relevant ones'. Notice that even if we have measured only one variable, i.e. we have a *scalar* series, we can construct a *vector* space whose axes represent all the (relevant) variables. The underlying idea is to make copies of the measured signal with fixed time delays and to consider these delayed values as coordinates of a phase space, retrieved from the data.

We have a time series with the following measured data:

$$s_1, s_2, \ldots, s_N$$

In the m-dimensional reconstructed phase space, a point at time t is given by a vector $\mathbf{y}_t$ with the following components:

first component	$s_t \equiv s(t)$
second component	$s_{t+d} \equiv s(t+d)$
third component	$s_{t+2d} \equiv s(t+2d)$
$\vdots$	$\vdots$
mth component	$s_{t+(m-1)d} \equiv s(t+(m-1)d)$

If $m = 2$, then $\mathbf{y}_t = (s_t, s_{t+d})$; if $m = 3$, then $\mathbf{y}_t = (s_t, s_{t+d}, s_{t+2d})$.

The vectors of the reconstructed phase space are also called *Takens vectors*.

Let us give a simple schematic example. Suppose that we have performed $N = 10$ measurements of a certain variable $x(t)$ and let $s(t)$ be the results:

measurements of $x(t) = s(t)$:   3, 5, 4, 2, 1, 1, 4, 6, 7, 2

The measurements are numbered as $1, 2, \ldots, 10$. Suppose that the measurements are taken every time interval, $\Delta = 30$ seconds. Then the first measure, taken at time $t = 1$, is $s_1 = 3$. After 30 seconds, at $t_2 = t + \Delta$, the second measure is $s_2 = 5$. The third measure is taken after 60 seconds at $t_3 = t + 2\Delta$ and is $s_3 = 4$. After 90 seconds, at $t_4 = t + 3\Delta$, the fourth measure is $s_4 = 2$, and so on.

For the embedding, we choose (following certain criteria that we shall describe shortly) $m = 3$ and $d = 1$. The variable m is called the *embedding dimension* and d

is called the *time delay* . To obtain the 'delayed' measurements $s(t+d)$, we move the $s(t)$ measurement to the left by an amount $d=1$ and to obtain the 'delayed' $s(t+2d)$, we move the previous measurement $s(t+d)$ to the left by a further amount $d=1$. In our example, we have the following:

number of measurements:	1	2	3	4	5	6	7	8	9	10
$s(t)$:	3	5	4	2	1	1	4	6	7	2
$s(t+d)$:	5	4	2	1	1	4	6	7	2	
$s(t+2d)$:	4	2	1	1	4	6	7	2		

We can see that at $t=1$, the point in the reconstructed phase space is defined by the vector $\mathbf{y}_1 = (3,5,4)$; at $t=2$, the point is $\mathbf{y}_2 = (5,4,2)$; up to $\mathbf{y}_8 = (6,7,2)$. Note that d is not a time, it is a number. In this case, it is 1. The eight points in the three-dimensional space are as follows:

$t=1$: $\mathbf{y}_1 = (3,5,4)$

$t=2$: $\mathbf{y}_2 = (5,4,2)$

$t=3$: $\mathbf{y}_3 = (4,2,1)$

$t=4$: $\mathbf{y}_4 = (2,1,1)$

$t=5$: $\mathbf{y}_5 = (1,1,4)$

$t=6$: $\mathbf{y}_6 = (1,4,6)$

$t=7$: $\mathbf{y}_7 = (4,6,7)$

$t=8$: $\mathbf{y}_8 = (6,7,2)$

Figure 3.5 shows the trajectory in the reconstructed phase space obtained from the 10 measurements.

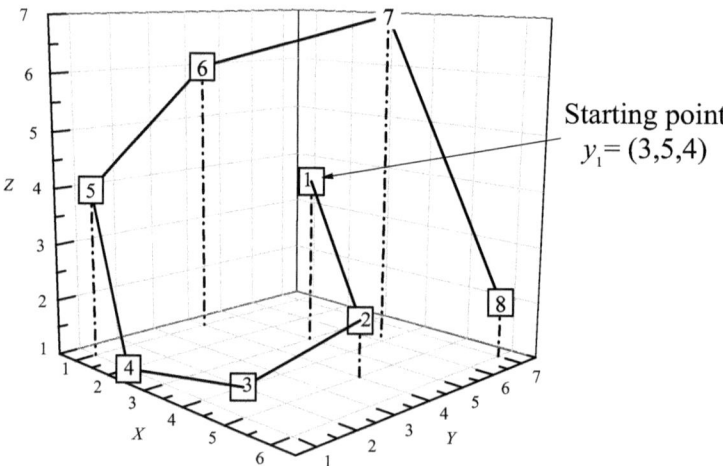

Fig. 3.5 Trajectory in the reconstructed phase space.

We can then write the trajectory joining the points $\mathbf{y}_1, \ldots, \mathbf{y}_8$ as

$$\mathbf{y}_1 \to \mathbf{y}_2 \to \mathbf{y}_3 \to \mathbf{y}_4 \to \mathbf{y}_5 \to \mathbf{y}_6 \to \mathbf{y}_7 \to \mathbf{y}_8$$

where

$$\mathbf{y}_t = [s_t, s_{t+d}, s_{t+2d}], \quad t = 1, \ldots, 8$$

In general, if N is the length of the series, the number n of embedding vectors obtained is

$$n = N - (m - 1)d$$

With the same measurements, but with $d = 2$, we have the following:

number of measurements:	1	2	3	4	5	6	7	8	9	10
$s(t)$:	3	5	4	2	1	1	4	6	7	2
$s(t+d)$:	4	2	1	1	4	6	7	2		
$s(t+2d)$:	1	1	4	6	7	2				

Now we have $n = 10 - (3 - 1) \times 2 = 6$. The six points in the three-dimensional space are as follows:

$t = 1$: $\mathbf{y}_1 = (3, 4, 1)$

$t = 2$: $\mathbf{y}_2 = (5, 2, 1)$

$t = 3$: $\mathbf{y}_3 = (4, 1, 4)$

$t = 4$: $\mathbf{y}_4 = (2, 1, 6)$

$t = 5$: $\mathbf{y}_5 = (1, 4, 7)$

$t = 6$: $\mathbf{y}_6 = (1, 6, 2)$

Embedding theorems (Takens, 1981; Mañé, 1981) ensure that, under certain circumstances and to a certain extent, the geometric object defined by the vectors $\mathbf{y}_t$ is 'equivalent' to the original trajectory defined by the variables $x(t), y(t), z(t)$. That is to say, the attractor defined by the vectors $\mathbf{y}_t$ is 'equivalent' to the original attractor living in an unknown phase space. 'Equivalent' here means that the dynamics is the same in the two spaces: true and reconstructed. In other words, certain properties of the original system, the so-called dynamical invariants, are preserved in the reconstructed space. For instance, the measure of trajectory divergence (the maximum Lyapunov exponent) must have approximately the same value in both the true and the reconstructed phase space. The crucial issue is that the reconstructed phase space of the system is constructed from only a single measured variable of the system. For a mathematical account of the embedding technique, see the paper by Sauer *et al.* (1991) on 'embedology'. The article by Packard *et al.* (1980) is widely quoted as the first containing the idea of reconstructing a phase space through delayed measurements. However, these authors themselves recognize that it was Ruelle who suggested, in a private communication, to consider delayed values of a single coordinate of a time series as coordinates of a faithful phase space.

Let us return to the ideal pendulum of eqn (3.1) and suppose that we have measured x (i.e. the angle θ). The R Code 3.2 embed_p_i.R is used to embed these measurements and reconstruct the phase space of Figure 3.2b.

```
#Code 3.2 Embedding pendulum: ideal small oscillations (embed_p_i.R)
#parms: 1, omega2       parameters
omega2<- 3
parms<- c(1,omega2)
tinit<-0
tfin<-10     # tfin<-1 if one wishes to write xy, otherwise 10 or 20
step<-0.01   # 0.01 is OK, but try also with  0.5 and 1 to see that they are bad values.
times<-seq(tinit,tfin,by=step)
funct<- function(t,integ,p){                      # the system of equations
x<-integ[1]
y<-integ[2]
dx<-  parms[1]*y     # that is: dx/dt = 1 y
dy<- -parms[2]*x     # that is: dy/dt = -omega2 x
list(c(dx,dy))
                            }   # end of funct
require(deSolve)
cinit<-c(1,1)
xy<-rk4(cinit,times,funct,parms)  # also xy<-lsoda(cinit,times,funct,parms)
#xy    # comment if you do not wish the xy values printed
plot(xy[,2],xy[,3],type="l",xlim=c(-3,3),ylim=c(-3,3),xlab="x(t)",ylab="y(t)",
cex.lab=1.7,cex.axis=1.3,lwd=3,lty=1)    # phase space portrait

# up to here it is the same as pend_id_s_o

# embedding
obss<-xy[,2]    # the 'observed' series is the second column of xy, that is x(t)
# m is the dimension of the reconstructed phase space (embedding dimension)
# d is the time delay
embedding<- function(x,m,d){
n<-length(x)
ne<- n-(m-1)*d  # number of vectors in the reconstructed phase space
y <- matrix(0,ne,m)  # ne rows, m column
for(i in 1:m) {y[,i] <- x[((i-1)*d+1):(ne+(i-1)*d)]}
y
    }
y<-embedding(obss,2,50) # if d is too small we have a bad reconstruction.
#To see it try d = 2
#y    # comment if you do not wish the y values printed
windows()
plot(y,type="l",xlim=c(-3,3),ylim=c(-3,3),xlab="x(t)",ylab="y(t)",
cex.lab=1.7,cex.axis=1.3,lwd=3,lty=1)
```

The first part of this code is the code pend_id_s_o.R presented earlier and repeated here for convenience. The variable $x(t)$ is then stored in the array obss, with the instruction obss<-xy[,2]. We have to consider the code as starting here: we do not know eqn (3.1), nor, of course, do we know how to integrate this ODE system. We have only an 'observed' time series – the quotation marks here mean that this series has to be considered as an experimental one in every sense, even though it has been created numerically. Therefore, with the function **embedding**, we embed the series obss, which is 1001 data long. This function has three arguments (x,m,d), the first of which is the series, the second the embedding dimension and the third the time delay. The result is the matrix y, with $ne = 951$ rows and 3 columns, the first of the latter, `y[,1]`, being the time. The first six and last three rows of the output matrix y are as follows:

```
       time          1           2
1      0.00   1.0000000    1.087662
2      0.01   1.0098495    1.080783
3      0.02   1.0193960    1.073581
4      0.03   1.0286368    1.066056
5      0.04   1.0375690    1.058212
6      0.05   1.0461899    1.050050
.....................................
949    9.48  -1.1342943   -0.570228
950    9.49  -1.1303809   -0.552752
951    9.50  -1.1261285   -0.535111
```

The reconstructed series is plotted in the reconstructed phase space in Figure 3.6a. The reconstructed trajectory, in general, is not identical to the original one, but appears as a more or less distorted reproduction. In this case, the reconstructed trajectory is still an ellipse, but with its long axis on the diagonal, rather than parallel to the y axis. This ellipse is 'equivalent' to the the original ellipse in Figure 3.2b; i.e. it is a *topologically faithful reconstruction* of the system's original dynamics. Note that the time delay d has to be chosen with care: Figure 3.6b shows what happens with an inappropriate choice of d.

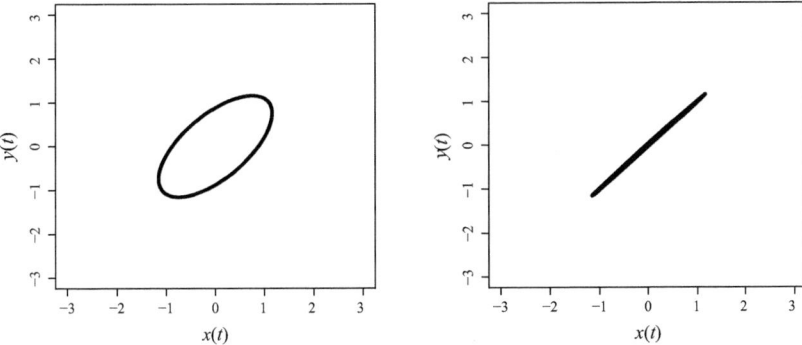

Fig. 3.6 Plot in the reconstructed phase space of the time evolution of the ideal pendulum (eqn 3.1). The 'true' phase space portrait is shown in Figure 3.2b. The embedding dimension is $m = 2$, and the time delay is $d = 50$ (a) and $d = 2$ (b).

3.4 Phase Space Reconstruction with R packages

The R package `tseriesChaos` offers a family of codes for the analysis of nonlinear time series. The author and maintainer is Di Narzo (2015). The work – as the author himself writes – is largely inspired by the TISEAN project, by Hegger *et al.* (1999), which is publicly available.

Our aim here is to teach the use of `tseriesChaos` and especially to highlight the underlying principles. The examples presented here is what is probably the most famous strange attractor, the *Lorenz attractor*. In his seminal paper *Deterministic nonperiodic flow*, Lorenz (1963) found that in forced dissipative hydrodynamic flow, nonperiodic solutions were unstable and highly dependent on initial conditions. He derived a set of equations that, in phase space, lead to the famous Lorenz strange attractor (Figure 2 of his paper), which described the evolution of the convective process. The two portions of the trajectories in the attractor represented the states of steady convection.

Lorenz derived these equations as a simplified model of atmospheric convection to study the unpredictable behaviour of weather. The *Lorenz equations* (Lorenz, 1963) are often presented as the quintessential example of a nonlinear system:

$$\frac{dx}{dt} = \sigma(y - x)$$

$$\frac{dy}{dt} = x(\rho - z) - y \tag{3.4}$$

$$\frac{dz}{dt} = xy - \beta z$$

The variables x, y and z describe the up and down motion of a fluid in convective flow. Lorenz discovered that the parameter values $\sigma = 10$, $\beta = \frac{8}{3}$ and $\rho = 28$, generate surprisingly complex aperiodic behaviour for a parsimonious deterministic model. With these parameter values, the system displays chaotic behaviour. To solve the system (3.4), we can proceed as we did for the ideal pendulum, namely with Code 3.3 `lorenz_attractor.R`:

```
#Code 3.3   Lorenz attractor (lorenz_attractor.R)
parms<- c(10,8/3,28) # parameters: sigma, beta, rho
tinit<- 0
tfin<- 100
step<- 0.01
times<-seq(tinit,tfin,by=step)
funct<- function(t,integ,parms){  # the system of equations
x<-integ[1]
y<-integ[2]
z<-integ[3]
sigma<- parms[1]
beta<- parms[2]
rho<- parms[3]
dx<- sigma*(y-x) # that is dx/dt = sigma(y-x)
dy<- x*(rho-z)-y # that is dy/dt = x(rho-z)-y
dz<- x*y-beta*z # that is dz/dt = xy -beta z
list(c(dx,dy,dz))
                          } # end of funct
require(deSolve)
cinit<-c(1,1,1)
xyz<-lsoda(cinit,times,funct,parms)
# xyz # comment if you do not wish the xyz values printed
par(mfrow=c(3,1))
par(mar = c(6.3, 4.8, 1., 3))
par(cex.lab=2,cex.axis=1.6,lwd=1,lty=1)
```

```
plot(xyz[,1],xyz[,2],type="l",xlab="t",ylab="x(t)",   # x(t) vs t
xlim=c(tinit,tfin),ylim=c(-30,30))
plot(xyz[,1],xyz[,3],type="l",xlab="t",ylab="y(t)",   # y(t) vs t
xlim=c(tinit,tfin),ylim=c(-30,30))
plot(xyz[,1],xyz[,4],type="l",xlab="t",ylab="z(t)",   # z(t) vs t
xlim=c(tinit,tfin),ylim=c(0 ,50))
windows()
# phase space portrait
require(scatterplot3d)
scatterplot3d(xyz[,2],xyz[,3],xyz[,4],type="l",xlim=c(-30,30),
cex.lab=1.4,cex.axis=1.2)
```

This code follows the same logic as `pend_id_s_o.R` for the ideal pendulum, but now there are three equations in the system. The integration step is 0.01, so the solution of the system of eqn (3.4) is represented by three series $x(t), y(t)$ and $z(t)$, with $N = 10001$ points each. In Code 3.3, the integration is performed with the function **lsoda**. The first six and last three rows of the output matrix **xyz** are as follows:

```
          time         1           2           3
1         0.00    1.0000000    1.000000    1.0000000
2         0.01    1.0125672    1.259918    0.9848912
3         0.02    1.0488232    1.523999    0.9731148
4         0.03    1.1072067    1.798314    0.9651593
5         0.04    1.1868657    2.088545    0.9617377
6         0.05    1.2875548    2.400161    0.9638068
..................................................
9999     99.98    2.2934869   -4.347472   29.7237663
10000    99.99    1.6621680   -4.329890   28.8571724
10001   100.00    1.0936143   -4.293252   28.0394308
```

The package `scatterplot3d` is required to plot $x(t), y(t)$ and $z(t)$ in a three-dimensional plot, that is, to display the phase space portrait of the attractor. If one wishes to try an interactive visualization, the package `rgl` must be employed with the instruction `rgl.linestrips(xyz[,2], xyz[,3], xyz[,4])`. Figure 3.7 shows the Lorenz strange attractor and time plots for each component of the system.

It is worth noting that the x and y time series are similar, but not identical. They oscillate between positive and negative values and the shift is fairly irregular. The trajectories displayed in the attractor never intersect and they are aperiodic, which is a feature of a chaotic system.

As we did for the logistic map in Chapter 2, we will show how to estimate the invariant density of the Lorenz system, by means of Code 3.4 **Lorenz_density**.

```
#Code 3.4 Lorenz system: density, 1 trajectory
#install.packages("tseriesChaos", dep = TRUE) # to install the package if not present
library(tseriesChaos)
lorenz.syst
a<- 10
b<- 28
c<- -8/3
x0<- 1
y0<- 1
z0<- 1
t_init<- 0
t_fin<- 1000
step.int<- .01 # integration step
x<-sim.cont(lorenz.syst,start=t_init,end=t_fin, dt=step.int,
```

(a) (b)

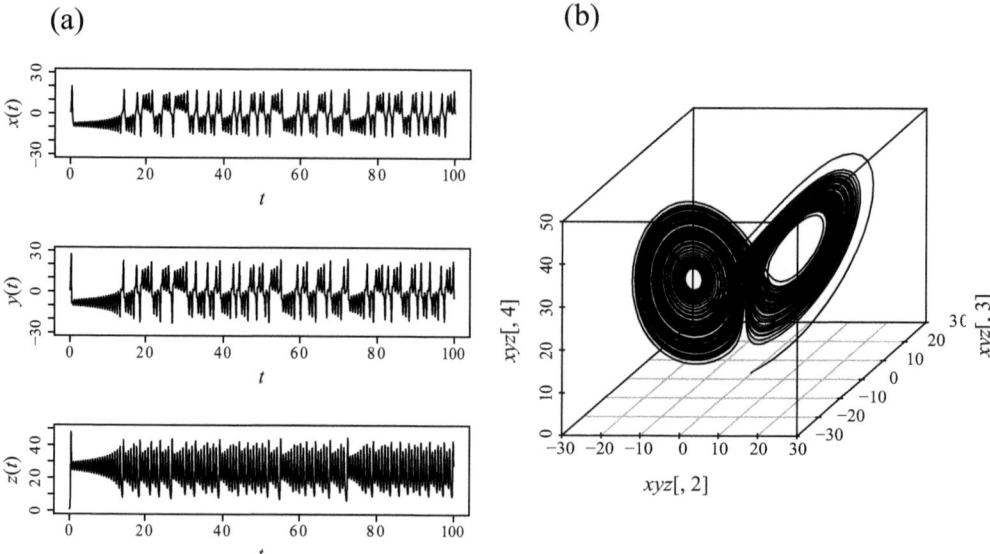

Fig. 3.7 Solutions of the Lorenz equations (a) and phase space portrait of the chaotic attractor (b).

```
start.x=c(x0,y0,z0),parms=c(a,b,c),obs.fun=function(x)x[1])
y<-sim.cont(lorenz.syst,start=t_init, end=t_fin,dt=step.int,
start.x=c(x0,y0,z0),parms=c(a,b,c),obs.fun=function(x)x[2])
z<-sim.cont(lorenz.syst,start=t_init,end=t_fin,dt=step.int,
start.x=c(x0,y0,z0),parms=c(a,b,c),obs.fun=function(x)x[3])

#install.packages("gplots", dep = TRUE) # to install the package if not present
require(gplots)
h2d <- hist2d(x,z,show=FALSE, same.scale=F,nbins=100)
h2d    # to summarize bins contents
h2d$counts<- h2d$counts/max(h2d$counts)   # normalization
par(mai=c(1.02,1.,0.82,0.42)+0.1,cex.axis=1.2,cex.lab=1.6)
filled.contour(h2d$x, h2d$y, h2d$counts,col=gray.colors(10,start=0,end=1), nlevels=10,
xlab="x(t)",ylab="z(t)", main = "",xlim=c(-20,20),ylim=c(0,50),las=0,
key.axes=axis(4,las=1) )
```

The system of equations are already defined in the function **lorenz.syst** and can then be directly integrated with the function **sim.cont**. The package `gplots` (Warnes *et al.*, 2016) contains several codes for plotting data. One of these is the function **hist2d(.)**, which produces a two-dimensional histogram. The number of counts in each bin `h2d$counts` is normalized to take values in [0,1]. Normalization is performed with respect to the maximum value `max(h2d$counts)`. The option `same.scale=F` prevents the use of the same range for the axes. The histogram is not directly plotted (`show=FALSE`), but its contents are passed to the function **filled.contour(.)** to produce the filled contour plot and its legend. The number of grey levels is `nlevel=10`, from `black` for bins containing no elements (`h2d$counts=0`) to `white` for `h2d$counts=1`. The line `key.axes=axis(4,las=1)` produces horizontal labelling in the legend of the

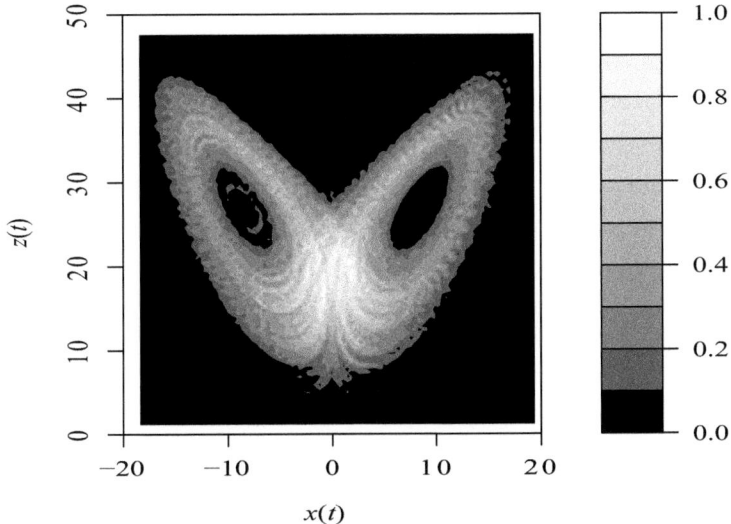

Fig. 3.8 Invariant density of the Lorenz system. Two-dimensional histogram of the positions $(x(t), z(t))$ obtained by 1000 iterations.

grey scale, but not on the y axis. One can change the function **filled.contour** to get colour maps, for instance as follows:

```
filled.contour(h2d$x,h2d$y,h2d$counts,col=topo.colors(10),nlevels=10, ...
filled.contour(h2d$x h2d$y,h2d$counts,col=terrain.colors(10),nlevels=10, ...
filled.contour(h2d$x,h2d$y,h2d$counts,col=rainbow(10,start=0,end=1),nlevels=10, ...
```

The result is shown in Figure 3.8. We have projected the three-dimensional scatter plot of Figure 3.7 onto the plane $(x(t), z(t))$.

As we did for the ideal pendulum, let us suppose that the the first component of the Lorenz system $x(t)$ is the 'observed' series, from which we wish to reconstruct the unknown phase space. The question is how to determine the values of the embedding dimension and of the time delay. We have already seen that with a bad choice of the time delay (Figure 3.6b), the reconstructed phase space portrait does not resemble the original one (Figure 3.2a), and therefore it is important to select the correct values.

3.4.1 Embedding Dimension

There are no mathematical rules for selecting the correct values of the embedding dimension and the time delay to achieve a faithful reconstruction of the 'true' phase space; however, some prescriptions have been proposed. It must be kept in mind that the choice of embedding parameters depends also on the specific problem to which the embedding procedure is applied. In general, this procedure is performed after the transient period has been discarded; in our case, we have considered as transient the time interval $t \lesssim 20$ (see Figure 3.9).

So, for each component $x(t), y(t), z(t)$, we delete the first 1999 integration time steps and start with the 2000th value. The length of the $x(t)$ series (and also of those for $y(t)$ and $z(t)$) is now 8002 points. To obtain the time plot of each component versus t as in Figure 3.7, but without the transient, we need a new initial time corresponding to $x(t = 2000)$. This new time is $t = 19.99$, named `t_start` in the following instructions that we add to Code 3.3 to remove the transient:

```
trans<- 2000                       # integration time step considered as transient
# discard  initial transient:
x <- window(xyz[,2],trans)
y <- window(xyz[,3],trans)
z <- window(xyz[,4],trans)
t_start<- trans*step - step        # new initial time
t_time<-seq(t_start,tfin,by=step)  # new time interval
windows()
par(mfrow=c(3,1))
par(mar = c(6.3, 4.8, 1., 3))
par(cex.lab=2,cex.axis=1.6,lwd=1,lty=1)
# x(t), y(t), z(t) after the transient:
plot(t_time,x,type="l",xlab="t",ylab="x(t)",
xlim=c(t_start,tfin),ylim=c(-30,30))
plot(t_time,y,type="l",xlab="t",ylab="y(t)",
xlim=c(t_start,tfin),ylim=c(-30,30))
plot(t_time,z,type="l",xlab="t",ylab="z(t)",
xlim=c(t_start,tfin),ylim=c(0  ,50))
```

We obtain Figure 3.9, in which the transient has been removed; that is, each component is plotted for `t_start` $= 19.99 \leqslant t \leqslant$ `tfin` $= 100$. Compare this with Figure 3.7a, in which `tiniz<- 0`. From here onwards, the series $x(t)$ is considered without the transient, with length 8002.

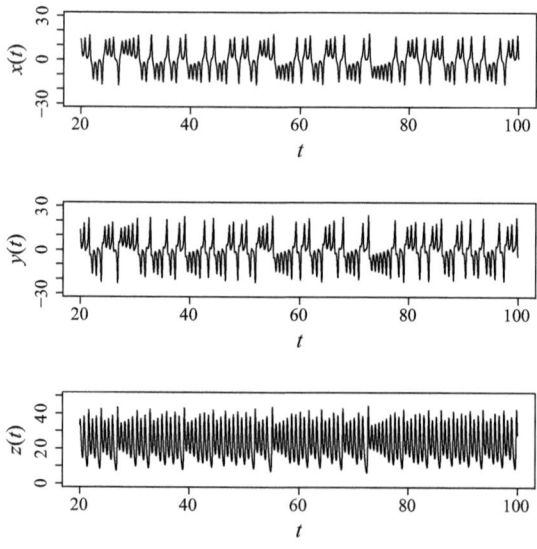

Fig. 3.9 As Figure 3.7a, but without the transient.

We now discuss the procedure to select the embedding parameters. Let us begin with the embedding dimension m.

As we have already underlined, for a dynamical system, self-intersections of trajectories are not possible. Takens (1981) showed that to avoid self-intersections of the reconstructed trajectory, it is *sufficient* that the dimension m be

$$m \geqslant 2d_A + 1 \tag{3.5}$$

where d_A is the attractor dimension (possibly fractal). Note that the condition (3.5) is a sufficient, but not necessary, condition, so m may be less than $2d_A + 1$; indeed, for practical purposes, it is advisable to reduce the value of m. A method that suggests the *minimal sufficient* embedding dimension m, called the *false nearest neighbour (FNN) method*, was proposed by Kennel *et al.* (1992). The concept of this method is shown schematically in Figure 3.10. The point C appears to be a neighbour of the point A when it is projected onto one dimension, that is, onto the line labelled $m-1$, but if the points are observed in two dimensions, only the point B is a real neighbour of A; the point C is a *false* neighbour.

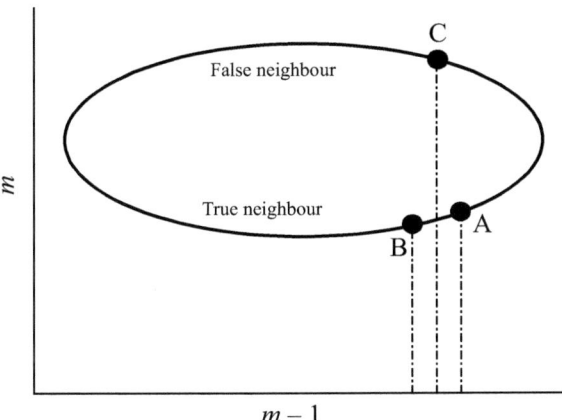

Fig. 3.10 Underlying principle of the false nearest neighbour method. The point C projected onto one dimension, the axis $m-1$, seems to be a neighbour of the point A, whereas in two dimensions, with the axis m included, it can be seen that the point C is a false neighbour of the point A, while the point B is a real neighbour.

We can illustrate false nearest neighbours with a couple of real-world examples. First, let us consider why ticket prices for sporting events, such as soccer, tend to be higher at midfield than behind the goals (Didonato *et al.*, 2014). Spectators at midfield are in a better position to accurately gauge the true distances between players, while spectators at the ends of the field tend to mistakenly perceive players towards the other end as closer to each other than they actually are. A larger percentage of players are false nearest neighbours from these seats, which makes it more difficult to watch attacks on the other goal unfold. As another example, consider cramming (not neatly folding) a jacket into a very small box. From the top of the box, we select two neighbouring points and mark them. We then unpack the jacket and lay it out flat. The two marked neighbours need not remain close together on the unfolded jacket, and may even be on opposite sleeves.

In the FNN method, one starts with a small m, even $m = 1$, and counts the numbers of neighbours of a point along the observed series. If the embedding dimension m is increased from m to $m + 1$, many of the previous neighbours will no longer be neighbours; instead, they will be *false* neighbours. In this way, one finds the estimated minimum embedding dimension m when (almost) all the neighbours are real. The FNN method is implemented in **tseriesChaos** with the code **false.nearest**. We add the following instructions to the code **lorenz_attractor.R**:

```
require(tseriesChaos)
m.max<- 6          # embedding dimensions: from 1 to m_max
d<- 18             # tentative time delay (see below)
tw<- 100           # Theiler window
rt<- 10            # escape factor
eps<- sd(x)/10     # neighbourhood diameter
fn <- false.nearest(x,m.max,d,tw,rt,eps)
fn
plot(fn)
```

The following parameters are selected:

- **m.max**: the embedding dimension ranges from **m** $= 1$ to **m.max** $= 6$.
- **d**: a tentative time delay d; we shall see how to estimate it in the following discussion.
- **tw**: Theiler window ; points separated by less than **tw** in the series **x** are excluded from the search for neighbouring points. For more on this subject, see Section 3.4.5
- **rt**: escape factor; this is a factor for escaping from the neighbourhood.
- **eps**: taking a point along the observed series as the centre of a ball of diameter **eps**, all the points inside this ball are considered neighbours, no matter whether they are real or false. Here **eps** is selected as **sd(x)/10**, where **sd(x)** is the standard deviation of the series **x**.

It should be noted that in the reference manual *tseriesChaos.pdf* that comes with the package, the mathematical symbols are slightly different from those used here.

The output of the code gives the following matrix:

	m1	m2	m3	m4	m5	m6
fraction	7.925454e-01	2.732335e-01	2.694656e-02	2.248345e-02	2.502960e-02	1.880152e-02
total	3.456910e+06	4.194800e+05	1.464380e+05	7.009600e+04	3.715600e+04	2.159400e+04

For each computed dimension, here **m1**, ..., **m6** (i.e. 1, ..., 6), the fraction of false neighbours (first row) and the total number of neighbours (second row) are reported. We can plot this matrix with the instruction **plot(fn)**. Figure 3.11 shows that $m = 3$ is the first embedding dimension that gives a small percentage of false nearest neighbours.

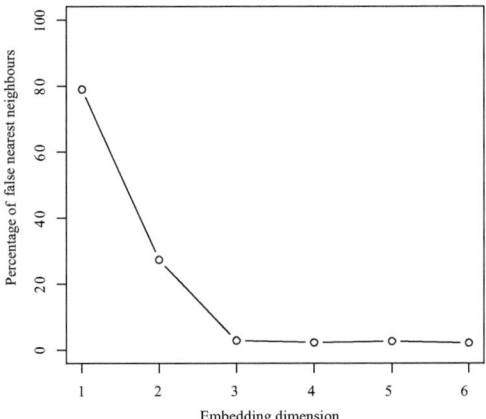

Fig. 3.11 Percentage of false nearest neighbours as a function of the embedding dimension, for the 'observed' series **x**.

3.4.2 Embedding Time Delay

The criterion for choosing the time delay d is that d must not be too small, but also not too large, compared with the internal timescales of the system. If the delays are small, then strongly correlated vectors are obtained; that is, $\mathbf{y}(t + d)$ adds no information to $\mathbf{y}(t)$. As shown in Figure 3.6b, the points are distributed close to the bisector of the reconstructed phase space, suggesting a linear dependence that, in contrast, is completely absent from the original trajectory. If the delays are too large, then $\mathbf{y}(t + d)$ and $\mathbf{y}(t)$ are practically measures performed at random, so the points in the embedding space appear to be randomly distributed. In fact, if we had to hand an infinite and perfectly clean (noise-free) series, then the time delay could be arbitrary. However, since in practice this is impossible, we will adopt some prescriptions.

One could think of choosing for d the first zero of the autocorrelation function (or the value of d at which this function is significantly reduced), but this criterion may lead to misleading results (Abarbanel, 1996). As we have already seen, with nonlinear series, the autocorrelation function may find no correlation. A suggestion for the time delay choice is given by the estimate of the *mutual information*, which is a quantity with no relation to the linearity or nonlinearity of the series. Let X and Y be two discrete random variables. Then the mutual information between them, $I(X;Y)$, is defined as

$$I(X;Y) = \sum_i \sum_j p(x_i, y_j) \log_2 \frac{p(x_i, y_j)}{p(x_i)p(y_j)}$$

where $p(x_i, y_j)$ is the joint probability distribution function of X and Y, and $p(x_i)$ and $p(y_j)$ are the marginal probability distribution functions of X and Y, respectively. Note that $I(X;Y)$ is in units of *bits*. The mutual information is a measure of how much the knowledge of X (Y) reduces uncertainty about Y (X). This implies that $I(X;Y) = 0$ if and only if X and Y are independent random variables, that is, if $p(x_i, y_j) = p(x_i)p(y_j)$.

If the measurements $s(t)$ are represented as a histogram, then $p(x_i)$ is the probability that $s(t)$ lies in the ith bin, and $p(x_i, y_j)$ is the joint probability that $s(t)$ lies in the ith bin *and* $s(t + d)$ lies in the jth bin. In other words, $I(X; Y)$ describes the information that the measurement $s(t)$ at time t brings to the measurement $s(t + d)$ at time $t + d$. In the series, the mutual information is averaged over all data $s(t)$ to yield the *average mutual information (AMI)*. If d is chosen as the value *around* the first minimum of the AMI, then $\mathbf{y}(t)$ and $\mathbf{y}(t+d)$ are partially but not totally independent.

The AMI estimate is implemented in `tseriesChaos` with the code `mutual(x, lag.max = lm)`, to yield the AMI as a function of the lag (from 1 to `lm`), i.e. as a function of d. We add the following instructions, to the code `lorenz_attractor.R`, after `plot(fn)`:

```
. . . . . . . . . . . . . . . . .
plot(fn)

lm<- 60                # largest lag
mutual(x,lag.max = lm) # average mutual information to suggest d
```

The result is displayed in Figure 3.12. It is clear that the first minimum of the AMI is for a lag equal to 18.

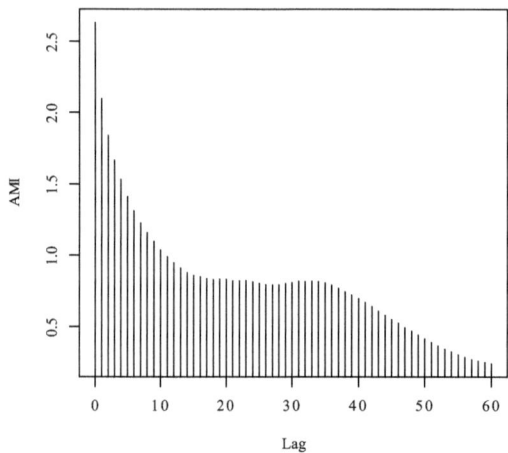

Fig. 3.12 The average mutual information as a function of the lag, for the 'observed' series x.

3.4.3   Embedding Procedure

We add the following instructions to the code `lorenz_attractor`:

```
m<- 3                         # choose embedding dimension
d<- 18                        # choose time delay; try also d<- 10
xyz <- embedd(x,m,d)          # embed the 'observed' series
windows()
scatterplot3d(xyz, type="l")
```

With $m = 3$ and $d = 18$, previously determined, the 'observed' series x is embedded into a reconstructed phase space. We also tried this with $d = 10$. The result of the embedding procedure is shown in Figure 3.13.

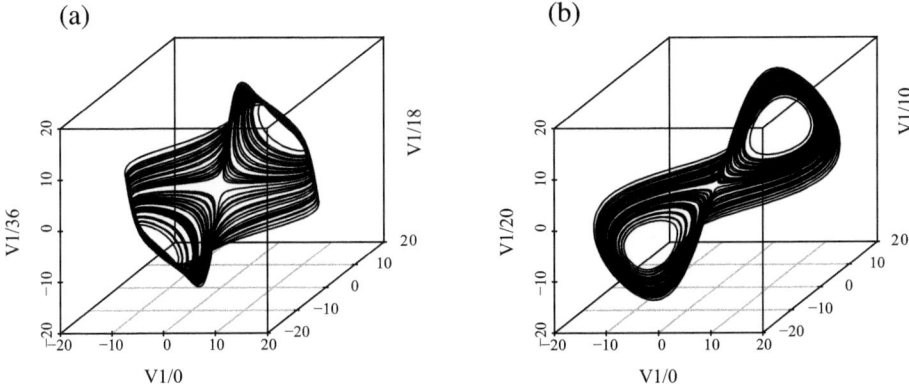

Fig. 3.13 Plot in the reconstructed phase space of the Lorenz strange attractor (3.4). The 'true' phase space portrait is shown in Figure 3.7b. The embedding dimension is $m = 3$ and the time delay is $d = 18$ (a) and $d = 10$ (b).

As we have already seen in the case of the ideal pendulum – although it is more evident here – reconstructed series, in general, appear as more or less faithful visual copies of the original ones. Note that with $d = 10$ the reconstructed attractor seems 'aesthetically' more satisfactory than the one with $d = 18$, even though $d = 18$ is the actual first minimum of the AMI. We recall, however, that the prescriptions for the choices of m and d are just that – *prescriptions*. In practice, a glance at reconstructions with various values d and m can provide important information allowing for a judicious choice of the embedding parameters.

Figure 3.14 shows what happens with a bad choice of d. In (a), the time delay is $d = 2$, and we can see that the phase space is virtually reduced to the main diagonal, as for the ideal pendulum in Figure 3.6b. On the other hand, if d is too large, as in (c), obtained with $d = 100$, the reconstructed phase space spreads throughout the available space.

A warning: the result of the command `xyz<-lsoda(cinit,times,funct,parms)` at the beginning of the code `lorenz_attractor.R` is the matrix `xyz` in which the first column `xyz[,1]` is the time, while the other three columns are the three series $x(t), y(t)$ and $z(t)$. After the embedding `xyz <- embedd(x,m,d)`, the matrix `xyz` has three columns `xyz[,1],xyz[,2],xyz[,3]` that describe the reconstructed trajectory.

3.4.4 Theiler Window

The so-called 'Theiler window', mentioned earlier, deserves further comment, since it will reappear several time in the following, in particular in codes for estimating invariants, as the maximum Lyapunov exponent or the correlation dimension. In fact, if the estimation procedures are based on the comparison between two or more Takens vectors, for instance their mutual distance, then the resulting estimates may be biased when the Takens vectors considered are not statistically independent. This may occur if points in phase space are close to each other because they are simply close in time,

(a)   (b)   (c)

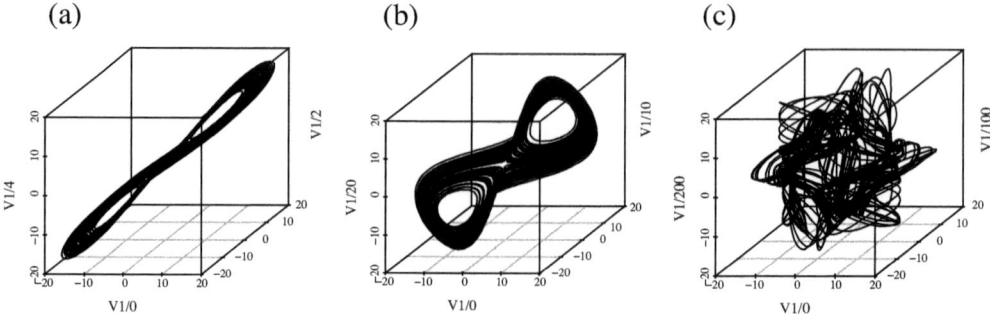

Fig. 3.14 Plot in the reconstructed phase space of the Lorenz strange attractor (3.4). The 'true' phase space portrait, shown in (b), is that in Figure 3.7. The embedding dimension is $m = 3$, and the time delay is $d = 2$ (a), 10 (b) (the best choice, as in Figure 3.13) or 100 (c).

and not because of the geometry of the phase space. To avoid such temporal nearness, the distance between pairs of points has to be greater than some time w (`tw` in the code `false.nearest`), called the *Theiler window* (Theiler, 1986). Provenzale *et al.* (1992) introduced the *space–time separation plot* method to estimate the Theiler window. The number of pairs of points is computed as a function of the spatial separation and the separation in time. The result is displayed as contours of constant probability for the distribution of the spatial separation as a function of the time separation. In the package `nonlinearTseries`, the function **spaceTimePlot(.)** yields the space–time separation plot, as shown in Figure 3.15, where the contour lines are at incremental values of 10% from the line below. The figure is obtained by adding the following lines to the code `lorenz_attractor.R`, after removing the transient:

```
require(nonlinearTseries)
spaceTimePlot(time.series=x,embedding.dim=3,time.lag=10,
time.step=1,max.radius=400,number.time.steps=400,numberPercentages=10,do.plot=TRUE,
main="",xlab="Separation in time",ylab="Separation in space")
```

where `max.radius` is the maximum radius of the sphere in which pairs of points are looked for. If the radius is not written, the code estimates it. `time.step` is the number of steps between two computations of the space–time plot, `number.time.steps` is the number of steps of the computed space–time plot and `numberPercentages` is the number of contour lines to be computed. The contour lines of constant probability at 10%, 20%,..., 100% are are shown in colour by **spaceTimePlot(.)**.

A very similar result obtained with TISEAN software is reported in Heathcote and Elliott (2011). As before, in the following, we set $w = 100$, since after this value almost-regular oscillations begin to appear. However, with $w = 50$ or 200, results are not significantly different.

3.4.5   Noisy Series

We have mentioned that in actual experiments, the state space is almost always neither 'observable' nor constructible through a system of differential equations. Moreover, in some cases, even the values of quantities defining the state of the system might be

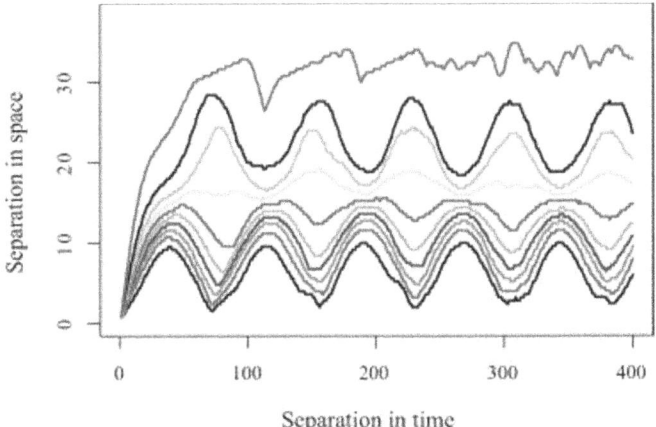

Fig. 3.15 Space–time separation plots for the Lorenz attractor. Contours of constant probability are shown for spatial separation as a function of time separation, with incremental values of 10% from the line below.

not directly measured. Suppose, for instance, we want to measure the angle $\theta(t)$ of a pendulum within a certain time interval. One way is to shine a light on the bob and measure the length of the projection of the shadow on a wall, from which, by a bit of trigonometry, the angle $\theta(t)$ can be estimated. We mean that the measure $s(t)$ is a *function* of the quantity $\theta(t)$. In general, let

$$\mathbf{x}(t) = \big[x_1(t), x_2(t), \dots, x_r(t)\big]$$

be the state of the system at time t in an r-dimensional physical phase space. If $r = 3$, we can also write $\mathbf{x}(t) = [x(t), y(t), z(t)]$. The time observed series $s_1, s_2, \dots, s_N$ represents the result of measurement of a certain quantity s_t, which is a function of the state $\mathbf{x}(t)$; that is,

$$s_t = s[\mathbf{x}(t)]$$

But measurement errors are unavoidable, so it is more correct to write

$$s_t = s[\mathbf{x}(t)] + \epsilon(t)$$

where $\epsilon(t)$ is the measurement error. There is also a further, more subtle, error, the so-called *dynamic noise*, due to sources of exogenous environmental noise, as already introduced in Chapter 2. Here we give only some indication about the measurement noise.

Let us turn to the first component of the Lorenz system, $x(t)$, and, as before, consider it as the 'observed' series. In this case, of course, it is just $s_t = x(t)$. Now, before applying the embedding procedure, let us simulate a measurement noise. To do so, we have to superpose on each clean observation s_t a noise ϵ_t, with $t = 1, \dots, N$. A normal noise model is adopted, and then at each step t we need a realization of the normal random variable.

R provides instructions to generates random deviates from some standard distributions. For the normal distribution, the appropriate instruction is `rnorm(n,mean,sd)`, where `n` is the number of required deviates, and `mean` and `sd` are the mean and standard deviation of the distribution, respectively. In the code `lorenz_attractor`, the instruction `xyz <- embedd(x,m,d)`, is preceded by the following instructions:

```
set.seed(3)
w<- 100
x<- x + rnorm(length(x),0,sd(x)/w)   # add noise to x
xyz <- embedd(x,m,d)                 # embed the noisy 'observed' series
scatterplot3d(xyz, type="l")
```

We require that `length(x)` deviates from a normal distribution with zero mean and with standard deviation equal to a factor of `1/w` times the standard deviation `sd(x)` of the series, which in our case is equal to 7.85. With $m = 3$ and $d = 10$, the results are shown in Figure 3.16 for two values of `w`: 100 (a) and 10 (b). We can see how, when the noise is low, as in (a), there is not a lot of difference with respect to the clean series, but high noise, as in (b), makes it impossible to study the features of the attractor. In practice, with very noisy real-word data, tools to reduce the noise are needed, and these will be described later.

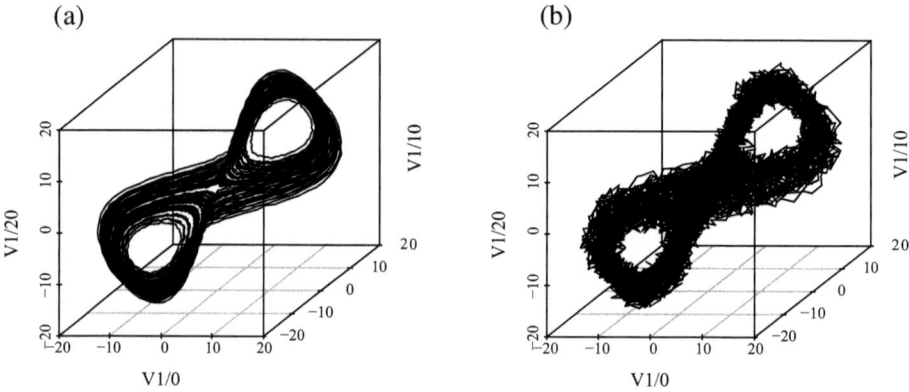

Fig. 3.16 Plot in the reconstructed phase space of the Lorenz strange attractor (3.4). The embedding parameters are $m = 3$ and $d = 10$. The 'observed' series $x(t)$ is corrupted by a measurement noise modelled by a normal distribution with standard deviation equal to $\frac{1}{100}$ (a) or $\frac{1}{10}$ (b) times the standard deviation of the series.

3.4.6 Alternative Embedding Procedure

An alternative procedure for embedding is based on *principal component analysis (PCA)*. Here we give a brief outline of PCA, limited to its application to the embedding process; for a comprehensive treatise, see for instance Jolliffe (2002) (although many multivariate statistics books deal with PCA). As the name suggests, PCA is a statistical technique to select the most important variables from a large set of (possibly correlated) variables. More technically, PCA is a linear transformation of the variables into a lower-dimensional space that retains the maximal amount of information about the variables. The earliest description of this technique was given by Pearson (1901), and it was later developed by Hotelling (1933).

Suppose we have to study a system. We start with a set of h variables; then, using PCA, we transform the original set into a new set of uncorrelated variables, called *principal components*. Each component is derived as linear function of the original variables. The transformation maximizes the variance and constrains each variable to be uncorrelated with all the others; thus, there are as many components as there are original variables. However, the PCA tells us whether most of the information is along the first few components, with a negligible gain of information being obtained by considering the remaining components.

Geometrically, this means that we create a new system of orthogonal axes that contains the greatest fraction of the total variance of the data. The principal directions are obtained as those eigenvectors of the symmetric autocovariance matrix that correspond to the largest eigenvalues. In the remaining orthogonal directions, the variance is small, providing no useful information about the data, so these directions can be ignored.

In fact, these components are sorted in decreasing order of their contribution to the total variance, with the first component having the maximum variance and the hth component the minimum. We can decide to neglect the components of lesser significance and retain only the relevant ones. In this manner, the k retained components can account for most of the information present in the data, with substantially fewer variables than those originally present.

There are several functions from different packages in the R environment for performing PCA. In the **stats** package, there are two functions: **princomp** and **prcomp**. The function **princomp** uses the spectral decomposition approach (in R, see the **spectral.methods** package), while the function **prcomp** uses the singular value decomposition (in R, see the **svd** package).

This latter function is known to be numerically more stable in some cases, and we shall employ it in the following. Code 3.5 **lorenz_PCA.R** shows how it is possible to execute the embedding of an observed series through PCA. The 'observed' series is the first component of the Lorenz system $x(t)$ after the transient, as was the case in the code **lorenz_attractor.R**.

```
#Code 3.5 Lorenz PCA (lorenz_PCA.R)
# embedding procedure with the Principal Component Analysis

parms<- c(10,8/3,28)   # parameters: sigma, beta, rho
tiniz<- 0
tfin<- 100
step<- 0.01
times<-seq(tiniz,tfin,by=step)
funct<- function(t,integ,parms){              # the system of equations
  x<-integ[1]
  y<-integ[2]
  z<-integ[3]
  sigma<- parms[1]
  beta<-  parms[2]
  rho<-   parms[3]
  dx<- sigma*(y-x)           # that is  dx/dt = sigma(y-x)
```

```
dy<- x*(rho-z)-y          # that is  dy/dt = x(rho-z)-y
dz<- x*y-beta*z           # that is  dz/dt = xy -beta z
list(c(dx,dy,dz))
}                         # end of funct
require(deSolve)
cinit<-c(1,1,1)
xyz<-lsoda(cinit,times,funct,parms)
trans<- 2000                    # integration time step considered as transient
x <- window(xyz[,2],trans)      # discard  initial transient
t_start<- trans*step - step     # new initial time
t_time<-seq(t_start,tfin,by=step) # new time interval
require(tseriesChaos)
options(digits=5)               # number of digits to print (only a suggestion)
m<- 11                          # embedding dimension
d<- 1                           # time delay
emb<- embedd(x,m,d)
######################           # up to here it is as in lorenz_attractor.R

# pc<- prcomp(emb)     # the simplified format of the function prcomp()
pc<- prcomp(emb,center=TRUE,scale.=TRUE)  # in a more explicit form
summary(pc)
sd<-pc$sdev
var <- sd^2
var.percent <- var/sum(var)*100
# the scree plot in bar plot form
barplot(var.percent,xlab="Principal Component",ylab="Percent Variance",
        names.arg=1:length(var.percent),            # to label the bars.
        ylim=c(0,100),col="gray",cex.axis=1.2,cex.lab=1.5)
# the scree plot in line plot form

plot(0,0,type="n", xlab="Principal Component", ylab="Percent Variance",ylim=c(0,100),
     xlim=c(1,m),cex.lab=1.5,cex.axis=1.2)
lines(var.percent,col="black",type="b",pch=19,cex=1.4)
# to plot the cumulative fraction of the total variance
cum.var<- cumsum(var.percent[1:m])
lines(cum.var, col="black",type="b",pch=21,cex=1.4)
# to plot the reconstructed attractor
xyz <- predict(pc)[ ,1:3]
require(scatterplot3d)
scatterplot3d(xyz,type="l",cex.lab=1.5,cex.axis=1.2,lab=c(4,4,7),lab.z=4)
# lab() and lab.z: parameters to control the number of tickmarks
```

We see that up to the instruction emb<- embedd(x,m,d), this code is the same as lorenz_attractor.R, but now the embedding dimension m is deliberately excessive and the time delay d is set very small. The result of the instruction emb<- embedd(x,m,d) is then a matrix with m = 11 columns and ne = 7992 rows, where ne is the number of embedding vectors; recall that the series $x(t)$ without the transient has length 8002, so ne = $8002 - (11 - 1) \times 1$. By default, with **prcomp**, the variables have zero mean, that is, center=TRUE is implied, but they are not scaled to have unit variance. Before the analysis is performed, it is advisable (and necessary if the range of variability is too large), to scale the variables with the logical option scale.=TRUE, so that their variances are equal to one. With raw data, the principal components are computed through the covariance matrix, while if the data are standardized, this is done through the correlation matrix, which in this case is the same as the covariance matrix. The command pc<- prcomp(emb) can then be written in more explicit form as pc<- prcomp(emb,center=TRUE,scale.=TRUE).

A summary of the PCA results is obtained using the **summary** function on the output of **prcomp**, that is, with the command `summary(pc)`:

```
Importance of components:
                       PC1    PC2     PC3      PC4     PC5      PC6       PC7
Standard deviation     3.27 0.5666 0.07883 0.00992 0.0012 0.000141 1.56e-05
Proportion of Variance 0.97 0.0292 0.00056 0.00001 0.0000 0.000000 0.00e+00
Cumulative Proportion  0.97 0.9994 0.99999 1.00000 1.0000 1.000000 1.00e+00
                          PC8      PC9    PC10    PC11
Standard deviation     1.62e-06 1.51e-07 1.9e-08 1.17e-08
Proportion of Variance 0.00e+00 0.00e+00 0.0e+00 0.00e+00
Cumulative Proportion  1.00e+00 1.00e+00 1.0e+00 1.00e+00
```

The **summary** function describes the importance of the principal components. The first row lists the standard deviation of each component, that is, the square root of the eigenvalues covariance matrix **emb**. The second row lists the proportion of the variance in the data explained by each component, and the third row lists the cumulative proportion of explained variance. The first principal component (**PC1**) explains the greatest percentage (97%) of the original total variance, while the remaining components explain a very small proportion of the variability. In any case, the first three components capture 100% of the original total variance, so the number of variables, that is, the embedding dimension m, is decreased from 11 to 3.

Further information can be extracted from **prcomp** with the command `pc$object`, where `object` is one of the following objects included in **prcomp**:

- `sdev`: the standard deviations of the principal components.
- `center`: the values used to centre the data (center=TRUE in prcomp()).
- `scale`: the values used to scale the data (scale=TRUE in prcomp()).
- `rotation`: the loadings matrix, which is a matrix whose columns are the loadings for the principal components. This matrix gives the position of each new axis relative to the original system of axes.
- `x`: the scores matrix (in prcomp(), add rtex=TRUE), which gives the location of each data point along each of the new principal component axes.

We will return to the loadings and scores matrices in the following. To decide the appropriate number of components, it may be useful in several cases to visualize the results of the PCA with a *scree plot*, in either bar plot or line plot form, as in Figure 3.17.

The name 'scree plot' comes from its appearance, which recall that of a hillside down which stones have rolled and accumulated at the bottom. Note that the command `sd<-pc$sdev` extracts the standard deviation of each principal component, so that what is plotted is the fraction of total variance as a percentage, `var.percent`, as explained by each component. With the command `screeplot(pc)` a bar plot is obtained directly, but with the y axis showing the variance: `(pc$sdev)^2`. To separate the 'most important' components from the 'least important', we look for a point, called the 'elbow' in the PCA literature, at which the variances become relatively small. In our example, the first three components explain the total variance, so we do not need to look for any elbow. There are other ways to decide how many components to retain.

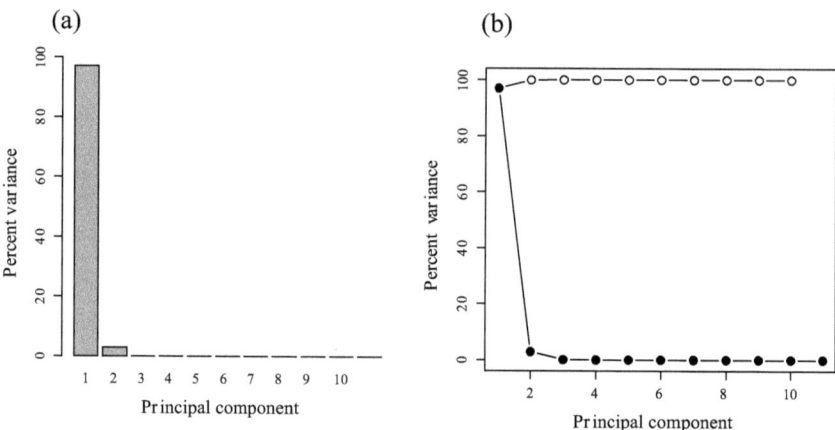

Fig. 3.17 Scree plot in bar plot form (a) and line plot form (b) as result of PCA applied to the 'observed' series $x(t)$ of the Lorenz strange attractor (3.4). The embedding parameters are $m = 11$ and $d = 1$. On the x axis are the principal components, sorted in decreasing order of variance, while on the y axis are the fractions of total variance (as percentages) as explained by each principal component. The cumulative fraction of total variance explained is also shown in (b), by the open circles.

Finally, we use the **predict** function, a generic R function for predicting the results of various models. Here, this function recognizes `pc` as the result of a `prcomp` instruction. From the **predict** point of view, each principal component is a new variable, and it then projects into the new space, described by the first three principal components, the values of these variables. The reconstructed attractor is shown in Figure 3.18.

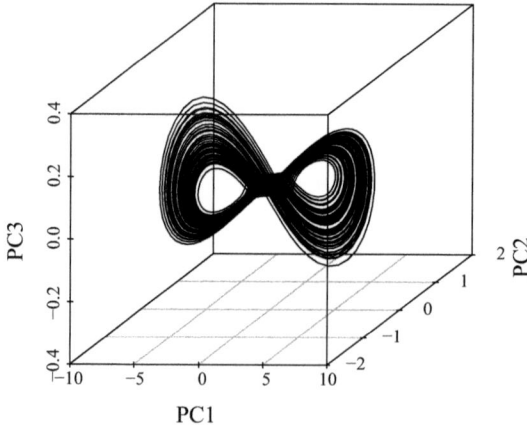

Fig. 3.18 Plot of the Lorenz strange attractor (3.4) in the reconstructed phase space obtained by PCA.

As before, we can rotate the graph interactively with the command

```
rgl.linestrips(xyz[,1], xyz[,2], xyz[,3])
```

It is useful to see in detail how one arrives at Figure 3.18, starting from the 'observed' series (the first component of the Lorenz system $x(t)$ after the transient), through PCA. With the command `xyz[1:6, ]` after the last line of the code `lorenz_attractor.R`, we print the retained three principal components for the first six steps, which are part of the plot of Figure 3.18:

```
        PC1     PC2        PC3
[1,] -5.3050 0.52102  0.05339045
[2,] -5.1346 0.60093  0.03659912
[3,] -4.9420 0.66719  0.01819302
[4,] -4.7311 0.71878 -0.00082834
[5,] -4.5063 0.75538 -0.01952906
[6,] -4.2718 0.77729 -0.03710483
```

However, we have said that there are 11 principal components in all, each of which is a linear combination of the original variables, for instance the first principal component PC1 can be written in our case as

$$PC1 = z_1 w_1 + z_2 w_2 + z_3 w_3 + \ldots + z_{11} w_{11}$$

where $z_1, \ldots, z_{11}$ are the elements of the first row of the standardized version of the matrix **emb** and $w_1, \ldots, w_{11}$ are the the elements of the first column of the loadings matrix (see below). The first six rows of **emb**, obtained with `emb[1:6, ]`, are

```
       V1/0    V1/1    V1/2    V1/3    V1/4     V1/5     V1/6     V1/7     V1/8     V1/9    V1/10
[1,] 13.594  13.557  13.426  13.204  12.896  12.5075  12.0498  11.5340  10.9727  10.3788   9.7649
[2,] 13.557  13.426  13.204  12.896  12.507  12.0498  11.5340  10.9727  10.3788   9.7649   9.1428
[3,] 13.426  13.204  12.896  12.507  12.050  11.5340  10.9727  10.3788   9.7649   9.1428   8.5234
[4,] 13.204  12.896  12.507  12.050  11.534  10.9727  10.3788   9.7649   9.1428   8.5234   7.9160
[5,] 12.896  12.507  12.050  11.534  10.973  10.3788   9.7649   9.1428   8.5234   7.9160   7.3283
[6,] 12.507  12.050  11.534  10.973  10.379   9.7649   9.1428   8.5234   7.9160   7.3283   6.7665
```

The matrix **emb** is standardized with the command `emb.st<- scale(emb)`, and the first six rows are

```
       V1/0    V1/1    V1/2    V1/3    V1/4    V1/5    V1/6    V1/7    V1/8     V1/9    V1/10
[1,] 1.7778  1.7733  1.7571  1.7294  1.6906  1.6418  1.5842  1.5192  1.4483  1.37335  1.29579
[2,] 1.7730  1.7568  1.7290  1.6902  1.6414  1.5838  1.5188  1.4480  1.3730  1.29549  1.21689
[3,] 1.7565  1.7287  1.6899  1.6411  1.5834  1.5184  1.4476  1.3727  1.2952  1.21661  1.13833
[4,] 1.7284  1.6896  1.6407  1.5831  1.5181  1.4473  1.3723  1.2949  1.2163  1.13805  1.06129
[5,] 1.6893  1.6404  1.5827  1.5177  1.4469  1.3720  1.2945  1.2160  1.1378  1.06102  0.98674
[6,] 1.6401  1.5824  1.5174  1.4466  1.3717  1.2942  1.2157  1.1375  1.0607  0.98648  0.91550
```

To ensure that the matrix **emb.st** is really standardized, we apply the command `apply(emb.st,2,mean)`, which returns

```
       V1/0        V1/1        V1/2        V1/3        V1/4        V1/5
 1.7773e-17  1.2006e-17  3.4712e-17  1.3665e-17 -1.4867e-17  1.2870e-17
       V1/6        V1/7        V1/8        V1/9       V1/10
-1.4421e-17  6.3550e-18 -1.7937e-17  2.0621e-17  2.0657e-17
```

and the command `apply(emb.st,2,sd)`, which returns

```
V1/0  V1/1  V1/2  V1/3  V1/4  V1/5  V1/6  V1/7  V1/8  V1/9 V1/10
  1     1     1     1     1     1     1     1     1     1     1
```

Then the means of the standardized variables are all substantially equal to 0 and their standard deviations are all equal to 1. The elements of the first row of the matrix emb.st are the variables $z_1, z_2, \ldots, z_{11}$.

The command pc$rotation returns the loadings matrix, that is, the matrix whose columns are the eigenvectors of the correlation matrix. Then, the columns are the loadings for the principal components, in the first column there are the loadings for PC1, in the second column the loadings for PC2, and so on. The first column is given with the command pc$rotation[,1] and is

```
-0.29483 -0.29877 -0.30190 -0.30416 -0.30553 -0.30599
-0.30553 -0.30416 -0.30190 -0.29877 -0.29480
```

which are the variables $w_1, w_2, \ldots, w_{11}$. Then, the first principal component PC1 is given by

$$
\begin{aligned}
\text{PC1} = &+ 1.7778 \times -0.29483 + 1.7733 \times -0.29877 + 1.7571 \times -0.30190 \\
&+ 1.7294 \times -0.30416 + 1.6906 \times -0.30553 + 1.6418 \times -0.30599 \\
&+ 1.5842 \times -0.30553 + 1.5192 \times -0.30416 + 1.4483 \times -0.30190 \\
&+ 1.3733 \times -0.29877 + 1.2958 \times -0.29480 \\
= &- 5.3050
\end{aligned}
$$

Then, -5.3050 is, as expected, the first element of the matrix xyz.

We can obtain the loadings matrix through direct computation of the the eigenvectors of the correlation matrix, that is, with the instructions

```
CorMatr<- cor(emb.st)    # correlation matrix
Eig <- eigen(CorMatr)    # eigenvalues and eigenvectors of correlation matrix
Eig$vectors              # eigenvectors (loadings)
```

A short code computes the principal components explicitly:

```
pc_comp<- function(v_st,loadings,nrow,m) {   # start function pc_comp
# nrow: number of row in the matrix v_st
# m (embedding dimension): number of principal components
pcc<- numeric()   # in this vector the components are stored
# compute the value of the components for each row
for(i in 1:nrow) {            # loop on the rows
cp.i<- 0
for (j in 1:m)   {cp.i<- cp.i + (v_st[i,j]*loadings[j])} # loop on the colums
pcc[i] <- cp.i
                }            # end loop on the rows
return(pcc)
                     }                          #  end function pc_comp
# to compute the values of the first principal component for the first 6 rows
nrow<- 6
pc_comp(scale(emb), pc$rotation[ ,1],nrow,m)
```

The result is

```
-5.3050 -5.1346 -4.9420 -4.7311 -4.5063 -4.2718
```

The second principal components for the first six rows are obtained by writing

```
pc_comp(scale(emb), pc$rotation[ ,2],nrow,m)
```

with the result

```
0.52102 0.60093 0.66719 0.71878 0.75538 0.77729
```

and so on.

All this computation is equivalent to the command **pc\$x** to yield the scores matrix x. For instance, with the command **pc\$x[1:6,1]**, we obtain again

```
-5.3050 -5.1346 -4.9420 -4.7311 -4.5063 -4.2718
```

Let W be the matrix by multiplication of the matrix **emb.st** and the loadings matrix resulting from the command **pc\$rotation**. We use the R function **all.equal()** to ascertain that W is equal to the scores matrix resulting from the command **pc\$x**:

```
W <- scale(emb) %*% pc$rotation
all.equal(W,pc$x)
### answer: TRUE
```

An alternative way to plot the reconstructed attractor is then

```
scatterplot3d(pc\$x[ ,1:3], ...)}.
```

The PCA can also be profitably exploited to filter noise, since most of the noise is presumably contained in the discarded orthogonal directions. In this manner, it may be possible (at least in theory) to separate noise, the components of lesser significance, from signal, the most important components. A similar method to reduce noise is the *singular spectrum analysis (SSA)* technique, which will be discussed in Chapter 6.

PCA offers further tools for data analysis in a great variety of disciplines, as testified by the wide literature on this technique. In the context of nonlinear time series analysis, see the clear introduction and discussion, with examples, in Kantz and Schreiber (1997).

3.5 Summary

The concept of 'phase space' has been introduced as a central topic in the analysis of time series, for both linear and nonlinear series. Historically, this concept arises in the context of classical and statistical mechanics and allows the behaviour of a system to be represented in geometric form. The axes of this space are all the variables that describe the state of the system at any instant. A point (*representative point*) in phase space corresponds to a unique state of the system.

The dynamics of the full system is described by the trajectory followed by the representative point in time.

In addition to 'phase space', we have introduced '*reconstructed* phase space'. To determine the phase space of a system, we require knowledge of all the variables describing that system, but, for a real process, we do not usually know all the state variables – indeed, in some cases, not all of them can be measured. However, there are mathematical theorems ensuring that, with measurements of *only one* variable, it is possible to reconstruct a phase space that is 'similar', in a *topological* sense, to the true phase space of the underlying system. This means that the characteristic invariants of the system (fractal dimensions, Lyapunov exponents, and others) are conserved in the reconstructed phase space. Then, the features of the system derived from studying the reconstructed dynamics also hold for the true dynamics. Delay-coordinate embedding can be considered as a standard method for this reconstruction. The *scalar* measured time series $s(1), s(2), \ldots, s(N)$ can be expand in an m-dimensional *vector* space, in which each point is represented by the vectors $\mathbf{y}_t = [s(t), s(t+d), s(t+2d), \ldots, s(t+(m-1)d)], t = 1, \ldots, n$, with $n = N - (m-1)d$. Here, m is the embedding dimension (i.e. the dimension of the reconstructed phase space) and and d is the time delay (i.e. the time shifts between the coordinates). A wide variety of methods and prescriptions have been proposed for the choice of the embedding parameters m and d. In this chapter, we have encountered the 'false nearest neighbour' method and the use of mutual information for the choice of m and d, respectively. We repeat here that the choice of embedding parameters depends also on the specific problem to which the embedding procedure is applied. The freely swinging *ideal* pendulum (with small swings, no air resistance, etc.) and the Lorenz attractor can act as guides when starting out on the study of dynamic systems.

We have illustrated a further method of embedding based on *principal component analysis* (PCA). In PCA, a new system of orthogonal axes is created that contains the largest fraction of the total variance of the data. The principal directions can be obtained as the eigenvectors of the symmetric autocovariance matrix that correspond to the largest eigenvalues. The remaining orthogonal directions presumably contain most of the noise, and PCA can thus be exploited to reduce noise.

Lastly, a warning is in order. In the literature, and indeed elsewhere in this book, reconstructed phase space is also called 'pseudo phase space' or 'shadow phase space'. The latter term could lead to confusion with the notion of 'shadowing trajectory'. The question arises as to the relation (if any) between a *computer-simulated* trajectory and a *real* trajectory. It has been proved that, in the real word, there exists a real trajectory that stays close to ('shadows') the computed one (Grebogi *et al.*, 1990). This is important since the fractal structures of chaotic attractors displayed by numerically simulations are actually real.

4

The Features of Chaos

'And, while Belbo's head followed the pull of the wire, his body at first in its final spasms, then with the disarticulated agility of a wooden marionette, arm here, leg there, described other arcs in the void, arcs independent of the head, the wire, and the sphere beneath.'

Umberto Eco, Foucault's Pendulum

4.1 Introduction

There is more than one strangeness in strange attractors. One strangeness is dynamical in character, the so-called 'sensitivity to initial conditions'. A quantitative measure of this sensitivity is given by the Lyapunov exponents, which reflect the average rate of divergence (if any) between two neighbouring trajectories. The dynamic 'strangeness' of the system has its counterpart in the 'strangeness' of the attractor's geometry and concerns with the texture woven by the system in phase space during its time evolution. Fractal dimensions are measures of such strange geometries.

Lyapunov exponents are among the most famous quantities employed to characterize chaotic phenomena. As we have already seen, the nonlinear logistic map has one Lyapunov exponent, which can be negative, zero or positive. For a continuous system $\mathbf{x}(t) = [x_1(t), x_2(t), \ldots, x_r(t)]$ in an r-dimensional physical state space, there are as many exponents as there are space axes. Suppose that the dimension of the phase space is 3 and that the space axes are (x, y, z). In this case, there are three Lyapunov exponents, say $\lambda_1, \lambda_2, \lambda_3$. It may happen that $\lambda_1 > 0, \lambda_2 = 0, \lambda_3 < 0$ (the convention is to order the exponents according to their magnitude, starting with the largest). This means that in the x direction there is an exponential divergence of initially close trajectories, in the z direction a convergence, and in the y direction neither a convergence nor a divergence, but asymptotic stability.

The set $\lambda_1, \lambda_2, \lambda_3$ is the so-called Lyapunov *spectrum*. The sum $\lambda_1 + \lambda_2 + \lambda_3$ is a measure of the contraction or expansion of the system in phase space. If one of the exponents is positive, then there are strong clues that the system is chaotic. Hence, from a practical point of view, it should be enough to test if the maximum Lyapunov exponent is positive. This is exactly what we will do in Section 4.2.

As far as the strange geometries of attractors are concerned, it must be said that there are several definitions of fractal dimension, for instance, box counting dimension, Hausdorff–Besicovitch dimension , information dimension, correlation dimension, and so on. Of all these types of dimensions, most attention has been paid to the so-called correlation dimension, as we will do in the following.

Nonlinear Time Series Analysis with R. Ray Huffaker, Marco Bittelli and Rodolfo Rosa, Oxford University Press (2017).
© Ray Huffaker, Marco Bittelli and Rodolfo Rosa. DOI: 10.1093/oso/9780198782933.001.0001

We will also see a further tool to explore the features of linear or nonlinear systems, the *recurrence plot*. This is a *topological* method (Gilmore, 1993) to analyse how strange attractors are structured, in particular to unveil hidden recurring patterns, and is especially suitable for relatively short time series, as are typical in economics and finance.

4.2   Lyapunov Exponent

Let us extend the procedure to compute the Lyapunov exponent for a one-dimensional map to a discrete (or continuous) r-dimensional system. Consider the map, without noise, $\mathbf{x}_{t+1} = f(\mathbf{x}_t)$. Let $\delta\mathbf{x}_t$ be a small perturbation on the orbit at step t; that is, the system at step $t + \Delta t$ is in the state $\mathbf{x}_t + \delta\mathbf{x}_t$. At step $t + 1$, expanding $f(\mathbf{x}_t)$ in a Taylor series, we have

$$\mathbf{x}_{t+1} + \delta\mathbf{x}_{t+1} = f(\mathbf{x}_t + \delta\mathbf{x}_t)$$

$$\approx \mathbf{D}f(\mathbf{x}_t) \cdot \delta\mathbf{x}_t + f(\mathbf{x}_t)$$

where $\mathbf{D}f(\mathbf{x}_t) = \mathbf{J}$ is the $(r \times r)$ Jacobian matrix, computed on the points $\mathbf{x}_t$ of the orbit. The perturbation $\delta\mathbf{x}_t$, provided it remains 'small', will evolve according to

$$\delta\mathbf{x}_{t+1} = \mathbf{D}f(\mathbf{x}_t) \cdot \delta\mathbf{x}_t$$

and, up to $t + k$ steps, $\delta\mathbf{x}_{t+k}$ is given by

$$\delta\mathbf{x}_{t+k} = \mathbf{D}f(\mathbf{x}_{t+k-1}) \cdot \mathbf{D}f(\mathbf{x}_{t+k-2}) \cdot \mathbf{D}f(\mathbf{x}_t)\,\delta\mathbf{x}_t = \mathbf{D}f^k(\mathbf{x}_t)\,\delta\mathbf{x}_t$$

where $\mathbf{D}f^k(\mathbf{x}_t)$ is the product of the Jacobian matrices along the trajectory. Let us consider the matrix (known as the Oseledec matrix)

$$\mathbf{O}(\mathbf{x}, k) = \left\{ \left[\mathbf{D}f^k(\mathbf{x}_t)\right]^{\mathrm{T}} \cdot \mathbf{D}f^k(\mathbf{x}_t)\right\}^{1/2k}$$

where T denotes the transpose. There is a theorem, due to Oseledec (1968), ensuring the convergence of $\mathbf{O}(\mathbf{x}, k)$ for $k \to \infty$, to the limit $\mathbf{O}(\mathbf{x})$. This limit is independent of $\mathbf{x}$ for almost all the vectors $\mathbf{x}$ belonging to the basin of attraction. The logarithms of the r eigenvalues of $\mathbf{O}(\mathbf{x})$ are the *global* Lyapunov exponents of the system. If the eigenvalues are computed for the matrix $\mathbf{O}(\mathbf{x}, k)$, they are the *local* Lyapunov exponents, measuring how fast a perturbation in the point $\mathbf{x}$ moves down the trajectory in a finite number k of steps. A variety of algorithms to estimate the complete Lyapunov spectrum have been proposed by, among others, Sano and Sawada (1985), Eckmann *et al.* (1986), Brown *et al.* (1991), McCaffrey *et al.* (1992) and Nychka *et al.* (1992). We will neither explore the mathematical details of the theory nor deal with these proposed approaches (for an extended discussion of this topic, see Chan and Tong (2001)), but in Section 4.2.1 we present a computer code to estimate the *maximum characteristic Lyapunov exponent (MCLE)* implemented in `tseriesChaos`.

It is worthwhile stressing that the theory of Lyapunov exponents is based entirely on the mathematical structure of *rational* (or analytical) mechanics. This deterministic discipline has developed through *logical* deduction from first principles, and now extends into the land of chaotic phenomena, where, while clarifying its own limits, it is able to give a sense and measure of unpredictability. It is still the *logos* that explains and quantifies the *clinamen* (Giannerini and Rosa, 2002).

4.2.1 Estimating the Maximum Characteristic Lyapunov Exponent

Let $\mathbf{x}_0$ and $\mathbf{x}_0'$ be two close initial conditions in an r-dimensional phase space. The MCLE may be then defined as

$$\lambda = \lim_{n \to \infty} \lim_{\delta \to 0} \frac{1}{n} \ln \left(\frac{\|\mathbf{x}_n - \mathbf{x}_n'\|}{\|\mathbf{x}_0 - \mathbf{x}_0'\|} \right)$$

where $\delta = \|\mathbf{x}_0 - \mathbf{x}_0'\|$ is the perturbation in the initial condition and $\|\cdot\|$ is an appropriate norm. This definition holds with probability equal to 1 for almost all initial conditions. In practice, however, we have only one measured scalar series available. We have learned that, thanks to the embedding procedure, we can construct a phase space equivalent to the physical one. Let $\mathbf{y}_t$ being the reconstructed trajectory in the m-dimensional phase space (d is the time delay):

$$\mathbf{y}_t = [s(t), s(t+d), s(t+2d), \dots, s(t+(m-1)d)], \quad t = 1, \dots, n$$

However, even after the embedding, the trajectory remains only a single one. How can we follow two trajectories starting at two close initial conditions? The strategy is to look at nearby points as they were on different trajectories. The method was first proposed by Wolf *et al.* (1985) and then revisited by Rosenstein *et al.* (1993), and by Kantz (1994). The underlying algorithm is explained in details in Kantz and Schreiber (1997) and is included in the TISEAN package (Hegger *et al.*, 1999). Suppose, for example, that $m = 2$ (Figure 4.1).

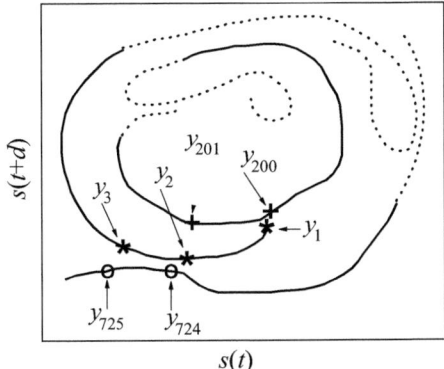

Fig. 4.1 Segments of reconstructed trajectory in two-dimensional phase space $(s(t), s(t+d))$, $t = 1, \dots, n$: scheme for estimation of the of the maximum characteristic Lyapunov exponent.

At time $t = 0$, we choose a point on the trajectory, for instance the initial point $\mathbf{y}_j$ with $j = 1$ (but it might be any other one), or also $\mathbf{y}_{j=1}$ ($\mathbf{y}_1$ in Figure 4.1):

$$\mathbf{y}_{j=1} = [s(0), s(0 + d)]$$

We locate its nearest neighbour $\mathbf{y}_i$. In the time domain, this may be the 10th, the 90th, the 200th, etc. – it is not necessarily the closest one in the time domain. In Figure 4.1, it is the 200th:

$$\mathbf{y}_i = [s(200), s(200 + d)]$$

We compute their distance $\delta_{j=1}(0)$:

$$\delta_{j=1}(0) = \min |\mathbf{y}_i - \mathbf{y}_j|$$

The subscript '$j = 1$' means 'point 1' (the first point considered) and the number 0 in parentheses refers to the time $t = 0$. The norm can be the Euclidean norm, the maximum norm or another one. In Figure 4.1, we have

$$\delta_1(0) = \min |\mathbf{y}_{200} - \mathbf{y}_1|$$

After one integral step, what happens to $\mathbf{y}_j$? We will have the point

$$\mathbf{y}_{j=2} = [s(1), s(1 + d)]$$

In Figure 4.1, this is $\mathbf{y}_2$. Also, $\mathbf{y}_i$ moves forwards one step and is the point $\mathbf{y}_{i+1}$ (in the figure, this is $\mathbf{y}_{201}$). We compute their distance $\delta_{j=1}(1)$. Notice that the subscript is still '$j = 1$', since we still consider the evolution of the point 1 with respect to its initial nearest neighbour. The number 1 in parentheses refers to the next time $t = 1$. In Figure 4.1, we have

$$\delta_1(1) = \min |\mathbf{y}_{201} - \mathbf{y}_2|$$

Following the same procedure, we obtain a series of distances:

$$\delta_{j=1}(0), \delta_{j=1}(1), \ldots, \delta_{j=1}(l)$$

In the example,

$$\delta_1(0), \delta_1(1), \ldots, \delta_1(l)$$

With this procedure, we have looked how the points 1 and 200 have moved away from each other in the reconstructed phase space, when at the beginning (time $t = 0$), they were very close to each other. The value of l (for 'last') will be discussed shortly.

If the divergence between two close points is exponential, then

$$\delta_j(\nu) \sim C_j \exp(\lambda \nu)$$

with $\nu = 0, 1, \ldots, l$. That is,

$$\log \delta_j(\nu) \sim C_j \lambda \nu$$

which, on a semilogarithmic scale, is a straight line, *if* and *where* there is exponential divergence. This line represents the evolution of the logarithm of the distance between the point 1 and its initial nearest neighbour. Notice that the straight line is an 'ideal' straight line – actually, two points sometimes move away from one another, sometimes more closely together, in which case we have a 'straight line' with oscillations. Figure 4.2 shows this schematically. Notice that the scale on the y axis is logarithmic and that the x axis is labelled by $\nu = 0, 1, 2 \ldots, l$.

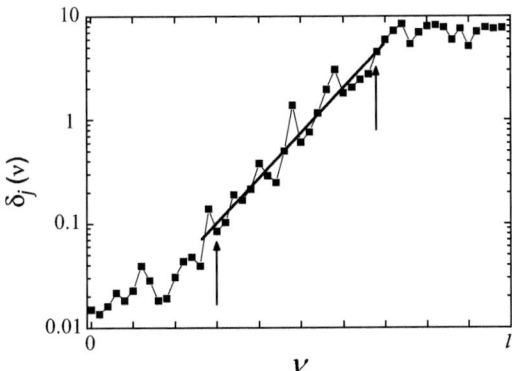

Fig. 4.2 Divergence $\delta_j(\nu)$ between two points on the same trajectory, starting very close to each other, as a function of the step ν. The arrows mark the approximately linear interval.

Notice also that to the left of the first arrow there is a initial transient and towards the end (to the right of the second arrow) a plateau is apparent since the maximum distance between the points is reached, (i.e. the diameter of the attractor). Therefore, it is completely useless to go beyond l in the computation of the divergence $\delta_j(\nu)$. The above procedure has to be repeated for all the points of the trajectory – or at least for a certain number of them.

After $\mathbf{y}_{j=1}$, we consider $\mathbf{y}_{j=2}$. We again locate its nearest neighbour, which need not be $\mathbf{y}_{i+1}$. In Figure 4.1, the nearest neighbour of $\mathbf{y}_2$ is not $\mathbf{y}_{201}$, but rather $\mathbf{y}_{724}$. In this manner, we now have the distances $\delta_{j=2}(0)$ and then $\delta_{j=2}(1)$, etc. In the example, $\delta_2(1)$ is the distance between $\mathbf{y}_3$ and $\mathbf{y}_{725}$. In this case also, we will obtain a 'straight line'. At the end of the whole procedure, assuming we have considered all the points n of the reconstructed trajectory, we know how the point 1 moves away from its nearest neighbour, how the point 2 does the same, ..., how the point n does it. We have computed

$$\delta_1(0), \ \delta_1(1), \ \ldots, \ \delta_1(l)$$
$$\delta_2(0), \ \delta_2(1), \ \ldots, \ \delta_2(l)$$
$$\vdots \qquad \vdots \qquad \quad \vdots$$
$$\delta_n(0), \ \delta_n(1), \ \ldots, \ \delta_n(l)$$

Hence we can obtain an 'average straight line' $S(\nu)$, $\nu = 0, 1, \ldots, l$ that, plotted as a function of ν, shows a linear trend in the interval where an exponential divergence exists (if it does exist). The slope of this 'average straight line' in the linear interval is an estimate of the MCLE; let us call it $\hat{\lambda}$. By 'average straight line', we mean the linear best fit of the points

$$\bar{\delta}(0), \bar{\delta}(1), \dots, \bar{\delta}(n)$$

where

$$\bar{\delta}(0) = \frac{1}{n} \sum_{i=1}^{n} \delta_i(0), \quad \bar{\delta}(1) = \frac{1}{n} \sum_{i=1}^{n} \delta_i(1), \quad \dots, \quad \bar{\delta}(n) = \frac{1}{n} \sum_{i=1}^{n} \delta_i(n)$$

In the scheme of the algorithm that we have outlined, the evolution of the distance between two points is considered: one reference point and its nearest neighbour. This approach is the same as that adopted in Rosenstein *et al.* (1993). In contrast, in Kantz (1994), the mean distance between the reference point and a set of neighbours is monitored. Actually, if the length n of the series suffices, it is suggested that, to save computer time, one should not average over the entire trajectory but rather consider only a number of reference points less than n.

There are advantages in using this type of estimators: (i) they do not involve any kind of modelling and thus (ii) their computation is easy compared with Jacobian estimators. Notice also that for non-chaotic systems, the scaling region is not linear and that for stochastic systems, it tends to become flatter with increasing embedding dimension. Further comments about MCLE estimation are given in Giannerini *et al.* (2007*b*)

4.2.2 Estimating the MCLE with `tseriesChaos`

The estimate of the MCLE is implemented in `tseriesChaos` with the code (the same notation as in `tseriesChaos` is adopted here)

```
ly <- lyap\_k(series,m,d,t,k,ref,s,eps)
```

where

- `series`: time series.
- `m`: embedding dimension.
- `d`: time delay.
- `t`: Theiler window. Points separated by less than `t` in the series `series` are excluded from the search for neighbouring points.
- `k`: number of neighbours considered.
- `ref`: number of points in `series` taken into account.
- `s`: iterations along which the neighbours of each point are followed.
- `eps`: radius of the ball inside which nearest neighbours are searched for.

In practice, for each of the `ref` reference points ($\text{ref} \leqslant n$) (where n is the length of the reconstructed trajectory), we look for `k` neighbours that are closer than `eps` and have a temporal separation greater than `t`. Then, we compute the log-average distance between the point and its `k` neighbours, and we follow it for a finite number `s` of steps ahead in time. The instruction `ly <- lyap_k()` is followed by the two lines

```
plot(ly)
locator()
```

With `plot(ly)`, the evolution of the logarithm of the mean distance $S(\nu)$ is plotted as a function of the step ν. With `locator`, it is possible to detect the mouse position inside the plot. One clicks the two points, say `lmin` and `lmax`, that by eye appear to delimit the linear region and then clicks the right button of the mouse to finish. The (x, y) values of the clicked positions are printed. The estimate of the MCLE is obtained by running the instruction

```
lyap(ly,lmin,lmax)
```

Turning now to the Lorenz attractor, we recall from Chapter 3 that the first component of the Lorenz system $x(t)$ can be considered, in every sense, as the 'observed' series. Noise is not taken into account and the transient is discarded. After the embedding, the series x is of length $n = 7982$. The parameters used are `series`: x, m = 3, d = 10, t = 20, k = 6, `ref` = 5000, s = 600, eps = 5. That is, we consider 5000 reference points, each with 6 neighbours inside a ball of radius 5. The evolution of the mean distance is monitored for 600 steps ahead in time and points separated by less than 20 from the reference point in the series x are not considered as neighbouring points. To plot $S(\nu)$ versus ν, we have slightly modified the instructions `ly <- lyap_k()` and `plot(ly)` as follows:

```
S_nu <- lyap_k(x,m=3,d=10,t=20,k=6,ref=5000,s=600,eps=5)
plot(S_nu,xlab = expression(paste(nu)),ylab=expression(paste("S",(nu))))
lmin<- 80
lmax<- 280
# add the vertical lines delimiting the linear region:
abline(v=lmin,lty=4,lwd=2,col="black")
abline(v=lmax,lty=4,lwd=2,col="black")
```

Note the use of `expression` to plot mathematical symbols and expressions in R graphics. The result is shown in Figure 4.3. We can see that there is a linear region ranging from $\nu \approx 80$ to $\nu \approx 280$, which allows the estimate $\hat{\lambda}$ of the MCLE.

If desired, it is possible to avoid the use of `locator` by computing the slope of the segment for $\nu \in [80, 280]$, through regression analysis, as follows:

```
lr<- S_nu[lmin:lmax]      # output of lyap_k() in [lmin, lmax]
l_lr<- length(lr)-1
lt<- seq(lmin*step,lmax*step,by=step)
lm(lr~lt)                 # regression analysis in R
abline(lyap(S_nu,lmin,lmax),lty=2,lwd=2,col="black")  # add the regression line
```

To perform linear regression, a linear model is created by using the `lm()` function. The regression line is is added to the plot with the instruction `abline()`. The output produced by R is

```
Call:
lm(formula = lr ~ lt)
Coefficients:
(Intercept)          lt
    -0.2442      0.9628
```

The estimate of the MCLE is thus $\hat{\lambda} = 0.96$. The reader can try to vary some parameters in the function `lyap_k()`. For instance, if we choose to find only one neighbour

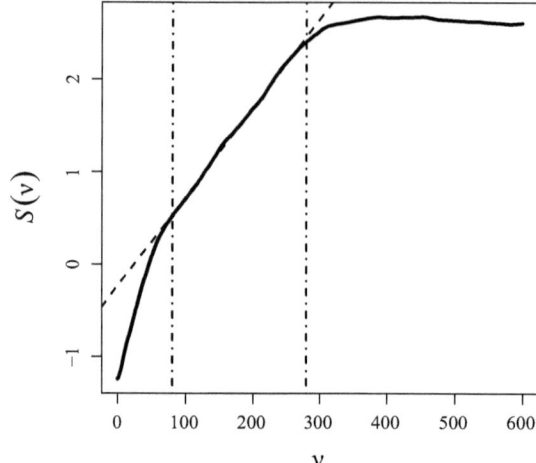

Fig. 4.3 Reconstructed Lorenz strange attractor: evolution of the logarithm of the mean distance $S(\nu)$ as a function of the time step ν. The dash–dot lines delimit the approximately linear region and the dashed line is the regression line.

for each reference point (k = 1), the result becomes $\hat{\lambda} = 0.88$. The choice of the linear region is somewhat subjective, and it is good practice to vary the extremes of this region slightly. If we choose the linear region as $[90, 270]$, we obtain $\hat{\lambda} = 0.97$, without any significant change.

It is interesting to analyse the algorithm more closely. Figure 4.4a shows the divergence $\delta(\nu)$ for three reference points, taken individually, with k = 1, as a function of the time step ν. It seems that it is very difficult to find a linear region – we see only some oscillations related to an increase or decrease of the distance. Figure 4.4b shows the mean distance $S(\nu)$ obtained with 200 reference points. Even though the oscillations are significantly reduced, a sharp linear region is still not apparent. This should provide a warning against the use of too short time series, when there are too few reference points.

In Chapter 6, estimates of the Lyapunov exponent will be used in user-defined functions for time series analysis. The appropriate codes will be shown there and described in detail.

4.3 Recurrence Plots

Henri Poincaré in an analysis of his own work on the three-body problem, writes (Poincaré, 1952, pp. 4–5):

Dans ce cas, si on laisse de côté certaines trajectoires exceptionnelles, dont la réalisation est infiniment peu probable, on peut démontrer que le système repassera une infinité de fois aussi près que l'on voudra de sa position initiale.

[In this case, neglecting some exceptional trajectories, the occurrence of which is infinitely improbable, it can be shown that the system will recur infinitely many times as close as one wishes to its initial state.]

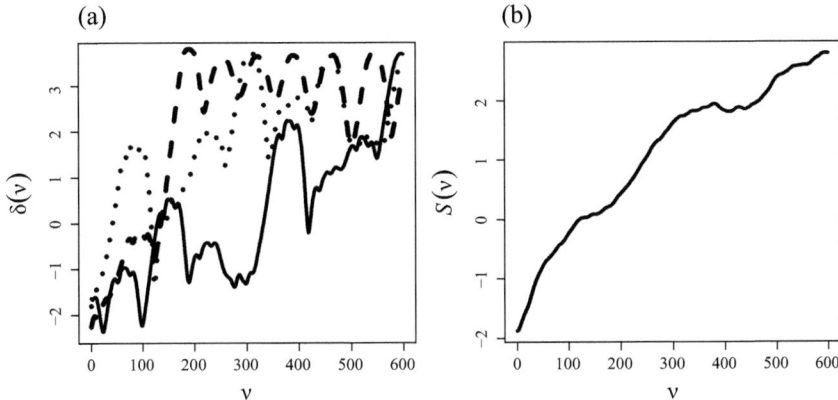

Fig. 4.4 (a) Divergence $\delta(\nu)$ for three reference points, taken individually. (b) Mean distance $S(\nu)$ obtained with 200 reference points

Poincaré refers here to his paper *Sur le probleme des trois corps et les équations de la dynamique* [*On the three-body problem and the equations of dynamics*] (Poincaré, 1890), for which he was awarded a prize by King Oscar II of Sweden and Norway on the occasion of his 60th birthday. This paper is famous not only because it introduced the concept of recurrences in conservative systems (known as Poincaré's *recurrence theorem*), but also because it was here that Poincaré sowed the seeds of chaos theory. Poincaré's recurrence theorem was at the basis of Zermelo's criticism of Boltzmann's *H*-theorem. For details of the debate between Boltzmann and Zermelo, as well as other historical controversies, we refer the reader to the classical book on the history of statistical mechanics by Brush (1986). Regarding recurrence plots, Marwan *et al.* (2007) provide a comprehensive overview of the subject, describing its application in various fields, while Marwan (2008) gives a historical review.

The essential question concerns the recurrence of chaotic dynamics. We know that in a deterministic system, a return at time j to the state x_i previously visited at time i can occur only if the state x_i is periodic, in which case the state recurs exactly every time period. Chaotic systems are deterministic but aperiodic. Can we again speak of recurrence here? The answer is affirmative – if we define recurrence within a certain threshold, that is, such that the state of a chaotic system is recurrent if it recurs, after a certain amount of time, not exactly, but in a sufficiently close way.

Let $\mathbf{y}_t$ be the reconstructed trajectory in the m-dimensional phase space (d is the time delay):

$$\mathbf{y}_t = [s(t), s(t+d), s(t+2d), \dots, s(t+(m-1)d)], \quad t = 1, \dots, n$$

Each $\mathbf{y}_t$ is a point in this space. We compute the distance δ_{ij} between two points at times i and j, that is, at $t = i$ and $t = j$:

$$\delta_{ij} = |\mathbf{y}_i - \mathbf{y}_j|$$

The distance as written here is the Euclidean distance, but it may be of a different type. R provides the function **dist** to compute the distances between the rows of a data matrix, using a distance measure that can be Euclidean, maximum, Manhattan, Canberra, binary or Minkowski. In the following, we refer to a reconstructed time series in the m-dimensional embedding phase space, but the discussion holds also for a true time series.

For a periodic time series, with period T, we have $\delta_{ij} = 0$ when $|i - j| = kT$, $k = 0, 1, 2, \ldots$ But, in general, we can ask ourselves when it is that two points are 'near enough', that is, when does $|\mathbf{y}_i - \mathbf{y}_j| < \epsilon$, where ϵ is a certain distance value. Suppose that the reconstructed time series is formed by n vectors:

$$\mathbf{y}_1, \mathbf{y}_2, \mathbf{y}_3, \ldots, \mathbf{y}_n$$

We can construct a new space i versus j, $i, j = 1, 2, \ldots, n$. When $\delta_{ij} = |\mathbf{y}_i - \mathbf{y}_j| < \epsilon$, the two points are considered recurrent points and we plot a black dot at the point (i, j); otherwise, we plot a small white circle or, better, we plot nothing. This kind of graph is called a *recurrence plot (RP)* and was introduced for the analysis of dynamical systems in 1987 by J.-P. Eckmann and co-workers (Eckmann *et al.*, 1987). RPs show how the trajectory repeats itself, and that it returns to visit the states already visited in the past, within the margin of a tolerance ϵ. It is clear that i and j are times, and therefore RPs visualize the time correlation between time series data. Note that an RP is a two-dimensional representation of a trajectory in a higher-dimensional phase space.

Formally, the RP can be written as the matrix

$$\mathbf{R}_{i,j} = \Theta(\epsilon - |\mathbf{y}_i - \mathbf{y}_j|), \quad i, j = 1, 2, \ldots, n \tag{4.1}$$

where $\Theta(\cdot)$ is the Heaviside function, that is, $\Theta(x) = 0$ if $x < 0$ and $\Theta(x) = 1$ otherwise. A black dot is always present on the main diagonal (the *line of identity, LOI*), that is, for $i = j$; indeed, $\mathbf{R}_{i,j} \equiv 1$. Moreover, the RP is symmetric with respect to the main diagonal; that is, $\mathbf{R}_{i,j} \equiv \mathbf{R}_{j,i}$ If the series is periodic, with period T, then, for fixed i, there is a black dot every $T - 1$ white circles; for example, for $T = 4$, we have ●○○○●○○○●... The R Code 4.1 logistic_RP.R gives the RP for the logistic map. This code allows us to see the main structures of the RPs in different dynamical regimes.

```
#Code 4.1 Logistic recurrence plot (logistic_RP.R)
x<- numeric()
xt<- numeric()
f.x<- function(x,r){
     r*x*(1-x)
                   }
##### function to compute the iterates #####
f.temp<-function(xinit,nstep,r){   # starting function f.temp
x<- xinit
xt[1]<- x
for(i in 2:nstep){
y<- f.x(x,r)
x<- y
xt[i]<- x
xt
```

```
                         }
return(xt)
                                   }   # ending function f.temp
### parameters and initial conditions
# r<- 3.52: period 4; r<- 4: chaos; r<- 3.835: periodic window
r<- 3.52
ntrans<-100      # transient
n<- 20          # iterations after the transient n <- 20, with r<- 4: n<- 100
nt<- ntrans+n   # total number of iterations
xinit<- 0.2
eps<- 0.01      #  with r<- 4:  do also eps<- 0.1, for comparison
xt<- f.temp(xinit,nt,r)                        # iterates
par(mai=c(1.02,1.,0.82,0.42)+0.1)              # to control the margin size
plot(0,0,type="n",xlab="i",ylab="j",xlim=c(0,n),ylim=c(0,n),
     cex.lab=1.7,cex.axis=1.3)
x[1:n]<- xt[(ntrans+1):nt]
for(i in 1:n){                                 # construction of the recurrence plot
for(j in 1:n){
distx<- abs(x[i]-x[j])                         # Euclidean distance between x(i) and x(j)
if(distx<eps) points(i,j,pch=19,cex=1.8)       # black point cex=1.8,  with r<- 4: cex=1
else points(i,j,pch=21,cex=1.8)                # white circle
                            # with r<- 4: no white circles, comment 'else points(.)'
    }
    }
```

This code is almost the same as the `logistic.R` code, but here it does not plot the
series in the time domain or its cobweb plot, but rather plots black or white circles
with the last instructions. Figure 4.5 shows the RPs for two periodic behaviours, (a)
when $r = 3.52$ (a cycle of period 4) and (b) when $r = 3.835$ (periodic window, a cycle
of period 3).

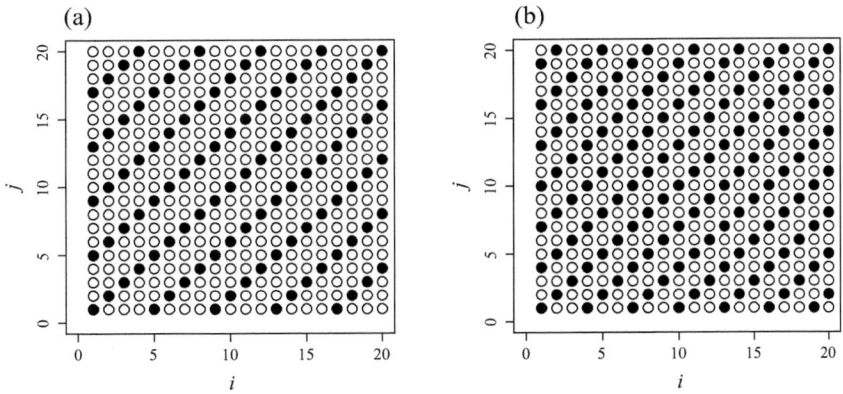

Fig. 4.5 RPs for the logistic map ($\epsilon = 0.01$): (a) $r = 3.52$ (a cycle of period 4); (b) $r = 3.835$
(periodic window, a cycle of period 3).

In a periodic orbit, the RP is characterized by uninterrupted parallel diagonal
lines separated by a constant distance corresponding to the period T. If we vary ϵ
(provided obviously that the variation is not excessive), the plots do not change. The
reason is that the points over a periodic orbit repeat themselves almost exactly and

are not affected by the presence of neighbours, as is shown in Figure 4.6, in which is a schematic representation of a period-4 trajectory in the two-dimensional phase space (x, y):

$$1 \to 2 \to 3 \to 4 \to 5 \, (\equiv 1) \to 6 \, (\equiv 2) \ldots$$

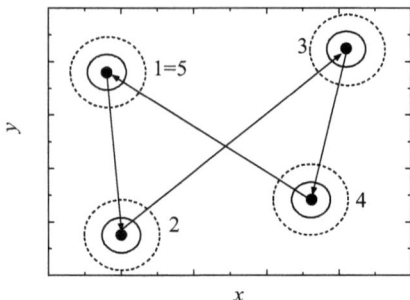

Fig. 4.6 Phase space with four periodic points. The radii of the circles around the points are for two different values of ϵ.

Things change if the (true or reconstructed) series is chaotic, as shown in Figure 4.7, in which only black dots are plotted. Now, apart from the main diagonal, where black dots are always present, we again note some diagonals, but these are much shorter than those of the periodic motion. Furthermore, the distance between diagonal lines is not constant, since there is no single defined period and the exponential divergence of nearby trajectories causes interruption of the lines. If ϵ is increased, the numbers of black dots and parallel lines also increase, but, as we shall see, increasing ϵ makes sense only up to a certain limit. It is important to note that the RP highlights an underlying structure. In particular, on the upper right, we can see a small rectangular cluster of points that resembles the RP of the periodic motion.

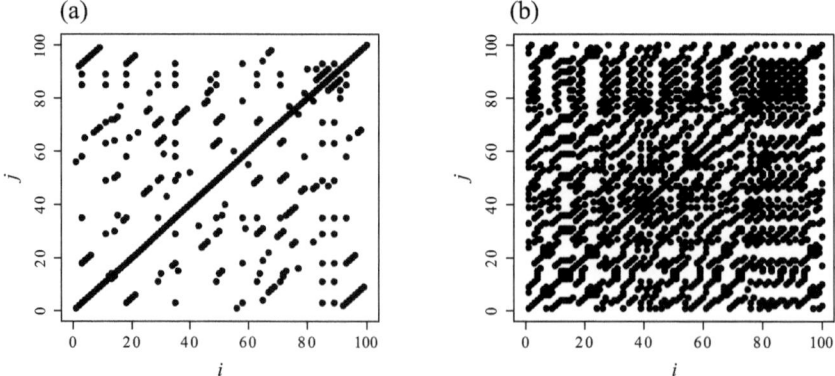

Fig. 4.7 RPs for the logistic map with $r = 4$ (chaotic regime): (a) $\epsilon = 0.01$; (b) $\epsilon = 0.1$.

With the code `random_RP.R`, we can examine the RP of a series obtained with the function **runif** (random deviates from the uniform distribution) with values uniformly distributed in the plane $[0, 1] \times [0, 1]$:

```
#Code 4.2 Random recurrence plot (random_RP.R)
# series with uniformly distributed points
x<- numeric()
n<- 100        # length of the series
eps<-  0.01    # do also eps<- 0.1, for comparison
plot(0,0,type="n",xlab="i",ylab="j",xlim=c(0,n),ylim=c(0,n),
     cex.lab=1.7,cex.axis=1.3)
set.seed(1)    # seed of the sequence of (pseudo) random numbers
x<- runif(n)   # n random deviates from the uniform distribution are generated
for(i in 1:n){ # construction of the recurrence plot
for(j in 1:n){
distx<- abs(x[i]-x[j])     # Euclidean distance between x(i) and x(j)
if(distx<eps) points(i,j,pch=19,cex=1.0)
           }
           }
```

Figure 4.8 shows the RPs obtained with this code for two values of ϵ.

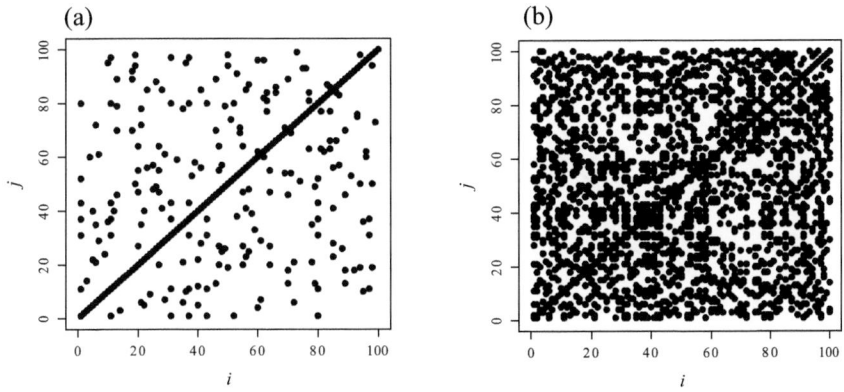

Fig. 4.8 RPs for a series with values uniformly distributed at random in the plane $[0, 1] \times [0, 1]$: (a) $\epsilon = 0.01$; (b) $\epsilon = 0.1$.

Certainly, increasing $\epsilon = 0.1$ leads to a greater number of black points, but in no case is any underlying structure visible, and instead we see rather homogeneous patterns. For a deeper analysis of the RPs for the logistic map, see Marwan *et al.* (2002).

We can add the following instructions to the code `lorenz_attractor.R` to obtain the RPs:

```
n<- 800
eps<- 5
distx<- matrix(,n,n)       # distances matrix
x<- xyz[,1]
y<- xyz[,2]
z<- xyz[,3]
for(i in 1:n){                      # construction of the thresholded recurrence plot
```

```
for(j in 1:n){
distx[i,j]<- sqrt( (x[i]-x[j])^2 + (y[i]-y[j])^2 + (z[i]-z[j])^2 ) # Euclidean distance
if(distx[i,j]<eps) points(i,j,pch=19,cex=0.01)
   }
   }
```

The matrix xyz, with the three columns $xyz[,1]$,$xyz[,2]$,$xyz[,3]$, describes the reconstructed trajectory. For clarity in the plots, we considered only the first 800 steps after the transient, so n represents the number of steps in which the Euclidean distance $distx[i,j]$ between two points i and j of the reconstructed trajectory is computed. Figure 4.9a shows the RP of the reconstructed Lorenz attractor. More precisely, such an RP is called a *thresholded recurrence plot (TRP)*, since it displays only those points that fall within a fixed distance, as in the previous figures in this chapter. Figure 4.9b shows the *unthresholded recurrence plot (UTRP)*, in which the distance between points is displayed by means of a greyscale. Small rectangular patches resembling the RP of periodic motion are apparent in both Figures 4.9a and 4.9b. In fact, such block-like structures reflect nearly periodic motions on the attractor, called *unstable periodic orbits (UPOs)*.

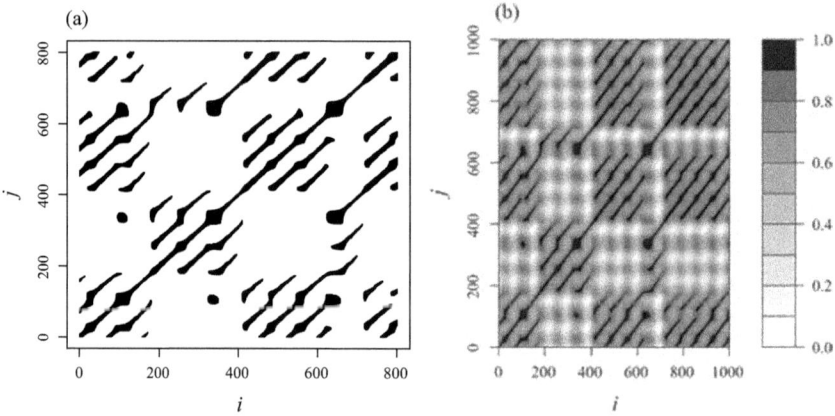

Fig. 4.9 Reconstructed Lorenz strange attractor: (a) thresholded RP with $\epsilon = 5$; (b) unthresholded RP.

In order to obtain the UTRP, we add the following instructions to the code `lorenz_attractor.R`:

```
time<- seq(1:n)
windows()
par(mai=c(1.02,1.,0.82,0.42)+0.1,cex.lab=1.6,cex.axis=1.2)
require(gplots)
distx<- distx/max(distx)        # normalized distances matrix previously computed
distx<- 1-distx
filled.contour(time,time,distx,col=gray.colors(10,start=1,end=0),nlevels=10,
xlab = "i", ylab = "j", main = "",xlim=c(0,n),ylim=c(0,n),las=0,
key.axes = axis(4, las=1) )
```

As in Code 3.3 `Lorenz system: density, 1 trajectory`, the number of counts

in each bin is normalized and the function **filled.contour** is used. The line `distx<-`
`1-distx` is a simple way to get white for empty bins and black for normalized `distx`
= 1. Compare with the TRP in Figure 4.9a. One can use different palettes, as already
written for the code `Lorenz system: density, 1 trajectory`.

In the R package `tseriesChaos`, the command

`recurr(series, m, d, start.time=start(series), end.time=end(series), ...)`

generates unthresholded RPs with greyscale as in Figure 4.9b. The '...' in `recurr()`
are to pass further parameters to **filled.contour**.

We stress that the choice of the threshold ϵ in the TRPs is quite critical when the
system is not periodic. If ϵ is too large, points that are not really neighbours may be
classified as recurrent. On the other hand, if ϵ is too small, almost no RPs may be
identified. Some prescriptions for the choice of ϵ have been suggested in the literature,
as reported in Marwan *et al.* (2007). For an application of a real-world TRP, see Greco
et al. (2011).

4.3.1 Recurrence Quantification Analysis

We have seen that lines parallel to the main diagonal mean that a segment of the
trajectory runs almost parallel to another segment at different times. We can imagine
a (hyper)tube in phase space with diameter ϵ through which the trajectory passes
at different times. The shorter the tube, the less predictable is the trajectory, and
therefore the length of the diagonals may be linked to the predictability of the system.
These and similar considerations led Zbilut and Webber (1992) to develop *recurrence
quantification analysis (RQA)* to quantify RPs; see also Webber and Zbilut (1994),
Marwan *et al.* (2007) and Marwan and Webber (2015). As we have already said for
the Lyapunov exponents, and as we will say for the correlation dimension, when we
have a finite observed series, we *estimate* such quantities; that is, they are *statistics*.
Several statistics have been proposed for the RPs, mainly based on the density of the
points and the conformation of the diagonals in the plots.

The Comprehensive R Archive Network (CRAN), the public clearing house for
R packages, contains a further two packages, in addition to `tseriesChaos`, for the
analysis of nonlinear time series, namely the package `fNonlinear` and the package
`nonlinearTseries`. This latter package offers, among several functions to compute
nonlinear statistics, the possibility of RQA. We will see how RQA works for a very
simple series, in order to introduce the most commonly used statistics of this technique.
The function **RQA** produces the following statistics:

- Recurrence rate (REC): percentage of black points on the TRP.
- Determinism (DET): percentage of recurrent points that form diagonal lines parallel to the main diagonal (LOI). Points on the LOI are, or are not, excluded from DET computation and only diagonals longer than or equal to a minimum length threshold $l_{\min}$ are considered. Typically, $l_{\min} = 2$.
- Ratio (RATIO): ratio of DET to REC. This can be useful in discovering dynamic transitions (Marwan *et al.*, 2007).
- Length of the longest diagonal line ($L_{\max}$). As already mentioned, the diagonal lines are linked to the divergence of the trajectory segments.

- Inverse of L_{max} (DIV). This quantity has been interpreted as an estimator of the MCLE (Trulla *et al.*, 1996); however, Marwan (2011) advises caution in relating the RQA measures to chaotic properties of the series.
- Average length of the diagonal lines (L_{mean}). This is the average time in which two segments of the trajectory are within the same 'tube' of phase space.
- Shannon entropy of the diagonal line length distribution (ENTR). This is a measure of the complexity of structure in the system.
- Trend (TREND): trend of the density of points depending on the distance to the LOI. TREND is employed to measure the degree of stationarity of the series.
- Laminarity (LAM): percentage of recurrent points that form vertical lines. This is analogous to DET, but refers to vertical, rather than diagonal, structures.
- Length of the longest vertical line (V_{max}). This is analogous to the diagonal length L_{max}. Only diagonals longer than or equal to a minimum length threshold v_{min} are considered.
- Average length of the vertical lines (V_{mean}), also called the trapping time (TT).

In addition to these statistics, the function **RQA** also computes the histogram of the length of the diagonals and *recurrenceRate*, the number of recurrent points depending on the distance to the LOI.

The 'toy series' to which the function **RQA** is applied here as an example is

$$0, 0, 1, 0, 0, 0, 0, 0, 1, 0$$

that is, it is a dichotomous series. The code is

```
y<- numeric()
y<- c(0,0,1,0,0,0,0,0,1,0)   # original time series
eps<- 0.01
require(nonlinearTseries)
rqa.analysis=rqa(time.series=y,embedding.dim=1,time.lag=1,
radius=eps,lmin=2,vmin=2,do.plot=TRUE,distanceToBorder=2,save.RM=TRUE)
rqa.analysis
```

For a dichotomous series, the threshold ϵ is useless, since two points are either equal or not. We can see that the code executes the embedding, if required. It is possible to give the Takens vectors directly. The arguments `lmin` and `vmin` are l_{min} and v_{min}, respectively. The argument `distanceToBorder` is introduced to avoid border effects which could yield distorted estimates. The recurrence matrix is saved with `save.RM=TRUE` and the RP is produced with `do.plot=TRUE`; this plot is shown in Figure 4.10a. Note that the LOI goes from the upper left vertex to the lower right vertex and that the scale on the y axis starts from 10 and ends at 1. With the instructions used to plot the RP of the logistic map in the `logistic_RP.R` code, the more familiar RP is shown in Figure 4.10b. The two plots are equivalent if the transpose of the matrix on the right is taken and thus the x axis (i) and the y axis (j) are swapped.

The RQA results using the function **RQA** are as follows:

REC	DET	RATIO	Lmax	DIV	Lmean	Lmean-LOI	ENTR	TREND	LAM	Vmax	Vmean
0.68	0.706	1.038	4	0.25	3.2	2.71	1.137	-0.013	0.823	5	3.5

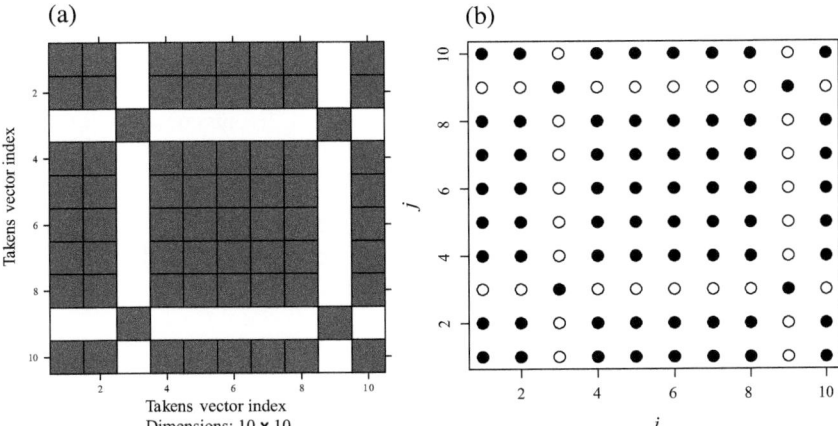

Fig. 4.10 (a) RP obtained with the package `nonlinearTseries`. (b) Plot instructions in the `logistic_RP.R` code.

We can see from Figure 4.10 that `REC` is computed without excluding the LOI, and the same has been done for `Lmean`, but the code also computes `Lmean` without the points on the LOI (`Lmean-LOI`). Examples of `DET` computation are (see Figure 4.10b) the line from $(i = 6, j = 1)$ to $(i = 7, j = 2)$, with length $= 2$, and the line from $(i = 7, j = 1)$ to $(i = 10, j = 4)$, with length $= 3$; that is each line is counted only once. We also have the histogram of the length of the diagonals:

```
diagonalHistogram: 0 8 2 4 0 0 0 0 0 1}
```

(there are eight line of length 2, two lines of length 3, etc.), and the number of recurrent points as a function of the distance to the LOI

```
recurrenceRate: 1.1111111 1.2500000 1.4285714 1.3333333 1.2000000 2.0000000
                0.6666667 1.0000000 2.0000000
```

4.3.2 Cross Recurrence Plots

The concept of RPs has been extended to the simultaneous analysis of the dynamics in two systems (Zbilut *et al.*, 1998; Marwan and Kurths, 2002). Besides the reconstructed time series $\mathbf{y}_1, \mathbf{y}_2, \mathbf{y}_3, \ldots, \mathbf{y}_n$ let us consider the reconstructed time series $\mathbf{z}_1, \mathbf{z}_2, \mathbf{z}_3, \ldots, \mathbf{z}_m$. Suppose that the two series are embedded in the same phase space, but that their lengths are not equal. Formally, eqn (4.1) can be rewritten as

$$\mathbf{CR}_{i,j} = \Theta(\epsilon - |\mathbf{y}_i - \mathbf{z}_j|), \quad i = 1, 2, \ldots, n, \quad j = 1, 2, \ldots, m \tag{4.2}$$

There are two trajectories in the same phase space, and it can happen that for certain time intervals the second trajectory is very close to the first. To ascertain whether this occurs, all points of the first trajectory are tested with all points of the second, giving rise to a *cross recurrence plot (CRP)*. In the CRP, there is a black dot at the point (i, j)

if the state $\mathbf{z}_j$ at time j is at a distance less than ϵ from the state $\mathbf{y}_i$ at time i. Since the lengths of the two series can differ, the CRP matrix can be a non-square matrix.

Marwan and Kurths (2004, p. 110) write (their italics):

This occurrence of neighbours in both trajectories is not a 'recurrence' of states, hence the matrix [**CR**] does not represent recurrences but the conjunctures of states of both systems. Therefore, this representation is not really a 'cross *recurrence* plot'.

Other authors (Coco and Dale, 2014) say that the CRP displays *co-visitation* patterns, that is (p. 4, their italics): 'one time series is *revisiting* states that the other time series has visited'. In any case, what the CRPs bring to light is a possible similarity of the dynamical behaviour of both processes.

In CRPs, in contrast to RPs, it is not always the case that $\mathbf{R}_{i,i} \equiv 1$; if $\mathbf{R}_{i,i} \not\equiv 1$, then the line of identity (LOI) is replaced by the *line of synchronization (LOS)*. Imagine that the second series $\mathbf{z}$ is identical to the first, $\mathbf{y}$, in which case the CRP becomes an RP, with its LOI. The more that $\mathbf{z}$ deviates from $\mathbf{y}$, the more often is the LOI interrupted. The LOI can be continuous, but bowed owing to a time dilatation or a time compression of one of the series (Marwan and Kurths, 2004). As with RQA in the case of RPs, examination of the diagonal lines in the CRPs is an important tool for *cross recurrence quantification analysis (CRQA)*. CRAN provides the package CRQA to execute CRQA. This package contains several functions, one of which, **crqa**, analogously to the function **RQA** in the package nonlinearTseries, computes the statistics characterizing the interaction between the two series. As an example, to the previous 'toy series' we now add a further one, $0, 1, 0, 1, 1, 0, 0, 0, 1, 1$, that is inserted in the code CRQA_example.R:

```
CRQA_example.R
ts1<- numeric()
ts2<- numeric()
ts1<- c(0,0,1,0,0,0,0,0,1,0)    # first time series
ts2<- c(0,1,0,1,1,0,0,0,1,1)    # second time series

#parameters initialization:
delay = 1; embed = 1; rescale = 1; radius = 0.001;
normalize = 0; mindiagline = 2; minvertline = 2;
tw = 0; whiteline = FALSE; recpt = FALSE; side = "both"
checkl = list(do = FALSE, thrshd = 3, datatype = "categorical",
pad = TRUE)

# install.packages("crqa", dep = TRUE)
# to install the package if not present require(crqa)

## call the function crqa
ans = crqa(ts2, ts1, delay, embed, rescale, radius, normalize,
mindiagline, minvertline, tw, whiteline, recpt, side, checkl)

print(ans[1:9]) ## last argument of list is the CRP
CRP = ans$RP ## take out RP
CRP
```

The parameters have the following meanings:

- **delay** and **embed**: time delay and embedding dimension if embedding of the two series is required.

- `rescale`: rescales the distance matrix. If `rescale` = 0, there is no rescaling.
- `radius`: threshold. as already remarked with regard to the code `RQA_example.R`, the threshold is useless for a dichotomous series.
- `normalize`: normalizes the time series. if `normalize` = 0, there is no normalization.
- `mindiagline` and `minvertline`: minimum diagonal and vertical lengths of recurrent points. In **RQA**, these are named l_{min} and v_{min}, respectively.
- `tw`: Theiler window.
- `whiteline`: a logical flag to calculate (`TRUE`) or not (`FALSE`) empty vertical lines.
- `recpt`: a logical flag to calculate the statistics directly from a recurrent plot (`TRUE`) or not (`FALSE`).
- `side`: can be 'upper', 'lower' or 'both', depending on whether the statistics are computed in the upper or lower triangle of the CRP, or in both.
- `checkl`: contains do: normalize (or not) the length of the time series.
- `datatype`: indicates the nature of the time series.
- `thrshd`: threshold difference between the number of points of the two series.
- `pad`: indicates whether (`TRUE`) or not (`FALSE`) the series have to be padded.

The CRP is as follows:

```
 [1,] | . | . . | | | . .
 [2,] | . | . . | | | . .
 [3,] . | . | | . . . | |
 [4,] | . | . . | | | . .
 [5,] | . | . . | | | . .
 [6,] | . | . . | | | . .
 [7,] | . | . . | | | . .
 [8,] | . | . . | | | . .
 [9,] . | . | | . . . | |
[10,] | . | . . | | | . .
```

where the small vertical dashes are the black points; the notation of the matrix is like that in Figure 4.10a. The following estimates are also computed:

RR	DET	NRLINE	maxL	L	ENTR	rENTR	LAM	TT
50	64	12	4	2.667	1.011	0.921	70	3.5

These statistics have the following meanings (the terms in parentheses are those used in the `nonlinearTseries` package):

- RR (REC): percentage of black points.
- DET (DET): percentage of recurrent points that form diagonal lines parallel to the main diagonal. Diagonal lines indicate segments of both trajectories visiting the same region of phase space.
- NRLINE: total number of diagonal lines, including the points on the main diagonal.
- maxL (Lmax): length of the longest diagonal line, excluding the main diagonal.
- L (L_{mean}): average length of the diagonal lines.
- ENTR (ENTR): Shannon entropy of the diagonal line length distribution.
- rENTR: entropy measure normalized by the number of lines in the CRP.
- LAM (LAM): percentage of recurrent points that form vertical lines.

- TT (V_{mean}): average length of the vertical lines.

It is worth noting that for categorical time series, as in this example, recurrence can be calculated by means of contingency tables with the function **CTcrqa(ts1, ts2, par)**. The code is CTcrqa_example.R:

```
CTcrqa_example.R
par = list(lags = seq(1, 40, 1), datatype = "categorical", thrshd = 3);
res = CTcrqa(ts1, ts2, par)
res
plot(seq(1,length(res),1), res, xlab = "Delays",
ylab = "Recurrence", type = "l", lwd = 3)
```

Before ending this section, we recall another bivariate approach to recurrence, namely the *joint recurrence plot (JRP)*, proposed by Romano *et al.* (2004). Suppose that we have two trajectories, **y** and **z**, in two separate phase spaces. By means of the JRP, we can explore possible joint recurrences, that is, the occurrence of two simultaneous events. In the first event, the system described by the first trajectory **y**, at time j, is in the state $\mathbf{y}_j$ and returns, at a distance less than $\epsilon_\mathbf{y}$, to the state $\mathbf{y}_i$, previously visited at time i. The second event is the same as the first, except that **y** is replaced by **z**. The probability that these events occur simultaneously, each in its own phase space, is the product of the probabilities of each single event. Formally, eqn (4.2) becomes

$$\mathbf{JR}_{i,j} = \Theta(\epsilon_\mathbf{y} - |\mathbf{y}_i - \mathbf{y}_j|) \times \Theta(\epsilon_\mathbf{z} - |\mathbf{z}_i - \mathbf{z}_j|), \quad i,j = 1, 2, \ldots, n$$

which is the product of the RPs of each series. The JRP can be also seen as the Hadamard product of the RP of the first system and that of the second system. For a comparison between CRPs and JRPs, see Marwan and Webber (2015).

4.4   Correlation Dimension

The correlation dimension measures the geometric dimension of the attractor and also gives the minimum number of variables required to construct the phase space, which is the nearest integer above the correlation dimension. The dimension may be integer or fractal. There are several ways to calculate a fractal dimension, the correlation dimension being one of these. We have already seen fractals in the context of the logistic map. Its bifurcation diagram (Figure 2.11) shows the self-similarity property. A fractal contains an infinite number of copies of itself; that is, it is similar to itself on all scales. Benoît Mandelbrot, in his book *The Fractal Geometry of Nature*, defines the term *fractal* (Mandelbrot, 1982, p. 15):

A fractal is by definition a set for which the Hausdorff–Besicovitch dimension strictly exceeds the topological dimension.

A brief digression on these concepts seems appropriate here.

4.4.1   Fractals

Let us begin with the *topological dimension*. In Liebovitch (1998, p. 46), we read (his italics):

The topological dimension describes how points within an object are connected to each other.
It tells us that an object is an edge, a surface, or a solid.
Its value is always an *integer* (for example: 1 or 2).

A line has topological dimension 1, a surface topological dimension 2 and a solid body topological dimension 3. Thus, in short, the topological dimension coincides with the Euclidean dimension for lines, planes and volumes.

Let us see another type of dimension. Consider a line segment of length L. We can cover it with another segment of the same length, or with 2 segments of length $L/2$ or with 4 segments of length $L/4$, etc., as shown schematically in Figure 4.11a, where, for the sake of clarity, the covering segments are sketched as small rectangles. Let $N(\epsilon)$ be the number of segments of length ϵ covering the segment of length L. Then

$$N(\epsilon) = L\frac{1}{\epsilon}$$

An analogous reasoning holds when the covering object is a square of side L, as shown in Figure 4.11b, in which the square of side L is covered by 4 squares each of side $L/2$ side and by 16 squares each of side $L/4$.

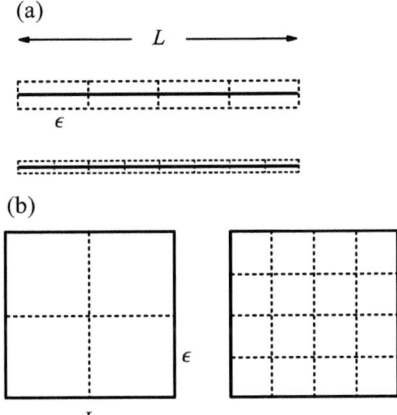

Fig. 4.11 Covering method for evaluating the Hausdorff–Besicovitch dimension.

In general, a square of side L can be covered by

$$N(\epsilon) = L^2\frac{1}{\epsilon^2}$$

squares of side ϵ. In d dimensions, we have

$$N(\epsilon) = L^d\frac{1}{\epsilon^d}$$

where ϵ is the side of the d-dimensional hypercube. Taking logarithms,

$$\log N(\epsilon) = d \log L + d \log \left(\frac{1}{\epsilon}\right)$$

Then

$$d = \frac{\log N(\epsilon)}{\log L + \log(1/\epsilon)}$$

If $\epsilon \to 0$, then $\log L$ becomes negligible compared with the other two terms. The *Hausdorff–Besicovitch (H–B) dimension* is defined as

$$D_0 = \lim_{\epsilon \to 0} \frac{\log N(\epsilon)}{\log(1/\epsilon)} \tag{4.3}$$

The subscript '0' is to distinguish this dimension from other dimensions D_q, $q = 0, 1, 2, \dots$. For $q = 1$, the dimension D_1 is the information dimension.

Actually, eqn (4.3) is not a rigorous definition of the Hausdorff–Besicovitch dimension (see e.g. Falconer (2014)), but this definition has become common in the dynamical systems literature. Sometimes, the H–B dimension is also called the *capacity dimension* or the *box-counting dimension*, although in fact these latter dimensions are similar but not identical to the H–B dimension. It is obvious for a line segment that $D_0(L) = 1$, for a surface that $D_0(L^2) = 2$, and analogously for a cube and a hypercube. In all these cases, the H–B dimension coincides with the usual Euclidean dimension, but let us see what happens for a fractal, namely the Cantor set, or more precisely *the middle-third Cantor set*, or more poetically *Cantor dust*, here denoted by $\mathcal{C}$.

The rules of the game to construct the Cantor set are sketched in Figure 4.12. At the beginning, the unit interval, which we shall denote by I_0, is divided into three equal parts and the middle part is deleted; that is, the open interval $\left(\frac{1}{3}, \frac{2}{3}\right)$ is cancelled, so there remains the set of points $I_1 = \left[0, \frac{1}{3}\right] \cup \left[\frac{2}{3}, 1\right]$. The same procedure is executed in successive iterations. Thus, I_2 consists of 4 segments, each of length $\frac{1}{9}$, and in general I_n consists of 2^n segments, each of length $\left(\frac{1}{3}\right)^n$. As the process proceeds, the number

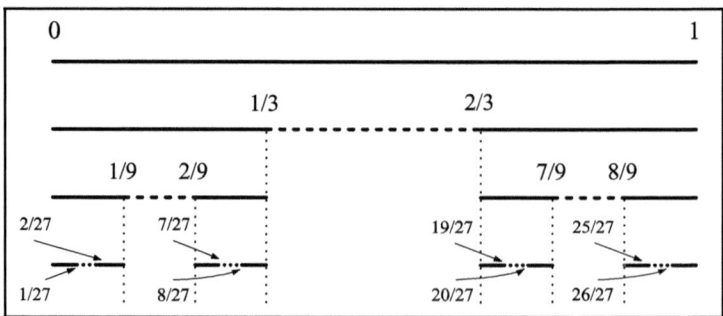

Fig. 4.12 Construction of the Cantor set. At the beginning, the middle third of the unit interval is removed. In each iteration, all of the remaining parts are divided into three and the middle parts are deleted. Here three iterations are shown.

of small segments increases, but their length decreases. The self-similarity property is clear: if at the nth iteration $(n \gg 1)$ we observe 'with a microscope' any tiny part of the figure, we always find the same scheme: two segments of equal length separated by a distance of the same length. The geometrical figure thus created, the middle-third Cantor set, is formally defined as

$$\mathcal{C} = \lim_{n \to \infty} I_n = \bigcap_{n=0}^{\infty} I_n$$

The set $\mathcal{C}$ is the set of points that belong to all I_n, $\forall n$.

At first sight, the Cantor set exhibits somewhat odd properties. What remains at 'the end' of all the iterations? Summing the lengths of the removed segments, we have

$$\frac{1}{3} + \frac{2}{9} + \frac{4}{27} + \ldots + \frac{2^{n-1}}{3^n} = \frac{1}{3} \sum_{n=1}^{\infty} \left(\frac{2}{3} \right)^{n-1}$$

$$= \frac{1}{3} \left(\frac{1}{1 - \frac{2}{3}} \right) = 1$$

Thus, the total length of the removed segments is equal to 1, which is just the same as the length of the starting interval $[0, 1]$, and therefore the length of $\mathcal{C}$ is zero: have we have taken everything away? No, something remains. Indeed, each time that I_{n+1} is created from I_n, two points are left in $\mathcal{C}$. The first are

$$0, \ 1, \ \tfrac{1}{3}, \ \tfrac{2}{3}, \ \tfrac{1}{9}, \ \tfrac{2}{9}, \ \tfrac{7}{9}, \ \tfrac{8}{9}, \ \tfrac{1}{27}, \ \ldots \in \mathcal{C}$$

They form a countably infinite set, all being rational numbers. Further points remain in $\mathcal{C}$. For instance, if $x_0 = \frac{1}{4}$, then $x_1 \in \left[0, \frac{1}{3}\right]$, and $x_2 \in \left[\frac{2}{9}, \frac{1}{3}\right]$, $x_3 \in \left[\frac{2}{9}, \frac{7}{27}\right]$, etc. If $x_0 = \frac{1}{4}$, then *all* the iterates remain in $\mathcal{C}$. It is possible to prove (see e.g. Alligood *et al.* (1997)) that, even though $\mathcal{C}$ has zero length, the set contains as many points as there are in the initial unit interval $[0, 1]$, namely an infinite number. To clarify this seeming paradox somewhat, let us calculate the H–B dimension of $\mathcal{C}$. From Figure 4.12, it is apparent that we can cover the initial interval $I_0 = [0, 1]$ by $N(\epsilon) = 1$ segment of length $\epsilon = 1$ and the interval $I_1 = \left[0, \frac{1}{3}\right] \cup \left[\frac{2}{3}, 1\right]$ by $N(\epsilon) = 2$ segments of length $\epsilon = \frac{1}{3}$. In this manner, we arrive at Table 4.1.

By applying eqn (4.3), we have

$$D_0 = \lim_{\epsilon \to 0} \frac{\log N(\epsilon)}{\log(1/\epsilon)} = \lim_{\epsilon \to 0} \frac{\log 2^n}{\log 3^n} = \frac{\log 2}{\log 3} \approx \frac{0.6931}{1.0986} \approx 0.63$$

This is not an integer number. D_0 is a fractional dimension – in other words, the Cantor set is a fractal. It is a geometric figure consisting of an infinite number of points, but its dimension is less than 1, so to speak, 'a little less' than that of a segment, whose topological dimension is 1, and 'a little greater' than that of a point, whose topological dimension is 0.

On the Internet and in books, the reader can find wonderful galleries of fractal

Table 4.1 Cantor set sketched in Figure 4.12 covered by $N(\epsilon)$ segments of length ϵ.

$N(\epsilon)$	ϵ
$2^0 = 1$	1
$2^1 = 2$	$\frac{1}{3}$
$2^2 = 4$	$\frac{1}{9}$
$2^3 = 8$	$\frac{1}{27}$
$\vdots$	$\vdots$
2^n	$\left(\frac{1}{3}\right)^n$

art, including also portraits of strange attractors, which have fractal dimension. To measure it, a further dimension is introduced: the correlation dimension.

4.4.2   Correlation Integral

Suppose we have, as usual, a reconstructed time series formed by n vectors:

$$\mathbf{y}_1, \mathbf{y}_2, \mathbf{y}_3, \dots, \mathbf{y}_n$$

We are interested in measuring their *spatial correlation*. Let us define the *correlation dimension* D_2 as the double limit

$$D_2 = \lim_{\epsilon \to 0} \lim_{n \to \infty} \frac{\log C(\epsilon)}{\log(\epsilon)} \tag{4.4}$$

where the *correlation integral* $C(\epsilon)$ is defined as the proportion of all pairs of reconstruction vectors closer than a distance ϵ in some norm and is given by

$$C(\epsilon) = \lim_{n \to \infty} \frac{1}{n^2} \times \{\text{number of times } |\mathbf{y}_i - \mathbf{y}_j| \le \epsilon\} \tag{4.5}$$

This is the classic definition of the correlation integral (Grassberger and Procaccia, 1983*a*; Grassberger and Procaccia, 1983*b*). It may be interpreted as a measure of the probability that two points chosen at random will be within a certain distance of each other. The simplest and most popular way to estimate the correlation integral is the empirical analogue:

$$\widehat{C}(\epsilon) = \frac{2}{n(n-1)} \sum_{i=1}^{n} \sum_{j=i+1}^{n} \Theta(\epsilon - |\mathbf{y}_i - \mathbf{y}_j|) \tag{4.6}$$

where $\Theta(\cdot)$ is the Heaviside function, which we have already encountered, with $\Theta(x) = 0$ if $x < 0$ and $\Theta(x) = 1$ otherwise. The denominator $n(n-1)$ is the maximum possible number of pairs $(\mathbf{y}_i, \mathbf{y}_j)$, excluding the cases where $i = j$. Note that eqn (4.6) is the same as that used to calculate the recurrence rate with the function **RQA**, except that in the recurrence rate the points on the the main diagonal ($i = j$) are not excluded.

Remember that different type of correlation integrals are used. An example is the so-called Gaussian correlation integral, in which two points $\mathbf{y}_i$ and $\mathbf{y}_j$ separated by a small distance are weighted more than those separated by large distances; in this case, a Gaussian weighting system is used. This algorithm is particularly suitable when the observations are contaminated by noise (Diks, 1996).

The essential point is that $C(\epsilon)$ is linked to the fractal dimension of the attractor, which we wish to estimate. Supposing that the phase space is a one-dimensional curve, let us take a reference point and count how many points are at a distance less than ϵ. If ϵ is increased, the number of points increases by an amount proportional to ϵ; that is, $C(\epsilon) = k\epsilon$. The same reasoning can be applied on a plane, considering how many points are within a circumference of radius ϵ. With increasing ϵ, the number of points increases by a factor $\pi\epsilon^2$; that is, $C(\epsilon) = k\epsilon^2$. In three dimensions, the corresponding factor is $(4/3)\pi r^3$. In general, if points are randomly distributed in a D_2-dimensional hyperspace, then the number of points at a distance less than ϵ from the reference point is proportional to ϵ^{D_2}:

$$C(\epsilon) = k\epsilon^{D_2} \tag{4.7}$$

Taking the logarithm of both sides gives

$$\log C(\epsilon) = D_2 \log \epsilon + \log k \tag{4.8}$$

The term $\log k$ is a constant, and then to obtain D_2 we have to plot $\log C(\epsilon)$ as a function of $\log \epsilon$ and evaluate the slope of the line in the linear region (called the *scaling region*) of the log–log plot. The quantity D_2 is the correlation dimension. The commonly used subscript '2' refers to the correlation dimension defined by Grassberger and Procaccia (1983a, 1983b). Hentschel and Procaccia (1983) show how it is possible to associate with fractals an infinite number of generalized dimensions D_q, $q > 0$. Note also the difference between the correlation dimension D_2 and the Hausdorff–Besicovitch dimension, or the capacity dimension, or the box-counting dimension, or whatever we call them. These latter dimensions are purely geometric concepts, and do not take into account the dynamic aspects of the system, namely the *seniority* of states, as it is termed by Grassberger and Procaccia (1983a). Certain states are visited more often than others and thus we say that they have higher seniority, but they are equally weighted in the estimation of the dimension. In contrast, the correlation integral is sensitive to the seniority effect, that is, to the visiting rate of states.

All of this reasoning holds for (reconstructed) series of infinite points. In practice, we have to use the estimate $\widehat{C}(\epsilon)$ (eqn 4.6) to obtain an estimate $\widehat{D}_2$ of the correlation dimension D_2. Figure 4.13 shows $\widehat{C}(\epsilon)$ as a function of ϵ for the reconstructed Lorenz attractor as we did for the estimate of the maximum characteristic Lyapunov exponent. The parameters used are as follows: length of the reconstructed series $n = 7982$, embedding dimension $m = 3$, time delay $d = 10$ and Theiler window $t = 100$. Note that the plot is a log–log plot (the code is given in Section 4.4.3: Code 4.3).

It is clear that over a certain value of ϵ, around the size of the attractor, all points are within a hypersphere of radius ϵ; thus, $\widehat{C}(\epsilon)$ saturates at 1, and does not change any further. On the other hand, when ϵ is very small, only very few points are counted (depopulation), and this leads to distorted estimates of $C(\epsilon)$. We choose the

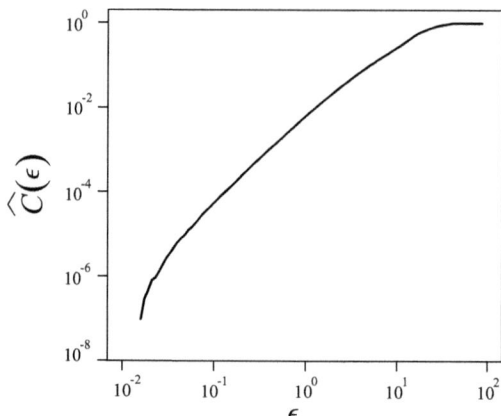

Fig. 4.13 Estimate $\widehat{C}(\epsilon)$ (eqn 4.6) of the correlation integral $C(\epsilon)$ (eqn 4.5) as a function of ϵ, for the reconstructed Lorenz attractor (Figure 3.13b), to obtain the estimate $\widehat{D}_2$ of the correlation dimension D_2 of the 'true' attractor.

scaling region to lie in $[0.1, 1.2]$, the slope of the line in this region is 2.046, which is considered an estimate $\widehat{D}_2$ of the correlation dimension of the Lorenz attractor. The nearest integer above 2.046 is 3, and thus to describe the system's dynamic in phase space, three independent variables are required.

Some notes of caution are in order here. As we will see, the slope of the line in the scaling region is fitted by unweighted least squares, even though the estimate $\widehat{C}(\epsilon)$ of $C(\epsilon)$ is more reliable when ϵ is large and so weighted fits should be more appropriate. As we said regarding the 'linear region' of the MCLE estimate, here also the choice of scaling region is somewhat subjective, and it is advisable to vary slightly the range of this region. For instance, with the scaling region in $[0.1, 1.0]$, we have $\widehat{D}_2 = 2.056$, while in $[0.1, 1.4]$, we have $\widehat{D}_2 = 2.041$. Needless to say, it would be a mistake to view $[2.041, 2.056]$ as something like an error interval of the estimate $\widehat{D}_2$ (we shall return to this further on).

The estimate of the correlation dimension can be exploited, at least in theory, for the classification of time series according to characteristics of the underlying dynamics. Given an experimental time series, the basic idea (Grassberger and Procaccia, 1983b), is to start with a low embedding dimension, for instance $m = 1$, and estimate the correlation dimension, denoted by $\widehat{D}_2^{(1)}$ (the superscript refers to $m = 1$). We increase the dimension of the reconstructed space by 1 ($m = 2$), compute the new estimate $\widehat{D}_2^{(2)}$, and so on. If the series is stochastic, then, as m grows in some range, the reconstruction vectors will always live on the whole of $\mathbb{R}^m$ (or an m-dimensional subset of it); that is, no saturation value is reached by $\widehat{D}_2^{(m)}$. But if the series is deterministic, then, no matter how much m is increased, $\widehat{D}_2^{(m)}$ will eventually become independent of m and saturate at $\widehat{D}_2$, the estimate of the correlation dimension of the reconstructed attractor. If the saturation value is a finite integer, then the system may be described by periodic deterministic dynamics, while for low-dimensional chaotic time series, the reconstruction vectors become concentrated on a subset of $\mathbb{R}^m$ of lower dimension, and $\widehat{D}_2$ has a fractal value. With this method, both the appropriate embedding dimension m and the estimate of the correlation dimension $\widehat{D}_2$ are always, in theory, simultaneously determined.

4.4.3 Estimating the Correlation Dimension with tseriesChaos

tseriesChaos provides a function **d2** to compute the sample correlation integral for embedding dimension $1, \ldots, m$ and return an estimate of the correlation dimension. As in Code 3.2 lorenz_attractor.R, we take the 'observed' series to be the first component of the Lorenz system $x(t)$, after the transient.

The code to estimate the correlation dimension is as follows:

```
#Code 4.3  Lorenz correlation dimension (lorenz_D2.R)
#
parms<- c(10,8/3,28)  # parameters: sigma, beta, rho
tiniz<- 0
tfin<- 100
step<- 0.01
times<-seq(tiniz,tfin,by=step)
funct<- function(t,integ,parms){          # the system of equations
x<-integ[1]
y<-integ[2]
z<-integ[3]
sigma<- parms[1]
beta<-  parms[2]
rho<-   parms[3]
dx<- sigma*(y-x)          # that is  dx/dt = sigma\(y-x)
dy<- x*(rho-z)-y          # that is  dy/dt = x(rho-z)-y
dz<- x*y-beta*z           # that is  dz/dt = xy -beta z
list(c(dx,dy,dz))
                         }                # end of funct
require(deSolve)
ciniz<-c(1,1,1)
xyz<-lsoda(ciniz,times,funct,parms)
trans<- 2000             # integral time step considered as transient
x <- window(xyz[,2],trans)    # discard  initial transient
t_start<- trans*step - step   # new initial time
t_time<-seq(t_start,tfin,by=step) # new time interval
require(tseriesChaos)
m<- 3                         # embedding dimension
d<- 10                        # time delay
xyz <- embedd(x,m,d)          # embed the 'observed' series
#library(scatterplot3d)       # here it is useless
#scatterplot3d(xyz, type="l")
################# up to here it is as in lorenz_attractor.R with the embedding

# estimate of D2
C.m <- d2(x,m=6,d=10,t=100,eps.min=0.01,neps=100)  # correlation integral, m = 1,...,6
C.m <- data.frame(unclass(C.m))                    # class attribute removed
C.3 <- subset(C.m,eps > 0.1 & eps < 1.2, select=c(eps,m3))      # eps in [0.1, 1.2]
lm(log(m3)~log(eps), data = C.3)       # D2 estimate with m=3
# windows()                            # do not comment if other plots are made before
par(mai=c(1.02,1.,0.82,0.42)+0.2)      # to control the margin size
plot(C.m[,1],C.m[,4], type="l",log="xy",main="",
xlab=expression(paste(epsilon)),ylab=expression(paste(widehat(C),(epsilon))),
lwd=2,lty=1,cex.axis=1.3,cex.lab=2.0, xlim=c(0.01,100),ylim=c(0.00000002,1))
# stop here to plot only C(m=3)
```

```
############################################
# add the lines below to plot C(m=1,...,6)
n.m<- ncol(C.m)
for(i in (n.m-0):2) lines(C.m[,c(1,i)],lwd=2)
```

Using this code without the two last lines gives Figure 4.13, that is, $\widehat{C}(\epsilon)$ (eqn 4.6) as a function of ϵ, with embedding dimension $m = 3$ and time delay $d = 10$. t=100 is the Theiler window and neps=100 means that 100 values of ϵ are used, starting from eps.min=0.01. The instruction lm(log(m3) log(eps), data = C.3) performs a regression analysis in the scaling region $[0.1, 1.2]$, with the following result:

```
Call:
lm(formula = log(m3) ~ log(eps), data = C.3)
Coefficients:
(Intercept)      log(eps)
     -5.019         2.046
```

The code yields the matrix C.3 with 7 columns and 100 rows. The first column contains the values of ϵ, while the remaining six columns contain the values of $\widehat{C}(\epsilon)$ for $m = 1, \ldots, 6$; thus, we can plot just one of these (as in Figure 4.13), or we can plot them all together (as in Figure 4.14a) by adding to the code the two last instructions. Note that after $m = 2$, the lines become parallel, which means that the slope, and hence $\widehat{D}_2$, saturates. Figure 4.14b shows the estimated correlation dimension $\widehat{D}_2$ as a function of the embedding dimension m derived from the scaling region of the estimated correlation integral in Figure 4.14a. The plateau after $m = 2$ is a signal of the saturation of the values of $\widehat{D}_2$.

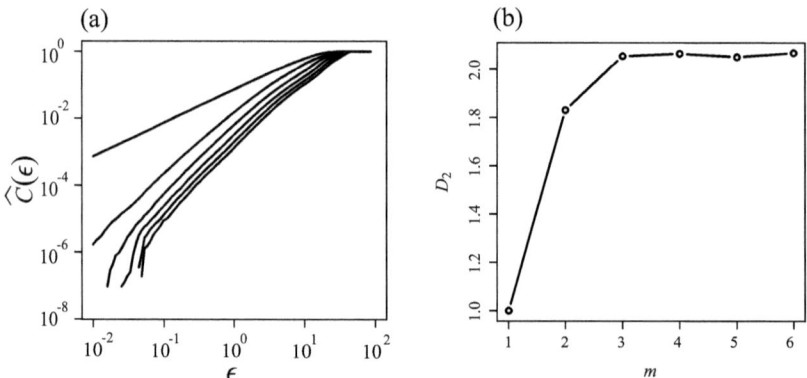

Fig. 4.14 (a) Estimated correlation integral $\widehat{C}(\epsilon)$ (eqn 4.6) as a function of ϵ as in Figure 4.13, but with the embedding dimension from $m = 1$ (uppermost curve) to $m = 6$ (lowest curve). (b) Estimated correlation dimension $\widehat{D}_2$ as a function of the embedding dimension m. For each curve, the limits of the scaling region are slightly adjusted.

The procedure that we have outlined is widely used for recognizing deterministic chaotic time series. However, there are many situations where this method may fail. For example, there are stochastic data sets for which the sequence of correlation dimension estimates also converges to a fractal saturation value (see e.g. Osborne and

Provenzale (1989)). On the other hand, there are deterministic dynamical systems for which time series appear stochastic, for instance when the noise level is high. A further consideration with regard to reliable estimation of the correlation dimension is the length of the embedded time series. Ding *et al.* (1993) showed that the estimated correlation dimension for rather short series falls systematically below that for longer series. The estimation of invariants is a somewhat sensitive matter – indeed, it has been said that *'[it] is often as much an art as it is a science'* (Diks, 1999).

To conclude this discussion of the estimation of invariants, we wish to draw attention to the problem of 'error bars'. For instance, suppose that we have measured a time series and, in the reconstructed space, estimated the MCLE as $\hat{\lambda} = 0.072$. May we believe that we have a chaotic system? This depends on how precise our estimate is. If the standard error were, say, 0.8, it would actually be difficult to be confident about the chaotic nature of the system – hence the need to assign measures of precision to estimates. In an imaginary world of Oz, if we had a measured series with one million observations, then we could divide this series into, say, 100 subseries, each 10 000 observations long, and estimate the MCLE and the correlation dimension in each of them. From this sample, it should be straightforward to derive the estimates of the whole original series and the confidence intervals. In our actual world, something similar can be done, even with only a single available series of reasonable length. Suppose that we have measured a series of length, say, 10 000 observations. Then, by means of so-called **bootstrap methods** (see Appendix C), we can generate 100 new series, each 10 000 observations long, since these are in all respects like 100 new measurements of the same process. We have shown that on the basis of such techniques, we can assign the standard deviation and confidence intervals to estimation of the MCLE and of the correlation dimension; the reader is referred to Giannerini *et al.* (2007*b*) and Borovkova *et al.* (2011), respectively. Incidentally, the estimate $\hat{\lambda} = 0.072$ refers to the Rössler attractor, introduced by Rössler (1976) in his studies on chemical kinetics, and its standard error is calculated in Giannerini and Rosa (2002) to be 0.006.

4.5 Poincaré Map

Imagine that the phase space is something real and so is the representative point. Suppose that the representative point follows the trajectory sketched in Figure 4.15a, namely a periodic orbit. Imagine now putting a piece of paper P parallel to the plane (x, z). The representative point intersecting the paper cuts a 'hole' in it. If the orbit has period 1, there is only one 'hole', A (Figure 4.15a). At every period, the representative point passes through the hole A. If the orbit is doubly periodic, there are two 'holes', A and B (Figure 4.15b), so the representative point passes through A and B alternately. The 'piece of paper' is the *Poincaré section*, and the 'holes' form the *Poincaré map*, also known as *first-return map*. In Figure 4.15, the 'holes' in the Poincaré section are mathematically points, but are enlarged for the sake of clarity, as will done also in the following. The process at the basis of the Poincaré section construction consists in converting the trajectory of a continuous dynamical system in an N-dimensional phase space into a discrete trajectory (i.e. a map) in an $(N - 1)$-dimensional phase space. We can look at this geometrical construction as a 'stroboscopic' representation

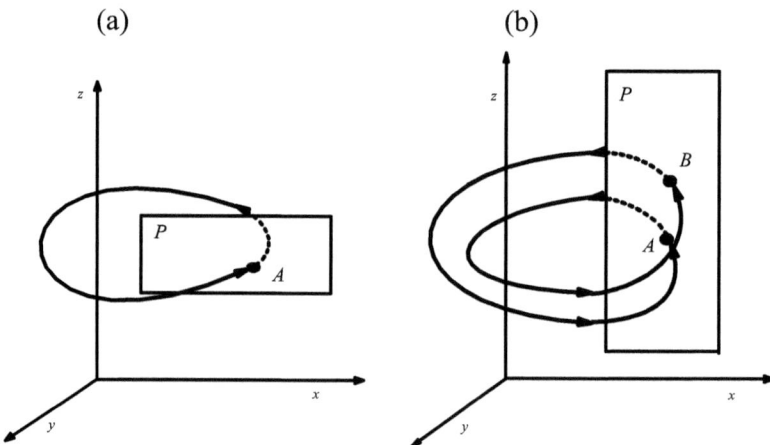

Fig. 4.15 Poincaré map for (a) a period-1 cycle and (b) a period-2 cycle.

of the dynamics of a system. We have seen that for a period-1 cycle, the Poincaré map consists of only one point in the Poincaré section; that is, at every period, the points are plotted upon each other. For a period-2 cycle, the Poincaré map consists of two points; for a period-n cycle, there are n points in the Poincaré section. Note that the intersection points that are considered are only those produced by trajectories in a specified direction, for instance those that cross the plane from the same side.

In the following, we will exploit the Poincaré map to show the dynamics of the *damped driven pendulum*. This pendulum derives from the ideal pendulum already discussed, but now it is more 'realistic', since, in addition to the force of gravity, two further forces acting on it are considered: the damping force and the driving force. The damping force takes into account possible friction with the suspension point and air resistance. The amplitude of oscillation will diminish gradually with time until eventually the system stops. However, an external force acts on the bob, supplying energy to the system, which can continue to oscillate. The interaction between the damping and driving forces makes the system exhibit an extraordinary variety of behaviour, including chaotic behaviour, as shown in Baker and Gollub (1996).

The damping force and the driving force are chosen with very simple forms: the former as being linearly dependent on the velocity of the bobs and the latter as periodic. The the equation of motion of the pendulum, in dimensionless form, is then

$$\frac{d^2x}{dt^2} + \gamma\frac{dx}{dt} + \sin x = F\cos(wt) \tag{4.9}$$

where γ is the *damping constant* (or damping coefficient). In the literature, the quantity $Q = 1/\gamma$, called the *quality factor*, or Q-factor, is also introduced Note that the small-angle approximation, $\sin x \approx x$, is *not* assumed. The parameters F and w are the amplitude and angular frequency of the periodic driving force. Do not confuse

w with ω in eqn (3.2) – the latter is the 'natural' angular frequency, that is, the pendulum frequency without the damping and driving forces. Note that ω does not appear explicitly in eqn (4.9), since it is taken equal to 1, so the driving angular frequency w is a multiple of ω.

Equation (4.9) can be rewritten as a system of three first-order differential equations:

$$\begin{cases} \dfrac{dx}{dt} = y \\[2mm] \dfrac{dy}{dt} = -\gamma y - \sin x + F \cos z \\[2mm] \dfrac{dz}{dt} = w \end{cases} \tag{4.10}$$

where $z = wt$ is the phase of the driving force. The phase space has three coordinate axes (x, y, z); recall that physically x is the angle $\theta(t)$ and y is the angular velocity $d\theta(t)/dt$ (see Figure 3.1). This system is clearly nonlinear owing to the presence of the terms $\sin x$ and $\cos z$.

For the system (4.10), the solutions depends on γ and ω (in our particular case, ω is fixed to be 1). Let $\gamma > 0$ and $\gamma^2 < 4\omega^2$. The solution is then a sine curve modulated by an exponential; in phase space, the orbits are spirals converging to the asymptotically stable fixed point, the origin $(x(0), y(0)) = (0, 0)$, which is an attractor, called the *focus* (see Figure 6.3b). The basin of attraction is the entire (x, y) plane. Recall that the term *basin of attraction* describes the set of initial points from which the trajectories start and will converge on the attractor at long times. The basin of attraction can be a region of phase space or the whole phase space. There are also trajectories that are not attracted by attractors and are called *separatrices* (singular *separatrix*). They separate regions of phase space inside which the trajectories have different qualitative and quantitative behaviours. In the chaotic regime, the basins of attraction have fractal boundaries.

Mathematically, we can consider also $\gamma < 0$. In this case, the origin $(0, 0)$ becomes an unstable focus, that is, a repeller. If $\gamma^2 < 4\omega^2$, with $\gamma > 0$, the attractor is a *node* and the trajectories approach it monotonically, without spiralling. We are in the presence of *overdamped* motion. See the discussion in Section 5.2. We can solve the system (4.10) by means of Code 4.4 `p_damp_driv_1.R`, which follows the same logic as `pend_id_s_o.R`, but now with three difference equations. As in Baker and Gollub (1996), the parameters γ (gamma) and w (w) are fixed ($\gamma = \frac{1}{2}$ and $w = \frac{2}{3}$), while F (F) assumes different values.

```
#Code 4.4      Damped driven pendulum (p_damp_driv_1.R)
# F=0.9        period-1 cycle
# F=1.07       period-2 cycle
# F=1.15       chaotic regime

parms<- c(1/2,0.9,2/3)      # F = 0.9 or 1.07 or 1.15

tinit<-0
tfin<-100
```

```
step<- 3*pi/100
times<-seq(tinit,tfin,by=step)
funct<- function(t,init,parms){
x<-init[1]
y<-init[2]
z<-init[3]

gamma<-parms[1]
F<-parms[2]
w<-parms[3]
dx<- y
dy<- -gamma*y - sin(x) + F*cos(z)
dz<- w
list(c(dx,dy,dz))
                            }       # end of funct
require(deSolve)
require(scatterplot3d)
cinit<-c(5,4,0)
xyz<-lsoda(cinit,times,funct,parms)
scatterplot3d(xyz[,2],xyz[,3],xyz[,4],type="l",cex.lab=1.5,cex.axis=1.2,lwd=3,
xlab = "x(t)",ylab = "y(t)",zlab = "z(t)",xlim=c(4,16),ylim=c(-2,4),zlim=c(0,50))
windows()
par(mai=c(1.02,1.,0.82,0.42)+0.1)
plot(xyz[,2],xyz[,3],type="p",pch=20,cex=0.8,xlab="x(t)",ylab="y(t)",
cex.lab=1.5,cex.axis=1.2,lwd=3,lty=1,xlim=c(5,15),ylim=c(-2,4))

## second initial point
cinit<-c(13,0,0)
xyz<-lsoda(cinit,times,funct,parms)
points(xyz[,2],xyz[,3],col="black",pch=1,cex=1)
```

The period T_D of one driving cycle is $T_D = 2\pi/w$, and the driving angular frequency w is taken equal to $\frac{2}{3}$. The time step (step) is $3\pi/100 \approx 0.09425$, so in 100 iterations the system makes one complete phase orbit.

With $F = 0.9$, the trajectory in the three-dimensional phase space starts from the point $(5, 4, 0)$ and, after a brief transient, the motion becomes periodic and we see the trajectory spiral along the z axis, which represents the flow of time, as shown in Figure 4.16a. For clarity, the three-dimensional trajectory is projected onto the (x, y) plane (as we will also do in the following) in Figure 4.16b (full circles), with a further trajectory (open circles) starting from the point $(13, 0, 0)$ also being shown. After some steps the two trajectories, independently of the initial conditions, converge on a unique closed curve, which is an attractor called the *limit cycle*, more specifically, the period-1 limit cycle. Physically, the pendulum, after the initial transient is extinguished, oscillates at a single frequency, exactly the driving angular frequency w.

Two improvements are introduced in the code. The angle x (i.e. θ) is restricted to the interval $[-\pi, \pi]$, so that if the trajectory exits at the right edge ($> \pi$), it is reported at the left edge, and vice versa. The second improvement concerns the Poincaré map. The R package **nonlinearTseries** contains the function **poincareMap** to compute the Poincaré map. So Code 4.4 becomes Code 4.5 **p_damp_driv_2.R**, which is run for $F = 0.9$ and $F = 1.07$.

```
#Code 4.5    Damped driven pendulum (p_damp_driv_2.R)
# Angle x restricted in [-pi, pi], nonperiodic transient discarded.
#Poincare map computed
```

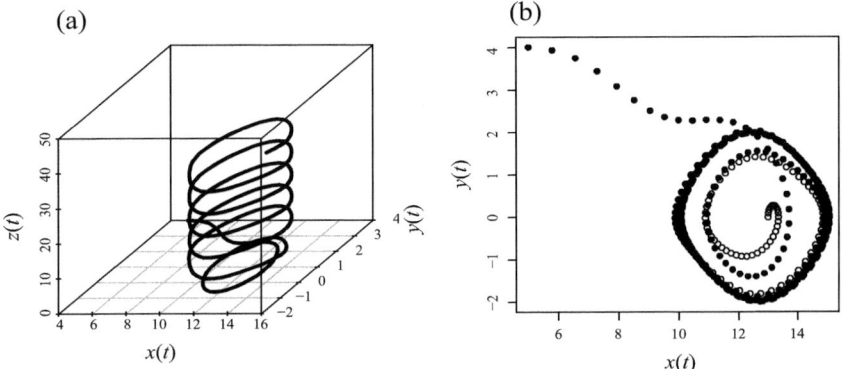

Fig. 4.16 (a) Trajectory in the three-dimensional phase space of the damped driven pendulum with parameters $\gamma = \frac{1}{2}$ and $w = \frac{2}{3}$ and $F = 0.9$ and with initial conditions $(5, 4, 0)$. (b) The same trajectory, but projected onto the (x, y) plane (full circles); a further trajectory starting from $(13, 0, 0)$ is also shown (open circles).

```
# F=0.9       period-1 cycle
# F=1.07      period-2 cycle
# F=1.15      chaotic regime

parms<- c(1/2,0.9,2/3)    # F = 0.9 or 1.07 or 1.15

tinit<-0
tfin<-100
step<- 3*pi/100
nstep<-tfin/step
itrans<-nstep*0.3   # the first 30% steps are discarted
times<-seq(tinit,tfin,by=step)
funct<- function(t,init,parms){
x<-init[1]
y<-init[2]
z<-init[3]

gamma<-parms[1]
F<-parms[2]
w<-parms[3]
dx<- y
dy<- -gamma*y - sin(x) + F*cos(z)
dz<- w
list(c(dx,dy,dz))
                         }       # end of funct
require(deSolve)
cinit<-c(5,4,0)
xyz<-lsoda(cinit,times,funct,parms)
# up to here it is the same as p\_damp\_driv\_2.R

axis.x<- numeric()
axis.y<- numeric()
axis.x<-xyz[itrans:nstep,2]
axis.y<-xyz[itrans:nstep,3]

for(i in 1:length(axis.x)){
```

```
if(axis.x[i]< -pi)    {              #  the trajectory is lesser than -pi ntimes
ntimes<-trunc(axis.x[i]/pi)
axis.x[i]<- axis.x[i]%%pi
if(ntimes%%2!=0)axis.x[i]<- axis.x[i]+pi
                         }
}


for(i in 1:length(axis.x)){
if(axis.x[i]> +pi)    {              #  the trajectory is greater than +pi ntimes
ntimes<-trunc(axis.x[i]/pi)
axis.x[i]<- axis.x[i]%%pi
if(ntimes%%2!=0)axis.x[i]<- axis.x[i]-pi
                         }
}
par(mai=c(1.02,1.,0.82,0.42)+0.1)
plot(axis.x,axis.y,type="p",pch=20,cex=0.3,xlab="x(t)",ylab="y(t)",
cex.lab=1.5,cex.axis=1.2,lwd=3,lty=1,xlim=c(-pi,pi),ylim=c(-pi,pi))

require(nonlinearTseries)
# compute poincare sections
takens=cbind(axis.x,axis.y)
pm=poincareMap(takens=takens,normal.hiperplane.vector=c(0,1),hiperplane.point=c(0,0))
points(pm$pm.pos,col="black",pch=19,cex=2.8,lwd=2.5)
```

To consider only the trajectory on the limit cycle, the nonperiodic transient is discarded. The line `takens=cbind(axis.x,axis.y)` means that the Takens vectors are known – that is, the real phase space, instead of the reconstructed phase space, is known to the user. If necessary, the embedding procedure can be performed with, for example, the instruction

```
pm=poincareMap(time.series=my.series, embedding.dim=3, time.lag=1,
takens=NULL, normal.hiperplane.vector=c(0,0,1), hiperplane.point=c(0,0,0))
```

The parameters `normal.hiperplane.vector=c(0,1)` and `hiperplane.point=c(0,0)` are used to select the Poincaré section. With `pm$pm.pos`, only intersection points produced by trajectories in the positive direction are considered. The result is shown in Figure 4.17 for $F = 0.9$ (a) and 1.07 (b).

Figure 4.17a shows the limit cycle in Figure 4.16b (initial condition $(5, 4, 0)$), but translated in $[-\pi, \pi]$. The result of the computer code is that the limit cycle in Figure 4.16b is inside $[10.06367, 15.07358]$. The right-hand limit is greater than $4 \times \pi$, so $x(t) = 15.07358$ has to be translated as $(15.07358 - 4 \times \pi = 2.507209)$. The Poincaré section is a plane perpendicular to the (x, y) plane and passing through the $y = 0$ axis. The Poincaré map appears as a single point – actually, single points are plotted in the same location on top of each other, at each integer multiple of T_D. When $F = 1.07$ (Figure 4.17b), the period doubles, $T_D = 6\pi$, so the Poincaré map consists of two points plotted within 3π of each other, signifying that the motion of the pendulum is characterized by two different periodic oscillations. Note that the small patch of trajectory on the right is due to the translation in $[-\pi, \pi]$ from left to right of the patch of trajectory less than $-\pi$.

Lastly, let us see what happens when $F = 1.15$. Figure 4.18a shows two trajectories starting from two very close points $A = (5, 5, 0)$ (continuous line) and $B = (5.001, 5.001, 0)$ (dashed line). Here the motion of the pendulum is chaotic: we see that, after a few steps, the trajectories separate themselves at an exponential rate and

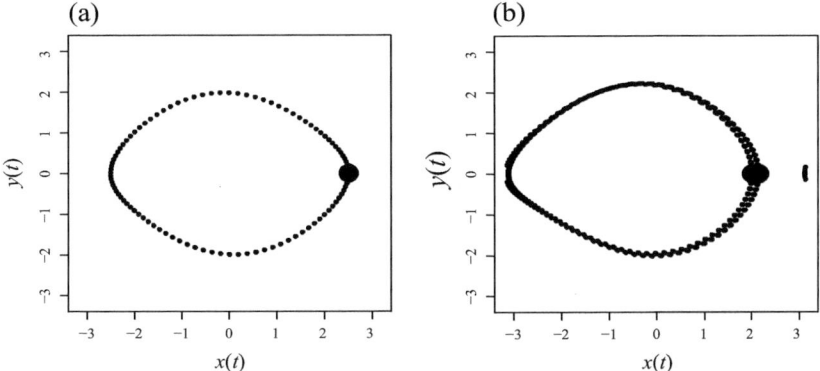

Fig. 4.17 Trajectory in the (x, y) plane of the damped driven pendulum with parameters $\gamma = \frac{1}{2}$ and $w = \frac{2}{3}$ and with initial conditions $(5, 4, 0)$. The trajectory is restricted to the interval $[-\pi, \pi]$ and the transient is discarded. The Poincaré map is also shown as full circles. (a) $F = 0.9$, (b) $F = 1.07$.

we again find a sensitivity to initial conditions, owing to the stretching and folding mechanism. Note that sometimes the single trajectory seems to intersect, although this is not possible for a deterministic system – actually, this impression is due to the projection of the trajectory onto the (x, y) plane. In Figure 4.18b, the trajectory starting from $(5, 5, 0)$ is 10 times longer and is translated in $[-\pi, \pi]$. The figure shows manifestly the 'strange' structure of the attractor, which can be seen in greater detail in the Poincaré map in Figure 4.19. There is no evidence of periodic behaviour, every cycle is different to every other and every point of phase space is visited by the system only once.

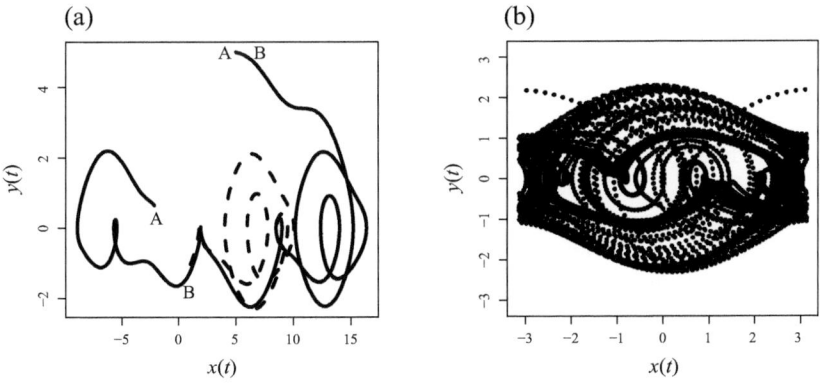

Fig. 4.18 (a) Trajectories in the (x, y) plane of the damped driven pendulum with parameters $\gamma = \frac{1}{2}$, $w = \frac{2}{3}$ and $F = 1.15$ and with initial conditions $A = (5, 5, 0)$ (continuous line) and $B = (5.001, 5.001, 0)$ (dashed line). (b) The trajectory starting from $A = (5, 5, 0)$ is restricted to the interval $[-\pi, \pi]$ and the transient is discarded. Compared with the trajectories in (a), the length of the trajectory here is 10 times greater.

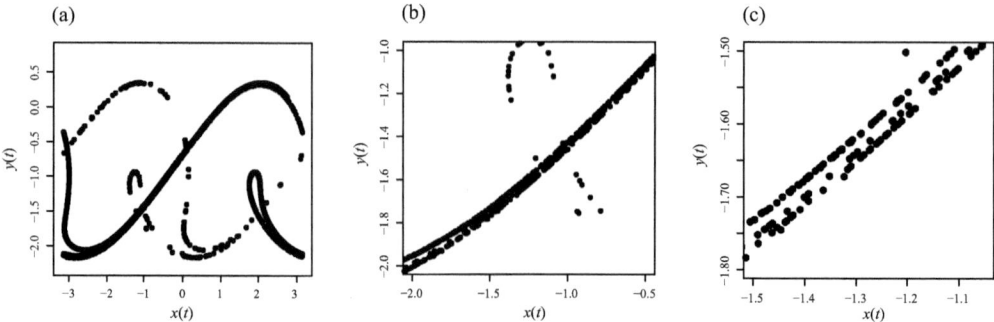

Fig. 4.19 (a) Poincaré map of the damped driven pendulum with parameters $\gamma = \frac{1}{2}$, $w = \frac{2}{3}$ and $F = 1.15$, and with initial conditions $(5, 5, 0)$, as in Figure 4.18b. (b, c) Zooming in on the Poincaré map highlights its fractal structure.

To compute the Poincaré map in this case, a 'stroboscopic picture' of the trajectory in Figure 4.18b is taken at every period of the driving force T_D. Figure 4.19a shows that the Poincaré map is no longer composed either of single points or of random points, but of a set of points with involute structure. If a 'line' on which the points are located is magnified, a group of similar lines emerges (Figure 4.19b, c; note the scales on the axes). Thus, the Poincaré map can be seen as an attractor having a fractal dimension. Looking at Figure 4.17, the cycle in (a), with $F = 0.9$, is symmetric with respect to the x and y axes, while cycle in (b), with $F = 1.07$, is asymmetric with respect to the x axis. If F is increased further, a period doubling occurs up to $F = 1.15$, when the system becomes chaotic. We will see such a route to chaos in Section 4.5.1.

4.5.1 Symmetry breaking

As we did for the logistic map, we construct a bifurcation diagram by computing the angular value $x(t)$ as a function of F, after the transient has been discarded, as shown in Figure 4.20. We have a cascade of bifurcations up to $F \approx 1.10$, even though only the first of these is clearly distinguishable. As F increases, a segment of period 3 becomes visible, around $F \approx 1.12$. For $F \approx 1.15$, a chaotic region begins, and, for $F \approx 1.30$, a period-1 window is born, which doubles for $F \approx 1.45$ and doubles again with a small increase in F. Around $F \approx 1.50$, the chaotic region reappears.

Let us examine in greater detail the first bifurcation occurring when F is very close to 1.066, as shown in Figure 4.21a and already seen in Figure 4.17b. It appears also that the period doubles again for $F \approx 1.08$. For Figure 4.21a, the initial conditions are $(5, 5, 0)$, but if the trajectory starts from $(5, 7, 0)$, we obtain Figure 4.21b. We can see that the arm between $F \approx 1.025$ and $F \approx 1.060$ is the same as that in Figure 4.21a, but as if it has been specularly 'reflected'.

To better understand what has happened, let us examine Figure 4.22a. It is immediately clear that the two innermost cycles are symmetric with respect to the x and y axes, while for the outermost cycles the symmetry with respect to the y axis is no longer apparent, as we have already seen in Figure 4.17. The orbit depicted by circles,

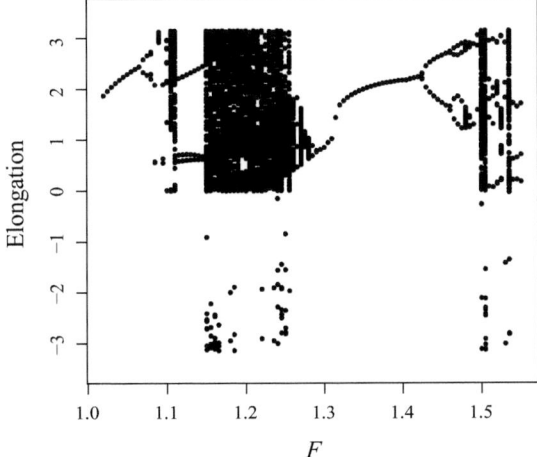

Fig. 4.20 Bifurcation diagram of the damped driven pendulum with parameters $\gamma = \frac{1}{2}$, $w = \frac{2}{3}$ and $F \in [1.020, 1.550]$.

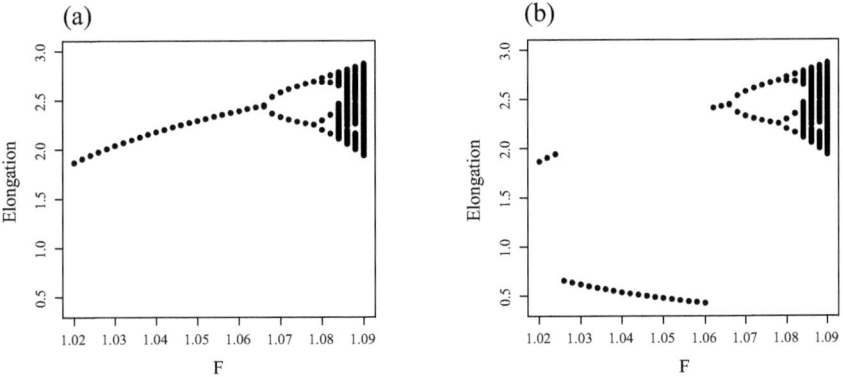

Fig. 4.21 Bifurcation diagram of the damped driven pendulum with parameters $\gamma = \frac{1}{2}$, $w = \frac{2}{3}$ and $F \in [1.020, 1.090]$, with initial conditions $(5, 5, 0)$ (a) and $(5, 7, 0)$ (b).

with initial conditions $(5, 7, 0)$, spends more time in the region with $x(t) < 0$; that is, the region to the left of the vertical line is visited more frequently than the region to the right. Specular behaviour occurs for the orbit depicted by open squares, with initial conditions $(5, 5, 0)$. Up to a value of $F \approx 1.024$, the trajectories converge on symmetric limit cycles, regardless of the initial conditions, but when F goes beyond this critical value, convergence occurs on two possible asymmetric limit cycles, each being the reflected image of the other. In other words, the trajectory may converge on a limit cycle or on its specular reflection, with both solutions being correct from a physical point of view and depending only on the initial conditions but not on the physical parameters. The existence of specular limit cycles explains the appearance of

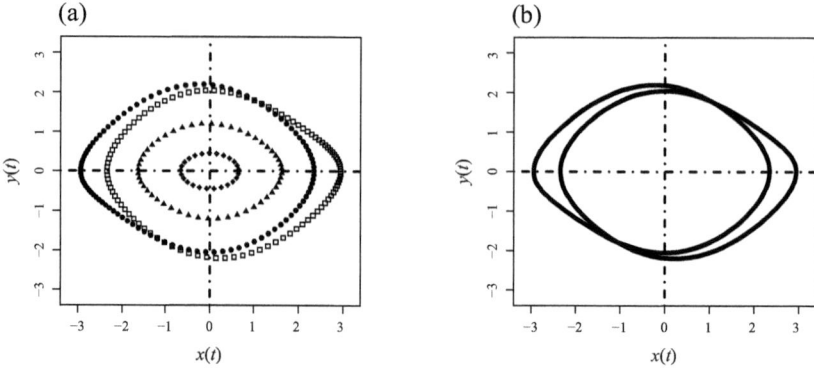

Fig. 4.22 Evolution in phase space of the damped driven pendulum with parameters $\gamma = \frac{1}{2}$ and $w = \frac{2}{3}$. (a) For the two innermost cycles, $F = 0.4$ (diamonds) and $F = 0.7$ (triangles), with initial conditions $(5, 7, 0)$. For the two outermost cycles, $F = 1.025$, with initial conditions $(5, 7, 0)$ (circles) and $(5, 5, 0)$ (open squares). (b) Only the two outermost cycles are plotted, with lines connecting the points.

the bifurcation plots in Figure 4.21, since the points on the specular arms are the co-ordinates x of the possible specular attractors. For instance, when $F = 1.026$, we find $x(t) = 0.657$ with initial conditions $(5, 7, 0)$ and $x(t) = 1.977$ with initial conditions $(5, 5, 0)$.

Nothing in the equations of the damped driven pendulum can reveal that the region to the right of the vertical line is more preferred than that on the left, or vice versa. This type of phenomenon is called *symmetry breaking*. Such symmetry breaking does not occur if the system evolution is described by linear equations, but is intimately connected with the nonlinear behaviour of the system. Note, however, that symmetry breaking of a single limit cycle leads to the emergence of a new symmetry if we imagine joining together the two specular attractors. We thereby obtain two limit cycles that are symmetric with respect to the vertical line. Figure 4.22b shows the two outermost cycles in Figure 4.22a, with the points now connected with a line, and the impression is of two symmetric limit cycles one inside the other.

The interconnected concepts of symmetry, symmetry breaking, conservation laws and degeneracy play a central role in the physical sciences (see e.g. Brading and Castellani (2013) and references therein). It is common knowledge that symmetry principles (or invariance principles) can be invoked to explain conservation laws. Examples from classical mechanics are energy conservation (related to invariance of the Hamiltonian under time translation) and momentum and angular momentum conservation (related to invariance of the Hamiltonian under spatial translation and rotation, respectively). In quantum mechanics, symmetry principles are even more powerful – recall, for instance, that so-called quantum statistics are related to a new kind of symmetry, the interchange of identical particles. A further consequence of symmetry is degeneracy of energy levels: greater symmetry in the Hamiltonian leads to more degeneracy in the energy spectrum. A further concept connected to symmetry is that of *symmetry breaking*. As David Gross (1996) writes: 'The secret of nature is symmetry, but much of the texture of the world is due to mechanisms of symmetry breaking.' As we have seen, the laws of mechanics describing the dynamics of the pendulum are symmetric, but when F is greater than a critical value, the states visited by the system are no longer symmetric in phase space.

4.6   Summary

The strangeness of strange attractors is exhibited in the dynamics of trajectories and in the fractal warp woven by the system in (reconstructed) phase space. One of the main features of deterministic chaotic systems is their well-known sensitive dependence on initial conditions. Recall here that historically von Mises, as far back as 1939, recognized sensitivity to initial conditions as the primary source of randomness even in stochastic systems.

The maximal characteristic Lyapunov exponent (MCLE) is an important measure of stability and one of the indicators of the presence of chaos. There are two classes of estimators for the MCLE. The first measures the exponential divergence of trajectories that are initially close through estimation of the Jacobian matrix along the trajectories. The second directly measures the evolution in time of the distance between initially close points. We have focused in this chapter on the performance of the estimator of the MCLE implemented in `tseriesChaos`, which belongs to the second class. For this type of estimator, computation is easy compared with that of Jacobian estimators, and they can also detect whether a system is non-chaotic. The presence of a sharp linear region in the plot of the evolution of the logarithm of the mean distance between points is a strong signal of chaos. For stochastic systems, in contrast, the linear region tends to become flatter with increasing embedding dimension. However, there are situations (the presence of a dynamic noise component and/or series of moderate size) in which these estimators can produce a linear scaling region giving a positive estimate of the MCLE even for non-chaotic systems.

The 'strangeness' of an attractor's geometry is measured by its fractal dimension. There are a number of definitions of fractal dimension, one of which is the correlation dimension D_2. This is estimated via the so-called correlation integral, which is the proportion of all pairs of reconstruction vectors closer than a given distance ϵ in some norm. In the Grassberger–Procaccia correlation integral, if a point is less than a threshold ϵ, it counts as 1; otherwise it counts as 0. In the Gaussian correlation integral, two points that are a small distances apart are weighted (using a Gaussian weighting) more than those separated by large distances. An estimate of the correlation dimension is given by the slope of the log–log plot of the correlation integral versus ϵ. If this estimate is repeated for increasing embedding dimensions m, for chaotic systems, it initially increases with m, but eventually – at least theoretically – it asymptotically reaches a plateau. The value m^* at which this plateau begins is the next integer above D_2. For instance, if the estimated correlation dimension is 2.056, as for the Lorenz attractor, then the minimum number of variables required to describe the system is 3. Other techniques exist to estimate D_2. In Chapter 7, the entire procedure to estimate D_2 for the Lorenz attractor will be repeated, except that, finally, it will be derived through a maximum likelihood procedure (the Takens estimator).

Jules Henri Poincaré makes two appearances this chapter: in *Poincaré's recurrence theorem* and in the *Poincaré section*. He introduced the concept of recurrence, namely that systems, under certain conditions and after a sufficiently long but finite time, will return to a state as close as you like to the initial state. A graphical analysis of the recurrence states of a dynamical system can be executed through recurrence plots (RPs), which can be quantified using recurrence quantification analysis (RQA). In this chapter, we have applied RPs to the logistic map and the Lorenz attractor. It is important to note that methods based on RPs allow trajectories in the m-dimensional phase space of a system to be analysed in terms of a two-dimensional representation of the system's recurrences.

For the study of the *damped driven pendulum*, we have exploited the Poincaré section. This is constructed by converting the trajectory of a continuous dynamical system in an N-dimensional phase space into a discrete trajectory (i.e. a map) in an $(N-1)$-dimensional phase space. We can view this geometric construction as a 'stroboscopic' representation of the dynamics of the system. The damped driven pendulum is the ideal pendulum, introduced in Chapter 3, made more realistic by the inclusion, in addition to the force of gravity, of two further forces acting on it: a damping force and a driving force. The damping force takes into account possible friction with the suspension point and air resistance. The driving force is an external force acting on the bob and supplying energy to the system. A wide variety of behaviours, including chaotic ones, are shown by the damped driven pendulum. Here, we have studied the bifurcation diagram, as for the logistic map, and the particular phenomenon called *symmetry breaking*.

5
Entropy and Surrogate Testing

'The program asks you how many lines you want in the
poem, and you decide: ten, twenty, a hundred. Then the program randomizes
the line numbers. In other words, a new arrangement each time. With ten
lines you can make thousands and thousands of random poems.'

Umberto Eco, Foucault's Pendulum

5.1 Introduction

The word 'entropy' was introduced by the German physicist Rudolf Clausius in the
middle of the nineteenth century, in his *Die Mechanische Wärmetheorie* ('The Me-
chanical Theory of Heat'). The term evokes 'energy' and derives from the Greek $\epsilon\nu$
(*en*: 'inside') and $\tau\rho o\pi\eta$ (*trope*: 'turning', 'change'). In thermodynamics, if a system
at an absolute temperature T (kelvin) receives an element dQ of heat, its entropy
S increases by an amount $dS = dQ/dT$. There is no interpretation involving lack of
knowledge, or randomness, or probability. The Second Law of Thermodynamics can
be expressed by saying that $dS/dt \geqslant 0$; that is, entropy increases with time: this
statement presupposes a direction of time, an 'arrow of time'.

In statistical mechanics, Ludwig Boltzmann linked the entropy to the concept of
probability, to tackle the problem of the Second Law of Thermodynamics, in particular
to determine the equilibrium state of a gas. A gas is in an equilibrium state if all
thermodynamics parameters (volume, temperature, pressure) have defined constant
values. The answer is carved on Boltzmann's grave at the Zentral Friedhof in Wien:
$S = k \log W$, the entropy S is proportional to the logarithm of the probability W (from
Wahrscheinlichkeit) and the constant k is called Boltzmann's constant. In an irony of
history, this formula was never written in this form by Boltzmann, but by Planck. The
formula says that the system spontaneously reaches the state of maximum entropy S
simply because all other states are extremely unlikely. For an extended discussion of
the concept of probability in classical statistical physics, see von Plato (1994).

Let X be a discrete random variable with support S_X, that is, the set of all possible
values of X. In the context of information theory, the support is called the 'alphabet'.
If I toss a coin, I am uncertain of the result. Is the coin fair or is it biased? Is it possible
to quantify this uncertainty? To this end, we need a number that, from the probability
distribution of the results, measures the uncertainty of the results themselves. Let this
number be denoted by $H(X)$:

$$\textit{probability distribution: } \{p(x_1), \dots, p(x_n)\}, x_i \in S_X \longrightarrow \textit{real number: } H(X)$$

Nonlinear Time Series Analysis with R. Ray Huffaker, Marco Bittelli and Rodolfo Rosa, Oxford University Press (2017).
© Ray Huffaker, Marco Bittelli and Rodolfo Rosa. DOI: 10.1093/oso/9780198782933.001.0001

In the following, we shall write p_i in place of $p(x_i)$. This $H(X)$ has to satisfy certain requirements:

1. H must be continuous in all its arguments p_i; that is, a small variation in the probability distribution $p_1, p_2, \ldots, p_n$ corresponds to a small variation in H.
2. If the probabilities p_i are all the same (a uniform distribution), that is, $p_i = 1/n$, then H is a monotonically increasing function of n.
3. H must be a function of the probability distribution $\{p_1, \ldots, p_n\}$, independently of how events are gathered within this distribution.

For the last point, consider the following example. Let X be a random variable with distribution

$$\mathsf{P}\{X = a\} = 0.5, \quad \mathsf{P}\{X = b\} = 0.2, \quad \mathsf{P}\{X = c\} = 0.3$$

where $\mathsf{P}\{\cdot\}$ denotes the probability. We can also say that the events $\{b\}$ and $\{c\}$ are realized half the time: when it happens, the event $\{b\}$ occurs with probability 0.4 and the event $\{c\}$ with probability 0.6. Then, the above distribution can be written as

$$\mathsf{P}\{X = a\} = 0.5, \quad \mathsf{P}\{X = Y\} = 0.5, \quad \mathsf{P}\{Y = b\} = 0.4, \quad \mathsf{P}\{Y = c\} = 0.6$$

H must be the same in the two cases. Shannon (1948) demonstrated that there is exactly one function H satisfying the above requirements, namely,

$$H = -k \sum_{x \in S} \mathsf{P}\{x\} \log \mathsf{P}\{x\}$$

$$= -k \sum_{i=1}^{n} p_i \log p_i$$

The constant k depends on the chosen unit of measurement, that is, on the basis of the logarithm. For $k = 1$ and logarithms to base 2, Shannon called the measure $H = -\sum_{i=1}^{n} p_i \log p_i$ the *entropy* of the set of probabilities $\{p_1, p_2, \ldots, p_n\}$. With the chosen basis 2, the unit of measurement is the *bit* ('binary digit'). Tribus (Tribus and McIrvine, 1971) recounts that Shannon told him that John von Neumann suggested the name 'entropy' for Shannon's measure, in the following words:

You should call it entropy, for two reasons. In the first place your uncertainty function has been used in statistical mechanics under that name, so it already has a name. In the second place, and more important, no one knows what entropy really is, so in a debate you will always have the advantage.

Turning to $H(X)$, it should be noted that the notation $H(X)$ for the entropy should not be taken to mean that H is a function of X (i.e. of the values taken by X), but rather that it is a function of the X distribution $\{p_1, p_2, \ldots, p_n\}$. Thus, X here is not the argument of a function, but instead is a label distinguishing, for instance, the entropy $H(X)$ of the random variable X from the entropy $H(Y)$ of the random variable Y.

The following are important properties of the Shannon entropy:

1. $H = 0$ if and only if X is known with certainty, that is, if all but one of the probabilities p_i are zero, with the remaining probability p_j being 1:

$$P\{X = j\} = 1 \quad \text{and} \quad P\{X = i\} = 0, \ \forall i \neq j$$

   This means that H takes the value 0 only when the result is certain; otherwise always $H > 0$.

2. Given a certain dimension n, H is maximal when all the probabilities p_i are equal, $p_i = 1/n$; this is a uniform distribution, the situation with the maximum uncertainty.

Suppose, for example, that we have an alphabet with only two symbols: H (heads) and T (tails). If we know that the coin has two H (or two T), then $H(X) = 0$, since $\log_2 1 = 0$. If $P\{X = T\} = P\{X = H\} = 0.5$, then we have

$$H(X) = -(0.50 \log_2 0.50 + 0.50 \log_2 0.50) = 1$$

($\log_2 0.50 = -1$). If, for instance, $P\{X = T\} = 0.75$ and $P\{X = H\} = 0.25$, then

$$H(X) = -(0.75 \log_2 0.75 + 0.25 \log_2 0.25) = 0.81128$$

The quantity $-\log_2 P\{X = i\}$, or $-\log_2 p(x_i)$, or $-\log_2 p_i$, is also called the *surprisal*. When p_i is small (in which case the surprisal, $-\log_2 p_i$, is large), we are 'surprised' if the result is that associated with x_i, whereas if p_i is large (and the surprisal is therefore small), our degree of surprise is low. All of this is in agreement with the interpretation of H as a measure of the amount of uncertainty: the more uncertain we are of a result, the more surprised we are if this result occurs.

We can also interpret H as a measure of the information content in a message. If H is large, then knowledge about the X distribution is highly informative. In other words, if an event is unexpected, it is very informative, whereas if it is easily predictable, it is not informative. Surprisal, uncertainty, randomness, and information are all related notions associated with the concept of entropy. As mentioned earlier, if a coin has two heads, then $H(X) = 0$, and no information is transmitted about the next toss, since the result is certain. However, if the coin is fair, then $H(X) = 1$, and to express the result, we need a one-bit message, denoting heads by 1 and tails by 0 (or the reverse). So, if we flip a fair coin 100 times, we need 100 numbers (1 or 0) to send a message with the results, but if the coin has two heads, there is no need for any sequence of digits to describe the outcomes.

We can also write $H(X) = -\sum_i p(x_i) \log_2 p(x_i)$ as

$$H(X) = \sum_i [-\log_2 p(x_i)] p(x_i)$$

$$= E[-\log_2 p(x_i)]$$

that is, as the expected value of the surprisal. From an informational perspective, we can say that entropy is a measure of the expected amount of information.

We now return to the definition of mutual information, given in Section 3.4.2, which we repeat here for the reader's convenience:

$$I(X;Y) = \sum_i \sum_j p(x_i, y_j) \log_2 \frac{p(x_i, y_j)}{p(x_i)p(y_j)}$$

Introducing the entropy $H(X)$, this equation can be rewritten as

$$I(X;Y) = \sum_i \sum_j p(x_i, y_j) \log \frac{p(x_i, y_j)}{p(x_i)p(y_j)}$$

$$= \sum_i \sum_j p(x_i, y_j) \log p(x_i, y_j) + \sum_i \sum_j p(x_i, y_j) \log \frac{1}{p(x_i)}$$

$$+ \sum_i \sum_j p(x_i, y_j) \log \frac{1}{p(y_j)}$$

$$= \sum_i \sum_j p(x_i, y_j) \log p(x_i, y_j) - \sum_i p(x_i) \log p(x_i) - \sum_j p(y_j) \log p(y_j)$$

$$= -H(X, Y) + H(X) + H(Y)$$

Thus,

$$\begin{aligned} I(X;Y) &= H(X) + H(Y) - H(X, Y) \\ &= H(X) - H(X|Y) \\ &= H(Y) - H(Y|X) \end{aligned} \tag{5.1}$$

As we anticipated in Section 3.4.2, the mutual information is a measure of the extent to which knowledge about X reduces uncertainty about Y. This implies that $I(X;Y) = 0$ if and only if X and Y are independent random variables or if one of the two variables has zero entropy. This measure is symmetric, that is, $I(X;Y) = I(Y;X)$, and it is always non-negative.

Historically, the idea that entropy, even in statistical mechanics, signifies something like 'lack of knowledge' has been put forward by a number of physicists and philosophers. In this view, entropy is a measure of incompleteness of human knowledge, so it becomes a subjective notion, an anthropomorphic concept. For a critical analysis of this viewpoint, see Denbigh and Denbigh (1985)

In the next section, the Shannon entropy is computed for the logistic map, and we see how it is related to the bifurcation diagram.

5.2 Shannon Entropy of the Logistic Map

Entropy can be used to characterize a sequence—in particular, whether it is chaotic. As in Code 2.7 `Logistic map: Density`, the interval $[0, 1]$ is divided into N subintervals I_k, given by

$$I_k = \left[(k-1)/N, k/N \right], \quad k = 1, \ldots, N$$

and the number of iterates in each interval I_k is computed. After the initial transient, if, for instance, $r = 2.8$, then all iterates lie in a single interval and therefore $H = 0$. It is maximal, $H = \log_2 N$, if all the intervals are equally probable. As we did for the Lyapunov exponent (Code 2.6 `logistic_Lyap.R`), we compute the entropy as a function of r with Code 5.1 `Logistic_Entr.R`.

```
#Code 5.1    Nonlinear logistic map: entropy (Logistic_Entr.R)

cell<- numeric()
entr<-  numeric()
f.x<-function(x,r){
      r*x*(1-x)    }
ncell<- 100       # number of cells
ntrans<- 1000     # transient
rin<-  2.8
rfin<- 4
n<- 400                   # number if iterations after the transient
nt<- ntrans+n             # total number of iterations
xinit<- 0.2
r<- seq(rin,rfin,by=0.005)    # 0.005: step of r
par(mar = c(6.3, 4.8, 1., 3))
plot(0,0,type="n",xlim=c(rin,rfin),ylim=c(0,7),xlab="r",ylab="H [bits]",   #ylim=c(0,4.7)
cex.lab=1.6,cex.axis=1.2)
ii<-0
for(i in r) {             # starting loop on r values
ii<- ii+1
x<-xinit
cell[1:ncell]<- 0
entr[ii]<- 0
for(j in 1:nt) {          # starting loop on the iterations
y<-f.x(x,i)
if(j > ntrans){k<-trunc(ncell*y)+1  #  only if j > ntrans
cell[k]<-cell[k]+1 }
x<-y
                }          # ending loop on the iterations

for(k in 1:ncell){    # loop to compute entropy
prob<-cell[k]/n
if(prob>0){entr[ii]<-entr[ii]-prob*log2(prob)}
                }         # ending loop to compute entropy
                }         # ending loop on r values
#r          # comment to not write r values
#entr       # comment to not write entr vs r
lines(r,entr,type='l',lwd=2)
emax<-log2(ncell)
emax
segments(3.569944,0,3.569944,emax,lty=4,lwd=2,col="black")    # or: col="red"
segments(rin,emax,rfin,emax,lwd=2,lty=8,col="black")          # or: col="blue"
```

The probability p_k that an iterate falls in the interval I_k is estimated by the frequency of occurrence in this interval (`prob<-cell[k]/n`). Figure 5.1, obtained using Code 5.1, shows H as a function of r. As expected, the entropy is zero before $r = 3$, when there is only a fixed point. Up to the beginning of the chaotic region, $r \approx 3.56994$, some stair-like steps are apparent, corresponding to the bifurcation points. In the chaotic region, H increases with r, except for those ranges of r corresponding to periodic windows. For an ideal random time series, partitioned into 100 intervals, $H = -\log_2 100 = 6.6439$ bits (the horizontal dashed line in Figure 5.1). Even for $r = 4$, H is less than this value, since, as we have already mentioned in Section 2.8, the iterates of the logistic map do not uniformly cover the interval $[0, 1]$.

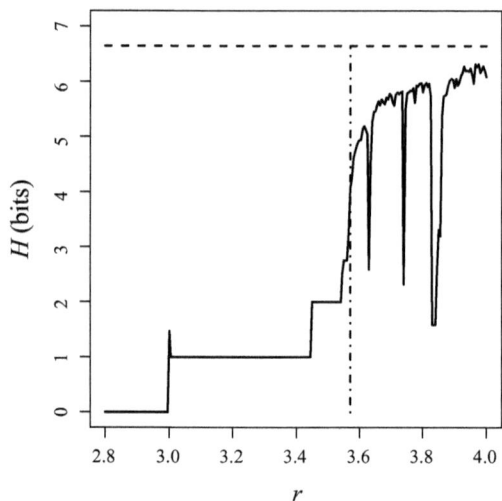

Fig. 5.1 Shannon entropy estimation for the logistic map. The vertical dash–dot line at $r \approx 3.56994$ denotes the beginning of the chaotic regime; the horizontal dashed line at $H = 6.6439$ denotes the maximum entropy, corresponding to 100 intervals of equal probability.

5.3 Entropy Test

There exists a vast literature on testing for nonlinearity in time series, but we shall focus on two approaches. In this section, we describe a method based on nonparametric entropy measures of dependence coming from information theory. In the next section, we introduce a simulated annealing method that yields surrogates constrained to take on prefixed values as the original data. The presentation in this section follows closely the fine review by Giannerini (2012) and the article by Giannerini *et al.* (2015), which describes the basis for the R package `tseriesEntropy` (Giannerini, 2015).

Let $x_1, x_2, \ldots, x_n$ be an observed time series. We know only that it has been generated by a stationary stochastic process X_t, which may be discrete or continuous. We ask ourselves whether the process X_t is linear or nonlinear, basing our decision only on the observed data. From a statistical perspective, we can formulate a null hypothesis

$H_0 : X_t$ *is a linear process* and test it. The alternative hypothesis is then $H_1 : X_t$ *is not a linear process*. However, what exactly does 'nonlinear' mean in this context? There are various kinds of nonlinearity, and therefore a test to determine linearity implies testing for specific nonlinear features that are absent in the case of a linear process. For instance, as we have already seen, linear processes show no sensitivity to initial conditions. Giannerini (2012) has classified and discussed various tests for nonlinearity; here, we mention only the BDS test (Brock *et al.*, 1986), based on the correlation integral (eqn 4.6), which has the form

$$\text{BDS}(m, \epsilon) = \frac{\sqrt{n}\{\widehat{C}_m(\epsilon) - [\widehat{C}_1(\epsilon)]^m\}}{\widehat{V}_m^{1/2}}$$

where m is the embedding dimension, $\widehat{C}_m(\epsilon)$ is given by eqn (4.6) (the subscript m is added here for clarity) and $\widehat{V}_m$ is an estimator of the asymptotic variance. Serial independence implies that $\widehat{C}_m(\epsilon) = [\widehat{C}_1(\epsilon)]^m$; note, however, that the reverse is not necessarily true.

Nonparametric entropy measures of dependence are at the basis of a class of tests for nonlinearity. Here we present the metric entropy measure due to Granger *et al.* (2004), often denoted by S_ρ, which has been exploited by Gonzalez *et al.* (2006) and Gonzalez *et al.* (2008) in studies of DNA sequences. This entropy S_ρ is a normalized version of the Bhattacharya–Hellinger–Matusita distance (Maasoumi and Racine, 2002) and is defined as

$$S_\rho(k) = \frac{1}{2} \int_{-\infty}^{+\infty} \int_{-\infty}^{+\infty} \left[\sqrt{f_{(X_t, X_{t+k})}(x_1, x_2)} - \sqrt{f_{X_t}(x_1) f_{X_{t+k}}(x_2)} \right]^2 dx_1 \, dx_2 \quad (5.2)$$

where $f_{X_t}(\cdot)$ and $f_{(X_t, X_{t+k})}(\cdot, \cdot)$ are the probability density functions of X_t and of the vector (X_t, X_{t+k}), respectively. This measure can be interpreted as a nonlinear autocorrelation function; that is, if S_ρ exceeds the confidence band at lag k, then there is a significant correlation between X_t and X_{t+k}, which are separated by k steps in the sequence. The measure satisfies the following properties:

- It is defined for both continuous and discrete variables.
- It is normalized and varies between 0 and 1.
- It takes the value 0 if X_t and X_{t+k} are independent
- It takes the value 1 if there is a measurable exact (nonlinear) relationship between the variables.
- It reduces to a function of the linear correlation coefficient in the case of Gaussian variables.
- It is a metric measure, not only a divergence measure; indeed, it obeys the triangle inequality and is a commutative operator.
- It is invariant with respect to continuous, strictly increasing transformations.

The measure S_ρ has been shown to have impressive and robust power for characterizing nonlinear processes (Giannerini *et al.*, 2007*a*). See also, for instance, Granger *et al.* (2004) for a detailed discussion of issues concerning the definition, implementation and estimation of $S_\rho(k)$.

The null hypothesis we test is that of independence, and we do this by resorting to a permutation scheme. The original sequence is randomly permuted, and the measure $S_\rho(k)$ is estimated on this new sequence. This procedure is repeated B times (say $B = 100$–1000) to obtain the bootstrap distribution of $S_\rho(k)$ under the null hypothesis and consequently the confidence intervals.

We end this section by mentioning that the *moving block bootstrap (MBB)* (see Appendix B) can be exploited as a test for independence. It has been applied to binary sequences of 0s and 1s by Gonzalez *et al.* (2006) on the basis of the following idea. Recall that, for independent data, for instance realizations of i.i.d. binomial variables, the standard error of the estimator of the proportion p of 0s in the sequences is given by $\hat{\sigma}_0 = \sqrt{\hat{p}(1 - \hat{p})/n}$, where $\hat{p}$ denotes an estimate of p. In this instance, both the i.i.d. bootstrap and the MBB give the same results. If there is some form of dependence, that is, if the observations can no longer be considered realizations of mutually independent random variables with the same distribution function, then the estimate given by the i.i.d. bootstrap is no longer valid, but the MBB is able not only to reveal the dependence but also to estimate the 'true' standard error σ.

5.4 Surrogate Test

Surrogate data tests are based on resampling methods: the original series is resampled, but with the constraint that one or more parameters are fixed. As a very simple example, let us assume that we wish to test the hypothesis of independence. We shuffle the data so that each surrogate consists of the same data as in the original series, but now in a random temporal order. All the means have the same values (fixed parameters) as before, but the serial correlations might be significantly different in the original data and in the shuffled surrogates.

The method of surrogate data was introduced by Theiler *et al.* (1992) and needs the following ingredients:

1. The null hypothesis H_0 regarding the process that has generated the observed data.
2. An ensemble of B resampled series, called surrogate series, consistent with H_0, that is, considered as different realizations of the hypothesized process.
3. A discriminating statistic that is computed on the surrogates and gives the distribution of this statistical ensemble.
4. The significance level of the test, which is derived by comparing the value of the parameter estimated on the original series with that obtained on the surrogate distribution.

For instance, suppose we have observed a time series that we presume has some degree of nonlinearity. We choose as the null hypothesis H_0: the time series has been generated by a Gaussian linear process. We generate B new series according to this hypothesis, namely B Gaussian linear series. Both for the original series and for each surrogate, we estimate a suitable quantity, which for the observed data could be significantly different from that of the surrogates. If this is the case, then the null hypothesis of linearity is rejected: it is very unlikely that the original time series has been generated by a Gaussian linear process. But this kind of inverse reasoning is not correct. If the statistic for the original series falls within the distribution of the results for the surrogates, we are not allowed to say that the null hypothesis is true. The only thing we can do at this stage is to not reject H_0. We could find a different discriminating statistic capable of distinguishing between the original data and surrogates. The use of different discriminating statistics is therefore suggested, since there may be situations where a single discriminating statistic is not sufficient (Kugiumtzis, 2001).

In the R package `nonlinearTseries`, the function **surrogateTest** executes surrogate data testing. The null hypothesis is that the observed series is generated by a linear stochastic process. Then it must be symmetric under time reversal. If the original data show a deviation from the distribution of the surrogates, then the hypothesis of linearity has to be rejected

Theiler *et al.* (1992) discuss different methods for generating surrogates. For instance, surrogates can be obtained by randomizing the phases of the Fourier transform of the original series. In this manner, the original mean and periodogram are preserved, but any further information is lost.

We must pay attention to the selected discriminating statistic when we wish to investigate whether the underlying dynamics is chaotic. There are several methods for the estimation of invariants, although not all of these are necessarily suitable to work as a discriminating statistic.

We now discuss a method to generate surrogates that is based on an algorithm developed by scientists involved in the construction of the first H-bomb: N. C. Metropolis, A. W. Rosenbluth, M. N. Rosenbluth, A. M. Harkanyi Teller and E. Teller. This algorithm is now called the *Metropolis algorithm* , or the $M(RT)^2$ algorithm with reference to the initial letters of the authors' names (Metropolis *et al.*, 1953).

Let $\mathbf{x}$ be a point of the phase space of the system. The Metropolis algorithm consists in the construction of a Markov process where each state $\mathbf{x}_j$ is derived from the previous state $\mathbf{x}_i$ via a suitable transition probability p_{ij}. This transition probability must be such that as the length of the sequence tends to infinity, the probability of occurrence of the state $\mathbf{x}$ is given by the so-called Boltzmann probability distribution $\mathsf{P}\{\mathbf{x}\} \propto \exp[-E(\mathbf{x})/kT]$, where $E(\mathbf{x})$ is the energy of the state $\mathbf{x}$ and k is Boltzmann's constant.

In practice, suppose that the system is in the state $\mathbf{x}_i$ with energy $E(\mathbf{x}_i)$. A new state $\mathbf{x}_j$ with energy $E(\mathbf{x}_j)$ is proposed by changing $\mathbf{x}_i$ via random numbers. The state $\mathbf{x}_j$ is accepted with probability $\alpha(\mathbf{x}_i, \mathbf{x}_j) = \min\{1, \exp\{-[E(\mathbf{x}_j) - E(\mathbf{x}_i)]/kT\}\}$. In other words, if $E(\mathbf{x}_j) \leq E(\mathbf{x}_i)$, the new state $\mathbf{x}_j$ is always accepted; if $E(\mathbf{x}_j) > E(\mathbf{x}_i)$, it is accepted with a nonzero probability. If $\mathbf{x}_j$ is rejected, then the current $\mathbf{x}_i$ is kept and counted as a 'new configuration'.

We now introduce a procedure, called *simulated annealing*, that consists in running the Metropolis algorithm more than once, lowering the temperature T each time. In metallurgy, annealing is a process in which metal is heated and then cooled very slowly, to bring it closer to its equilibrium state. Such a process can be simulated by choosing one or more constraints in terms of a cost function C, which is interpreted as the energy E in the thermodynamic system. Minimizing C can be considered as equivalent to searching for the equilibrium state of the system by looking at all possible permutations of the data. The Metropolis algorithm decides which changes to accept and which to reject. The aim is to reach a *global* minimum in which the constraints are satisfied; therefore, as in real annealing, the temperature has to decrease slowly in order to avoid local minima.

5.5   Tests for Nonlinear Serial Dependence with R Packages

The R package `tseriesEntropy` (Giannerini, 2015) implements the entropy measure S_ρ (eqn (5.2)). This measure is exploited to construct tests for *nonlinear* serial dependence for continuous and categorical time series. The null hypothesis is not of independence but of general *linear* dependence. Note that some routines have a parallel version that can be used in a multicore/cluster environment.

Here we apply the function **Srho.test.ts** to the Lorenz attractor. This function is for continuous time series; for categorical ones, there is the function **Srho.test**. Remember that in Code 3.2 `lorenz_attractor.R`, the 'observed' series is the first component of the Lorenz system $x(t)$, after the transient. The following lines are added to this code:

```
#install.packages("tseriesEntropy", dep = TRUE) # to install the package if not present
library(tseriesEntropy)
set.seed(1)
S<- Srho.test.ts(x,B=40,lag.max=10,ci.type=c("perm"),quant=0.95)
S
plot(S,cex.lab=1.2,cex.axis=1.2,lwd=2)
```

This function can be run with a univariate series (as here) or a numerical vector. These types of codes are very time-consuming, and therefore, where possible, the parallel version is suggested. Here, to speed execution, some parameters in Code 3.2 `lorenz_attractor.R` are modified: `tfin<- 50`, `step<- 0.05` and `trans<- 200`. Lastly, only `B=40` permutations are chosen. One can select the maximum lag at which to calculate S_ρ; the default is $n/4$, where n is the number of observations; here `lag.max=10` only. The parameter `ci.type` allows one to select how the distribution under the null hypothesis of independence is obtained. Here, each resampled series is a random permutation of the original one; in the bivariate case, the MBB can be used. The parameter `quant` is used for computation of the significant lags and to plot confidence bands. The output consists of a plot of S_ρ as a function of the lags (see Figure 5.2), the lags at which S_ρ exceeds the chosen confidence bands under the null hypothesis and the bootstrap p-value for each lag. It is clear that, as expected, the S_ρ is well beyond the 95% confidence band for all lags.

In the R package `tseriesEntropy`, the function **surrogate.AR** generates surrogate series in this way. Thorough the Akaike information criterion (AIC), the best autoregressive (AR) model is chosen and the residuals are resampled with replacement by means of the *sieve bootstrap* (Bühlmann, 1997). Each resampled series is considered as a surrogate series. We apply the function **surrogate.AR** to the Lorenz

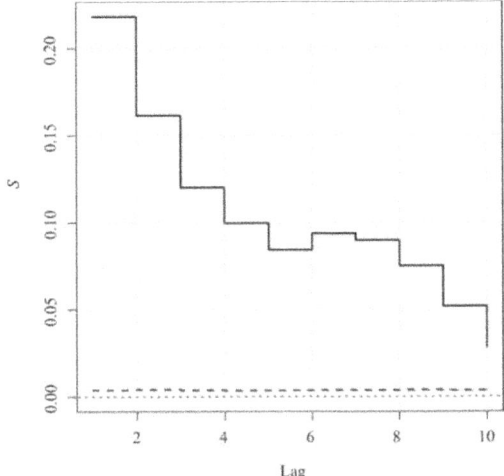

Fig. 5.2 Plot of S_ρ for the Lorenz attractor at lags 1–10 with rejection bands at 95% (dashed).

system and, as was done in Theiler *et al.* (1992), we use the correlation dimension D_2 as the discriminating statistic. The 'observed' series is the first component of the Lorenz system $x(t)$, after the transient (Code 3.2 `lorenz_attractor.R`), and, under the null hypothesis, is generated by an AR model. For the sake of completeness, we give the entire code here:

```
#Code 5.2   Function surrogate.AR (Lorenz_SurrAR.R)
# From Code 3.2   lorenz_attractor.R; transient discarded;
# x(t) is the "observed" series to be embedded
# D2 as discriminating statistics

parms<- c(10,8/3,28)  # parameters: sigma, beta, rho
tiniz<- 0
tfin<-  100
step<-  0.01
times<-seq(tiniz,tfin,by=step)
funct<- function(t,integ,parms){          # the system of equations
x<-integ[1]
y<-integ[2]
z<-integ[3]
sigma<- parms[1]
beta<-  parms[2]
rho<-   parms[3]
dx<- sigma*(y-x)          # that is  dx/dt = sigma\(y-x)
dy<- x*(rho-z)-y          # that is  dy/dt = x(rho-z)-y
dz<- x*y-beta*z           # that is  dz/dt = xy -beta z
list(c(dx,dy,dz))
                          }               # end of funct
require(deSolve)
ciniz<-c(1,1,1)
xyz<-lsoda(ciniz,times,funct,parms)
trans<- 2000                  # integral time step considered as transient
x <- window(xyz[,2],trans)    # discard  initial transient
```

```
t_start<- trans*step - step       # new initial time
t_time<-seq(t_start,tfin,by=step) # new time interval

# the plot(x,...) is skipped
#par(mar = c(6.3, 4.8, 1., 3))
#par(cex.lab=1.4,cex.axis=1.4,lwd=1,lty=1)
#plot(x,type="l",xlab="time",ylab="x(t)",xlim=c(1,length(x)),ylim=c(-30,30))
#windows()
require(tseriesChaos)
m<- 3                             # embedding dimension
d<- 10                            # time delay
xyz <- embedd(x,m,d) # embed the 'observed' series
library(scatterplot3d)
#scatterplot3d(xyz, type="l")  # to skip the scatterplot
# estimate of D2
C.m <- d2(x,m=6,d=10,t=100,eps.min=0.01,neps=100) # correlation integral, m = 1,...,6
C.m <- data.frame(unclass(C.m))          # class attribute removed
C.3 <- subset(C.m,eps > 0.1 & eps < 1.0, select=c(eps,m3)) # eps in [0.1, 1.0]
D2<-lm(formula=log(m3) ~ log(eps), data = C.3)       # D2 estimate with m=3
est_d2<- summary(D2)$coefficients[2]
est_d2
#windows()                               # if other plots are made before
par(mai=c(1.02,1.,0.82,0.42)+0.2)        # to control the margin size
plot(C.m[,1],C.m[,4], type="l",log="xy",main="",
xlab=expression(paste(epsilon)),ylab=expression(paste(widehat(C),(epsilon))),
lwd=2,lty=1,cex.axis=1.3,cex.lab=2.0, xlim=c(0.01,100),ylim=c(0.00000002,1))

# up to here it is as in Code 4.3 lorenz_D2.R without 2 plots

#install.packages("tseriesEntropy", dep = TRUE) # to install the package if not present
library(tseriesEntropy)
set.seed(1)
est_d2s<- numeric()
x.surr<- surrogate.AR(x, order.max=10, nsurr=100)
# order.max is the maximum order of the AR model to fit
for(i in 1:100){
# estimate  D2 surrogates
C.ms <- d2(x.surr$surr[,i],m=6,d=10,t=100,eps.min=0.01,neps=100)
C.ms <- data.frame(unclass(C.ms))
C.3s <- subset(C.ms,eps > 0.1 & eps < 1.0, select=c(eps,m3)) # eps in [0.1, 1.0]
# add 3 correlation integral surrogates
if(i<=3) lines(C.ms[,1],C.ms[,4],col="black",lwd=2,lty=2)
D2s<-lm(formula=log(m3) ~ log(eps), data = C.3s)             # D2 estimate surrogates
est_d2s[i]<- summary(D2s)$coefficients[2]
            }

# surrogates distribution
windows()
hist(est_d2s,freq=F,xlab="D2_surr",ylab="density",main="",cex.axis=1.3,cex.lab=1.5)
# to plot 3 surrogates remove the comments #
#windows()
#plot.ts(x.surr$surr[1:length(x),1:3],col="black",type="l",xlab="time",ylab="x(t)",
#      cex.lab=1.1,cex.axis=1.1,lwd=2,main="")
```

We can add the following lines to check that the surrogates have the same autocorrelation function (`acf`) as the original series:

```
corig <- acf(x,10,plot=FALSE)$acf[,,1];
csurr <- acf(x.surr$surr[,1],10,plot=FALSE)$acf[,,1];
round(cbind(corig,csurr,"abs(difference)"=abs(corig-csurr)),3)
```

The instruction `x.surr<- surrogate.AR(x, order.max=10, nsurr=100)`, where `order.max` is the maximum order of the AR model to fit, generates 100 surrogates, and the estimation of the correlation dimension of the surrogates is performed as for the original series. In the output, we can see first the autocorrelation function for the original (`corig`) and then that for the surrogate (`csurr`) series:

```
       corig csurr abs(difference)
 [1,]  1.000 1.000           0.000
 [2,]  0.998 0.998           0.000
 [3,]  0.993 0.992           0.001
 [4,]  0.985 0.984           0.002
 [5,]  0.974 0.972           0.003
 [6,]  0.961 0.956           0.004
 [7,]  0.944 0.938           0.006
 [8,]  0.926 0.918           0.008
 [9,]  0.905 0.895           0.010
[10,]  0.882 0.869           0.012
[11,]  0.857 0.843           0.014
```

Figure 5.3a shows the estimate $\widehat{C}(\epsilon)$ of the correlation integral $C(\epsilon)$ (eqn 4.5) as a function of ϵ as in Figure 4.13, with three further $\widehat{C}(\epsilon)$ computed on three surrogates (dashed line). A scaling region is also apparent for the surrogates; however, it is rather steeper. Figure 5.3b shows the distribution of the estimate $\widehat{D}_2$ of the correlation dimension D_2 for 100 surrogates. When we recall that for the original series $\widehat{D}_2$ was about 2.056, which falls clearly outside the distribution for the surrogates, we can see that the null hypothesis that the original series is linear can be rejected.

We compare also the estimated correlation dimension $\widehat{D}_2$ as a function of the embedding dimension m for the original series and for the surrogates. We add to Figure 4.14b the $\widehat{D}_2$ of the surrogates, obtained as the mean value of 20 surrogates for each m value; see Figure 5.4.

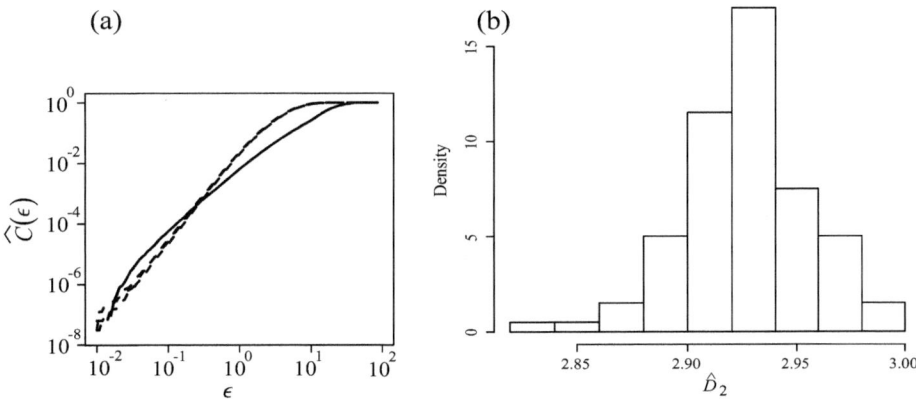

Fig. 5.3 (a) Estimate $\widehat{C}(\epsilon)$ of the correlation integral $C(\epsilon)$ (eqn 4.5) for the original $x(t)$ component of the Lorenz system and for three surrogates derived from it. (b) Distribution of the estimate $\widehat{D}_2$ of the correlation dimension D_2 for 100 surrogates.

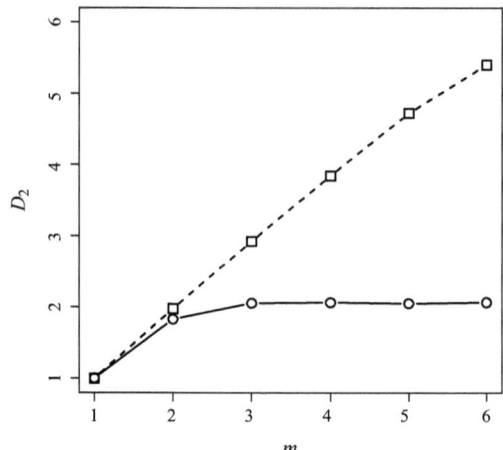

Fig. 5.4 Estimated correlation dimension $\widehat{D}_2$ as a function of the embedding dimension m for the original series, as in Figure 4.14b, and for the surrogates (dashed line).

The nonlinearity is very evident: for the original data, the sequence of correlation dimension estimates converges to a fractal saturation value, while no plateau is reached for the surrogates. A similar result for a different chaotic attractor is shown in Figure 7a of Theiler *et al.* (1992).

Giannerini *et al.* (2015) discuss tests of nonlinearity based on the entropy S_ρ and surrogate data. The null hypotheses are (1) a linear Gaussian process and (2) generic linearity. Two test statistics are proposed:

$$
\widehat{T}_k = \begin{cases} \left(\widehat{S}_k^{\mathrm{u}} - \widehat{S}_k^{\mathrm{p}}\right)^2 & \text{for Gaussian linearity} \\[2mm] \widehat{S}_k^{\mathrm{u}} & \text{for generic linearity} \end{cases}
$$

where S_k is $S_\rho(k)$, and the superscripts 'p' and 'u' stand for 'parametric' and 'unrestricted nonparametric', respectively. If the null hypothesis is of Gaussian linearity, then $\widehat{S}_k^{\mathrm{p}}$ can be derived analytically, while $\widehat{S}_k^{\mathrm{u}}$ requires numerical computation. The package `tseriesEntropy` contains two functions performing entropy tests of nonlinearity with the statistic $\widehat{T}_k$: **Trho.test.AR** and **Trho.test.SA** (together with their respective parallel versions). Both functions generate surrogates with the same autocorrelation function and the same sample mean as the original series. With the first function, the surrogates are obtained by means of the sieve bootstrap (see **surrogate.AR**); with the second function, simulated annealing is adopted. Here, we shall use **Trho.test.SA** to test Gaussian linearity for the logistic map in the chaotic regime ($r = 4$). For the sake of completeness, we give the code in full:

```
#Code 5.3   Function Trho.test.SA (LogisticMap_Trho.R)

f.x<- function(x,r){
    r*x*(1-x)
```

```
                         }
xt<- numeric()
sr<- numeric()
ntrans<- 200        # transient
n<- 100             # number if iterations after the transient
nt<- ntrans+n
nw<- 50
f.temp<-function(xiniz,nt,r){   #  starting function f.temp
x<- xinit
xt[1]<- x
for(i in 2:nt){
y<- f.x(x,r)
x<- y
xt[i]<- x
                }
return(xt)
                         }   # ending function f.temp
###  parameters and initial conditions
r<- 4
xinit<- 0.2
ss<-f.temp(xinit,nt,r)
ss<- ss[ntrans+1:n]
# comment the following lines if the time plot is not requested
plot(ss[1:nw],type="b",xlab="time",ylab="x(t)",
     cex.lab=1.1,cex.axis=1.1,lwd=2,main="")
#install.packages("tseriesEntropy", dep = TRUE) # to install the package if not present
library(tseriesEntropy)
set.seed(1)
windows()       # if the time plot has been executed
# with the simulated annealing
Tr <- Trho.test.SA(ss,lag.max=10,B=20,bw='reference',method='integral',quant=0.95,plot=F)
# with the sieve bootstrap
#Tr <- Trho.test.AR(ss,lag.max=10,B=20,bw='reference',method='integral',quant=0.95,plot=F)
Tr
plot(Tr,cex.lab=1.2,cex.axis=1.2,lwd=2,col="black",main="",xlab="lag",ylab="Trho")
```

Both `Trho.test.` algorithms compute T_k (`Trho`) up to $k =$`lag.max`. Here, `lag.max` $=$ 10, and $B = 20$ surrogates are generated. The result is displayed in Figure 5.5.

The code also yields the following p-values:

```
p-values:
   1    2    3    4    5    6    7    8    9   10
0.00 0.00 0.55 0.20 0.55 0.45 0.75 0.90 0.80 0.70
```

As stressed by Giannerini *et al.* (2015), the null hypothesis of linearity is rejected if the test rejects for at least one of $k = 5$ lags. In our case, for the first two lags, T_1 and T_2 are beyond the 95% confidence band, and thus the null hypothesis of Gaussian linearity is rejected, as expected.

5.6 Summary

From statistical mechanics to information theory, the word 'entropy' has received many interpretations: disorder, uncertainty, randomness, amount of information, surprisal, etc. In this chapter, we have first focused on the Shannon entropy, which can be interpreted as the degree of disorder in a system or the amount of information needed to specify the state of the system, We have computed this entropy for the logistic map as a

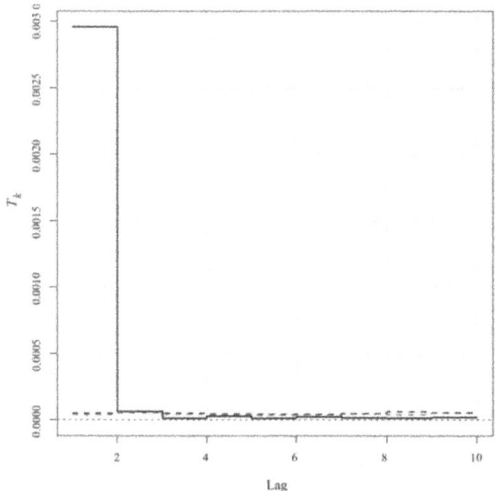

Fig. 5.5 Plot of T_k for the logistic map in the chaotic regime at lags 1 to 10 with rejection bands at 95% (dashed). The surrogates are obtained through simulated annealing.

function of r, and we have seen a strong correspondence with the bifurcation diagram. In particular, in the chaotic region, the entropy increases with r, but impressive dips are apparent, corresponding to periodicity windows. Note that even for $r = 4$, the value of the entropy is less than that of a purely random process, since the iterates of the logistic map do not uniformly cover the interval $[0, 1]$ (see Section 2.8).

We have also introduced another type of entropy, the *metric* entropy measure, known as S_ρ, which possesses some desirable properties. This measure can be interpreted as a nonlinear autocorrelation function, and it has been applied as a test statistic with the surrogate data method.

The method of surrogate data tests consists of the following steps. First, we formulate a null hypothesis H_0 on the type of process that has produced the observed data – for instance that the process is linear and Gaussian. Second, surrogate series that are consistent with H_0 are generated by means of bootstrap methods (see Appendix B). These surrogate series are considered as different realizations of the hypothesized process. Then, a discriminating statistic, computed for each surrogate series, allows us to obtain the distribution of this ensemble of statistics. Finally, by comparing the value of the parameter estimated on the original series with that estimated on the surrogate distribution, we can decide whether or not to reject the null hypothesis. There is a danger here, however: there may be situations where one discriminating statistic leads us to reject H_0, but according to another statistic we cannot reject H_0. The use of different discriminating statistics is therefore suggested.

The chapter ends with a test of nonlinearity based on the entropy S_ρ and surrogate data. The package `tseriesEntropy` generates surrogates with the same autocorrelation function and the same sample mean as the original series. We can choose the way in which we obtain the surrogates – either through bootstrap methods or through simulated annealing. It worth recalling here that the latter technique is based on the so-called Metropolis algorithm, which originated in the 1950s in a statistical mechanics context.

6
Data Preprocessing

'Eliminate all other factors, and the one which remains must be the truth.'
Arthur Conan Doyle, The Sign of the Four

6.1 Introduction

Reconstructing real-world system dynamics from a single time series is challenging owing to the broad menu of possible reconstructions that are available. Consider, for example, the four time series in Figure 6.1. The periodic cycling in Figure 6.1a is the regular behaviour of a linear system of ordinary differential equations (ODEs). The aperiodic (nonrepeated) cycling in Figures 6.1b and 6.1c is induced exogenously by subjecting the linear model to uniform random shocks (i.e. *noise*) each period (b), and to nonstationarity generated by a slowly evolving parameter through time (c). Finally, the aperiodic cycling in Figure 6.1d is generated deterministically with the Lorenz nonlinear system of ODEs (Lorenz, 1963) .

In practice, we face a quandary over whether irregular dynamics observed in real-world data should be attributed to noisy or nonstationary linear behaviour or to deterministic nonlinear dynamics. Real-world data are often corrupted by *noise*, meaning that they exhibit unstructured variation that does not evolve systematically with time. Noise is typically attributed to measurement, processing or recording errors in collecting data or to inherently random environmental shocks (Feder, 1979; Uusitalo *et al.*, 2015) . Noise obscures detection of systematic dynamic structure in data by decreasing the resolution of reconstructed attractors, and increasing the embedding dimensions required to capture noisy phase space dynamics (Kot *et al.*, 1988).

In addition, real-world data are often of short duration. Noise may be due to short data length that does not adequately capture structural variation in slow-moving variables. For example, 100 years of precipitation data are insufficient to detect patterns in the occurrence of a 100-year flood that might appear as noise in a single century. Increasing the sample time to 400 years allows for four repetitions of the 100-year flood, but this may still be too short to establish a reliable pattern. A 100-year flood might appear as a slow-moving linear or nonlinear trend violating the standard of *stationarity* requiring that the duration of the measurement be long compared with the timescales of the system (Schreiber, 1999). In general, a stationary time series exhibits similar behaviour throughout its duration (Kaplan and Glass, 1995).

Nonlinear Time Series Analysis with R. Ray Huffaker, Marco Bittelli and Rodolfo Rosa, Oxford University Press (2017).
© Ray Huffaker, Marco Bittelli and Rodolfo Rosa. DOI: 10.1093/oso/9780198782933.001.0001

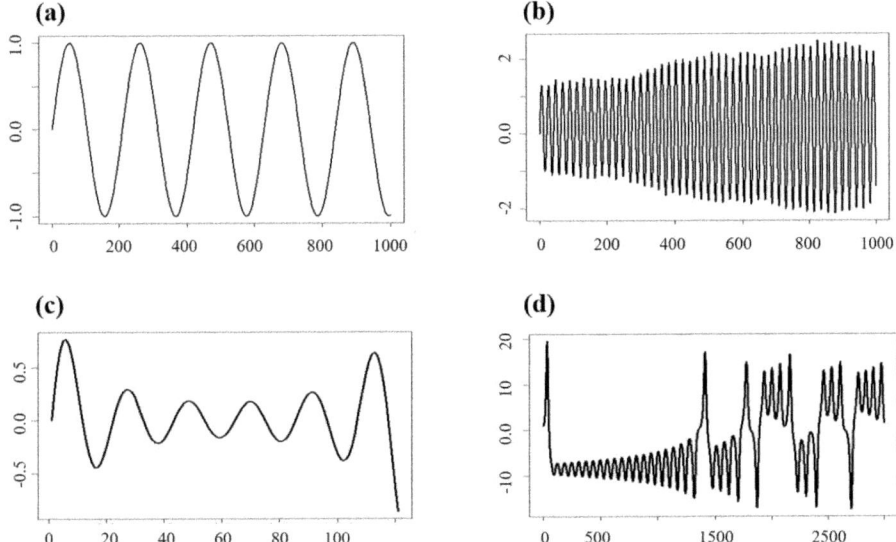

Fig. 6.1 Time series exemplifying the range of possible explanations for volatile dynamics: (a) periodic cycling in the linear model; (b) aperiodic (nonrepeated) cycling created exogenously by randomly shocking the linear model; (c) aperiodic behaviour created by introducing nonstationarity into the linear model; (d) aperiodic cycling endogenous to nonlinear Lorenz equations.

Nonlinear time series (NLTS) methods search for system dynamics likely to have generated observed irregularity in data. Like Sherlock Holmes, we progressively eliminate factors, and the surviving factor remains a possible reconstruction of the system dynamics. In this chapter, we get to know factors potentially contributing to real-world dynamics, and investigate how we can preprocess time series data to begin eliminating them as possibilities.

We start by reviewing why the regular behaviour of linear systems of ordinary differential equations is limited to exponential and periodic dynamics. We next illustrate how linear ODE models can be induced to mimic irregular behaviour exogenously with the introduction of noise and nonstationarity, and how data can be preprocessed to reduce their impact. In particular, we discuss a signal processing technique, *singular spectrum analysis (SSA)*, that separates unstructured variation (*noise*) from structural variation (*signal*) in real-world data. We also present a test based on *nonlinear cross prediction* methods to detect nonstationarity in the signal separated from observed data (Schreiber, 1997). Beyond removing noise from the data, SSA may also eliminate detected nonstationarity by further purging the signal of slow-moving trend components possibly causing the problem. Finally, we see how irregular dynamic behaviour can be generated deterministically with nonlinear ODE systems composed of at least three equations.

After we have preprocessed the data, we have ideally removed nonstationarity as a possible driver of observed irregular dynamics and mitigated the impact of noise in obscuring dynamic structure. We can proceed to the remaining NLTS procedures, which require stationary data (Schreiber, 1999), to distinguish whether irregularity in the isolated signal is most likely governed by deterministic nonlinear dynamics or linear stochastic dynamics due to lingering unstructured variation.

6.2   Regular Behaviour of Linear ODE Models

We begin by exploring the regular behaviour of two-dimensional linear ODE systems with the following model:

$$\frac{dx}{dt} = \dot{x} = ax_t + by_t, \qquad\qquad x_{t=0} = x_0$$

$$\frac{dy}{dt} = \dot{y} = cx_t + dy_t, \qquad\qquad y_{t=0} = y_0$$

or, in matrix form,

$$\begin{bmatrix} \dot{x}_t \\ \dot{y}_t \end{bmatrix} = \underbrace{\begin{bmatrix} a & b \\ c & d \end{bmatrix}}_{A} \begin{bmatrix} x_t \\ y_t \end{bmatrix} \tag{6.1}$$

The system parameters are the constants a, b, c and d. The system is autonomous, because it is not explicitly a function of time. The solution comprises the time trajectories of the variables x_t and y_t satisfying the differential equations from given initial levels x_0 and y_0. The *nullclines* set each ODE equal to zero:

$$y\big|_{\dot{x}=0} = -\frac{a}{b}x_t$$

$$y\big|_{\dot{y}=0} = -\frac{c}{d}x_t$$

The system is at equilibrium when the nullclines intersect. This occurs at the origin, so equilibrium is $x^* = y^* = 0$.

The general solution to the system (6.1) is

$$\begin{bmatrix} x_t \\ y_t \end{bmatrix} = k_1(x_0, y_0)e^{\lambda_1 t} EV_{\lambda_1} + k_2(x_0, y_0)e^{\lambda_2 t} EV_{\lambda_2} \tag{6.2}$$

where λ_1 and λ_2 are real *eigenvalues* of the coefficient matrix A, EV_{λ_1} and EV_{λ_2} are their corresponding *eigenvectors*, and k_1 and k_2 are constants tying the general solution to the initial values x_0 and y_0. Appendix C presents a brief primer on computation of eigenvalues and eigenvectors. Casual observation of eqn (6.2) demonstrates that stability depends on the signs of the eigenvalues of A, because they are exponentiated in the solution. When both eigenvalues are negative, each term in the solution dampens to a *stable node* equilibrium at $x^* = y^* = 0$. Alternatively, when both eigenvalues are positive, each term explodes away from an *unstable node* equilibrium. When the eigenvalues have opposite sign, the term associated with the positive eigenvalue explodes away from, and the term associated with the negative eigenvalue decays towards, a *saddle point* equilibrium. In all of these cases, stability depends on the system eigenvalues, which in turn depend on the system parameters.

When the eigenvalues of A are complex numbers, $\lambda_1 = \alpha + \beta i$ and $\lambda_2 = \alpha - \beta i$, the general solution becomes

$$\begin{bmatrix} x_t \\ y_t \end{bmatrix} = e^{\alpha t} \left\{ k_1 \left(\begin{bmatrix} u_1 \\ u_2 \end{bmatrix} \cos \beta t - \begin{bmatrix} w_1 \\ w_2 \end{bmatrix} \sin \beta t \right) \right.$$

$$\left. + k_2 \left(-\begin{bmatrix} w_1 \\ w_2 \end{bmatrix} \cos \beta t - \begin{bmatrix} u_1 \\ u_2 \end{bmatrix} \sin \beta t \right) \right\} \tag{6.3}$$

where α and β are respectively the real and imaginary parts of the complex eigenvalues and w_1 and w_2 are the real and imaginary parts of the corresponding complex eigenvectors. The constants k_1 and k_2 continue to tie the solution to initial conditions. The stability of eqn (6.3) is controlled by the sign of the real part of the complex eigenvalue, α, since it is exponentiated. The presence of the cosine and sine functions gives rise to oscillatory behaviour. Consequently, when α is negative, the solution follows dampened oscillations towards a *stable focus* equilibrium. When α is positive, the solution undergoes explosive oscillations away from an *unstable focus* equilibrium. When $\alpha = 0$, the system orbits along an ellipse around a *centre point*.

Solutions to the system (6.1) can be portrayed graphically in **phase diagrams** . Each point in phase space shows the state of the system at time t, and rests on a unique trajectory that records how the variables co-evolve from initial values (x_0, y_0) through time. Appendix C works through the analytical construction of phase space for a sample linear system. Code 6.1 uses the R package **phaseR** (Grayling, 2015) to numerically generate phase diagram solutions illustrating the linear dynamic behaviours that we have just discussed.

```
#Code 6.1  Phase diagrams for two-dimensional linear ODE system
rm(list=ls(all=TRUE))   #clear values between runs

library(phaseR)

#Set parameters for integration
t.end<-10 #integrate from t=0 to t=10
delta<-0.01 #default value in phaseR is t.step = 0.01

#Define linear Model
ode_linear<-function(t,y,parameters){
x<-y[1]
y<-y[2]
a<-parameters[1]
b<-parameters[2]
c<-parameters[3]
d<-parameters[4]
dy<-numeric(2)
dy[1]<-a*x+b*y
dy[2]<-c*x+d*y
list(dy)
}
```

```
#Equilibrium Type
 #Stable node attractor
parms.sn<-c(-3,2,2,-3)  #parameter values
x0.sn<-c(-0.5,0,0.5,-1,1,-0.5,0,0.5)
y0.sn<-c(1,1,1,0,0,-1,-1,-1)
xy0.sn<-cbind(x0.sn,y0.sn)  #initial conditions
 #Unstable node
parms.un<-c(3,2,2,3)
x0.un<-c(0.4,0.6,-0.4,-0.6,-0.1,0.1)
y0.un<-c(-0.5,-0.5,0.5,0.5,0,0)
xy0.un<-cbind(x0.un,y0.un)
 #Stable focus attractor
parms.sf<-c(-1,3,-3,-1)
x0.sf<-c(0.5,-0.5)
y0.sf<-c(-1,1)
xy0.sf<-cbind(x0.sf,y0.sf)
 #Unstable focus
parms.uf<-c(1,3,-3,1)
x0.uf<-c(0,0)
y0.uf<-c(-0.01,0.01)
xy0.uf<-cbind(x0.uf,y0.uf)
 #Saddle point
parms.sp<-c(1,3,3,1)
x0.sp<-c(-0.9,-0.8,-0.7,0.7,0.8,0.9,-0.1,0.1)
y0.sp<-c(0.8,0.8,0.8,-0.8,-0.8,-0.8,-0.1,0.1)
xy0.sp<-cbind(x0.sp,y0.sp)
 #centre
parms.cn<-c(0,3,-3,0)
x0.cn<-c(0)
y0.cn<-c(1)
xy0.cn<-cbind(x0.cn,y0.cn)

#Select equilibrium type for which to draw phase space diagram
parms=parms.cn
xy0<-xy0.cn    #initial conditions

#Compute time series solutions x (x.soln) and y (y.soln) from selected
  #initial values and plot
y0<-c(0,1) #initial values
ts<-numericalSolution(ode_linear,y0=y0,t.end=t.end,t.step=delta,type="one",
   ylim=c(-1.5,1.5),parameters=parms,colour=c("black","black"),ylab="x,y",lwd=3,
   cex.axis=1.1,font.axis=2)
x.soln<-ts$x  #time series solution of x
y.soln<-ts$y  #time series solution of y

#Phase Diagram: Vector-field arrows, nullclines, trajectories
arrows<-flowField(ode_linear,x.lim=c(-1,1),y.lim=c(-1,1),lwd=2,
   parameters=parms,points=19,add=FALSE,cex.axis=1.1,font.axis=2)

null<-nullclines(ode_linear,x.lim=c(-1,1),y.lim=c(-1,1),parameters=parms,
   col="black",lwd=3,points=500)

flow<-trajectory(ode_linear,y0=xy0,t.end=t.end,t.step=delta,parameters=parms,
   colour=rep("black",3),lwd=3,lty=1)

#Add eigenvectors to phase diagram
parms1<-matrix(parms,ncol=2,nrow=2,byrow=TRUE) #coefficient matrix
eigens<-eigen(parms1)   #compute eigensystem
values<-eigens$values  #eigenvalues
```

```
#Plot eigenvectors only for real eigenvalues
if(is.numeric(values[1])){
x.ev<-seq(-1,1,0.1)
y_ev1<--((parms1[1,1]-values[1])/parms1[1,2])*x.ev  #eigenvector equation
y_ev2<--((parms1[1,1]-values[2])/parms1[1,2])*x.ev  #eigenvector equation
par(new=TRUE)  #add eigenvectors to phase plot
plot(x.ev,y_ev1,ylab="",xlab="",axes=FALSE,col="black",type='l',lty="dashed",lwd=2)
lines(x.ev,y_ev2,ylab="",xlab="",col="black",type='l',lty="dashed",lwd=2)
par(new=FALSE)} #end if statement

#Stability Check for x*=y*=0
stability<-stability(ode_linear,y.star=c(0,0),parameters=parms)
eigenvalues<-values

#Export time series solutions to *.csv output file
ts.solns<-cbind(x.soln,y.soln)
dir.results<-as.character("C:/Users/Ray/Documents/R_files/Results")
setwd(dir.results)
write.table(ts.solns,sep=",",col.names=c("x.soln","y.soln"),"ts.solns.csv",
  row.names=FALSE)
```

In Code 6.1, we first set the parameters for integration. We integrate 10 periods (t.end $= 10$) with an integration step of $\delta = 0.01$. Consequently, $1000 = 10/0.01$ iterations beyond the initial time period ($t = 0$) are required to generate the 10 periods (for a total of 1001 iterations). We next set the parameters and initial conditions for six *nondegenerate* stability cases including stable nodes and foci, unstable nodes and foci, saddle points, and centre points ('nondegenerate' means that the system eigenvalues are nonzero). After we have selected one of these cases for simulation, the code computes and plots the time series solutions of x_t and y_t from given initial values. It subsequently generates a phase diagram of y_t plotted against x_t with vector-field arrows, trajectories from specified initial conditions, nullclines, and eigenvectors (when the eigenvalues are real). The code runs a stability check by computing system eigenvalues. Finally, it exports the time series solutions **xsoln** and **ysoln** to a *.csv* output file.

Figure 6.2 shows the time series plots and phase diagram for node equilibria. For a *stable node* (the top row of the figure), the time series plots show that the variables (x_t, y_t) are drawn towards it from their initial levels (Figure 6.2a). For example, when we set the initial levels at $(x_0 = 0, y_0 = 1)$, x_t first increases and then decreases towards the stable node at $(0, 0)$, while y_t steadily decreases in its approach. The phase diagram solution (Figure 6.2b) plots the vector field, showing directions of motion for the two variables (grey arrows pointing towards equilibrium), the nullclines (black lines intersecting at equilibrium), the eigenvectors (dashed lines intersecting at equilibrium) and solution trajectories from selected initial conditions (denoted by dots at the front of the trajectories). In a deterministic system, solution trajectories do not intersect (Kantz and Schreiber, 1997). Variables co-evolve towards the stable node from all initial conditions. Trajectories cross the two nullclines perpendicularly to the corresponding axis. For example, the flatter-sloped nullcline shows the locus of variable combinations, setting $\dot{y} = 0$, so that a crossing trajectory holds y_t constant along the vertical axis. The two eigenvectors direct the trajectories in their journey towards equilibrium. For example, the positively sloped eigenvector directs trajectories coming in from the lower left corner (low initial levels of both variables) and upper right

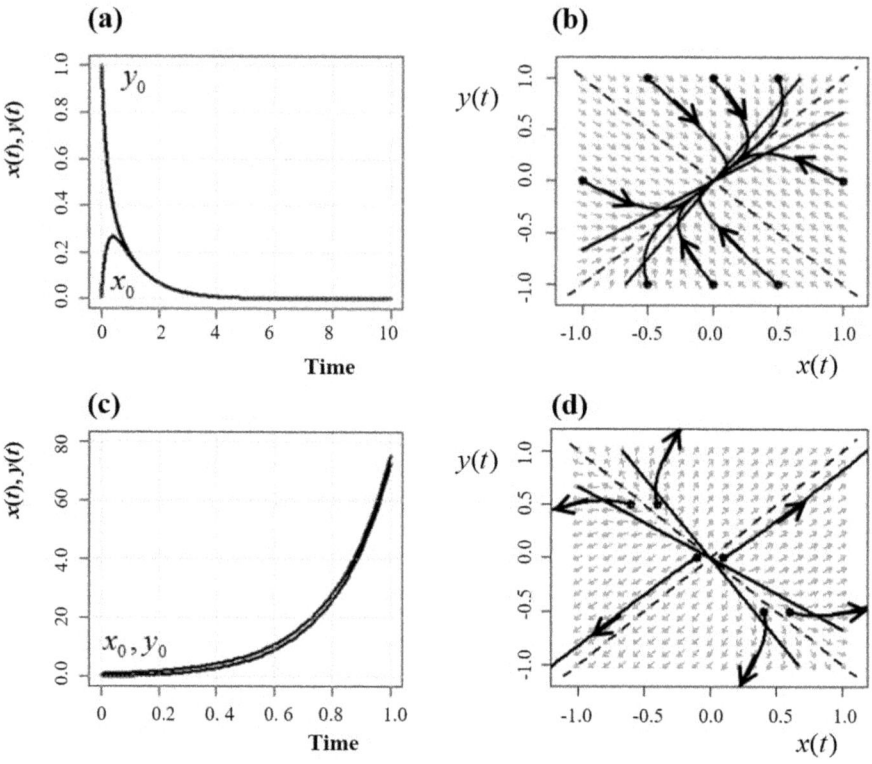

Fig. 6.2 Linear dynamic behaviour: nodes. The first row shows (a) the time series plots and (b) the phase diagram for a stable node equilibrium. The second row shows (c) the time series plots and (d) the phase diagram for an unstable node equilibrium.

corner (high initial levels). Trajectories converge towards an eigenvector and then run parallel to it towards equilibrium. The stability check results in negative real eigenvalues, consistent with a stable node: $\lambda_1 = -1, \lambda_2 = -5$. Alternatively, for an *unstable node* (the second row of Figure 6.2), the variables are repelled away from equilibrium, and the eigenvectors direct the trajectories in their explosive journeys towards positive and negative infinity. As expected, computed eigenvalues are real and positive: $\lambda_1 = 5, \lambda_2 = 1$.

Focus equilibria are shown in Figure 6.3. For a *stable focus*, the time series solutions follow dampened oscillations towards a steady state (the first row of the figure), and trajectories spiral towards a steady state in phase space. For an *unstable focus* (the second row of the figure), the time series solutions follow oscillations of increasing amplitude away from the steady state, and trajectories spiral away from the steady state in phase space. The computed eigenvalues are a complex conjugate pair with a negative real part for the stable focus ($\lambda_{1,2} = -1 \pm 3i$) and a positive real part for the unstable focus ($\lambda_{1,2} = 1 \pm 3i$).

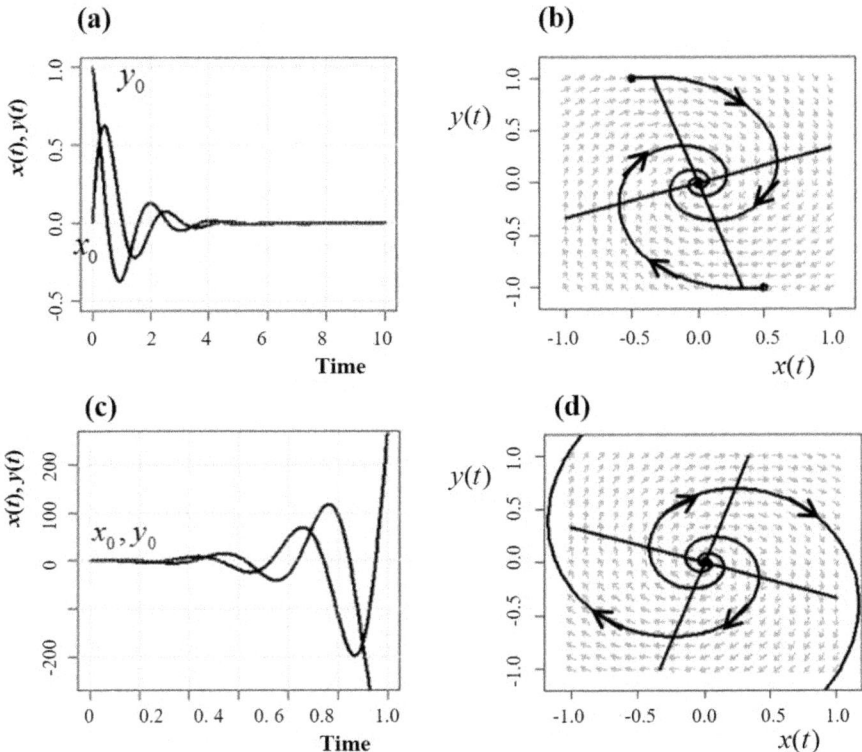

Fig. 6.3 Linear dynamic behaviour: foci. The first row shows (a) the time series plots and (b) the phase diagram for a stable focus equilibrium. The second row shows (c) the time series plots and (d) the phase diagram for an unstable focus equilibrium.

Finally, Figure 6.4 depicts the dynamics for the special cases of a *saddle point* equilibrium and a *centre point*. A saddle point (the first row of the figure) has real eigenvalues of opposite sign, which for this example are $\lambda_1 = -2, \lambda_2 = 4$. The eigenvector corresponding to the negative eigenvalue (the negatively sloped trajectory in Figure 6.4b) directs other trajectories from the upper left and lower right regions of the phase diagram towards equilibrium. Alternatively, the eigenvector corresponding to the positive eigenvalue (the positively sloped trajectory) leads other trajectories away from equilibrium towards the lower left corner (negative infinity) and upper right corner (positive infinity). Initial points resting on the stable eigenvector gravitate towards equilibrium (Figure 6.4a). Initial points away from either eigenvector initially follow the stable eigenvector towards equilibrium and then eventually veer away following the unstable eigenvector. A centre point (the second row of the figure) has the property that initial conditions rest on a unique elliptical trajectory, so variables cycle through time at constant amplitude. The centre point is not a stable equilibrium, because it does not attract initial conditions resting on other ellipses. Centre points have complex

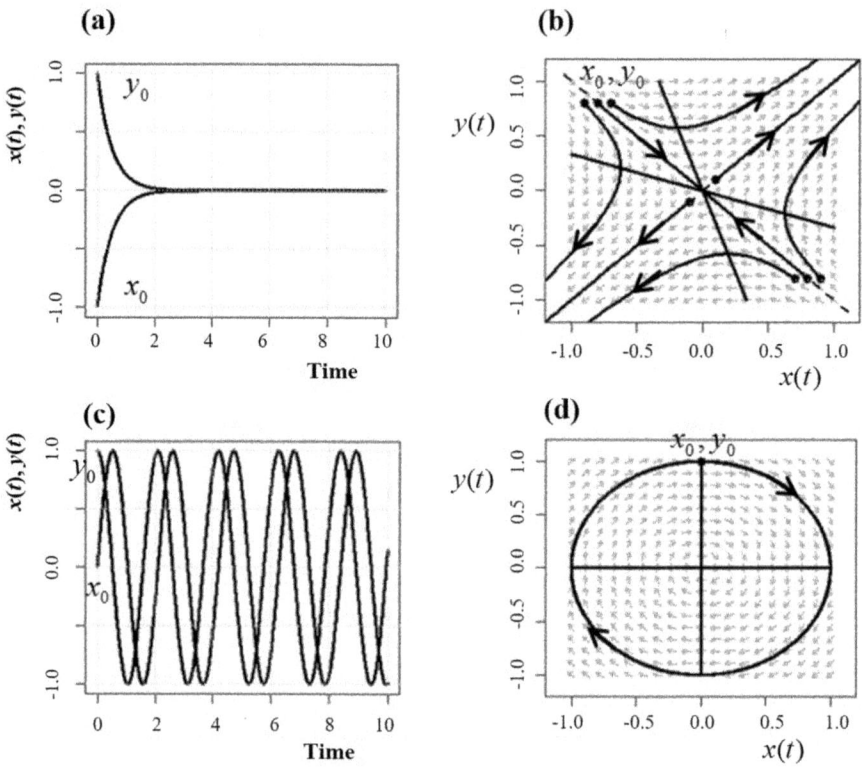

Fig. 6.4 Linear dynamic behaviour: saddle point and centre point. The first row shows (a) the time series plots and (b) the phase diagram for a saddle point equilibrium. The second row shows (c) the time series plots and (d) the phase diagram for a centre point.

eigenvalues with a zero real part. The eigenvalues for the centre point in this example are $\lambda_{1,2} = \pm 3i$.

We next explore what regular linear dynamics looks like in three dimensions by solving the following model:

$$
\begin{bmatrix} \dot{x} \\ \dot{y} \\ \dot{z} \end{bmatrix} = \begin{bmatrix} a_{11} & a_{12} & a_{13} \\ a_{21} & a_{22} & a_{23} \\ a_{31} & a_{32} & a_{33} \end{bmatrix} \begin{bmatrix} x \\ y \\ z \end{bmatrix} \tag{6.4}
$$

Code 6.2 uses the R package deSolve (Soetaert *et al.*, 2016) to solve the system (6.4) for the cases of a three-dimensional spiral sink and a spiral centre (Blanchard *et al.*, 2012; Hirsch *et al.*, 2004). We first define the parameters giving each case and report the corresponding eigenvalues. We next enter the information required by deSolve, including the parameters for the case we want to run, the initial conditions, the integration parameters and the model equations. Finally, we run `deSolve`, plot the time series, and animate the phase space trajectory with a user-defined function `animat3d_udf`.

```
#Code 6.2  Phase diagrams for three-dimensional linear ODE system
rm(list=ls(all=TRUE))  #clear values between runs

#Load user-defined functions
setwd("C:/User Defined Functions")
dump("animat3d_udf", file="animat3d_udf.R")
source("animat3d_udf.R")

#3D linear model parameters
#Case 1: Spiral sink
parms.ss<-c(-0.1,100,0,-100,-0.6,0,-1,0,-0.2)
#eigenvalues= -0.35+99.99969i -0.35-99.99969i -0.20

#Case 2: Spiral centre
parms.sc<-c(0,200,0,-1,0,0,0,0,-1)
#eigenvalues= 0+14.14214i, 0-14.14214i, -1

#Parameters for deSolve
parameters<-parms.sc  #Parameters for selected case
state<-c(x=1,y=-0.1,z=-0.1) #initial conditions
t.end<-5  #number of integration periods
delta<-0.001  #integration time step

#ODE model
model<-function(t,state,parameters){
  with(as.list(c(state,parameters)),{
  dx<-parameters[1]*x+parameters[2]*y+parameters[3]*z
  dy<-parameters[4]*x+parameters[5]*y+parameters[6]*z
  dz<-parameters[7]*x+parameters[8]*y+parameters[9]*z
  #With noise
  #dx<-parameters[1]*x+parameters[2]*y+parameters[3]*z+runif(1,-0.1,1)
  #dy<-parameters[4]*x+parameters[5]*y+parameters[6]*z+runif(1,-0.1,1)
  #dz<-parameters[7]*x+parameters[8]*y+parameters[9]*z+runif(1,-0.1,1)
  list(c(dx,dy,dz))
  }) #end with (as.list...
}  #end model function

#deSolve run statement
times<-seq(0,t.end,by=delta)  #integration step
library(deSolve)  #ODE solver package
out<-ode(y=state,times=times,func=model,parms=parameters,method="lsoda")

#Solution variables
x<-out[,"x"];y<-out[,"y"];z<-out[,"z"]
Mx<-cbind(x,y,z)

results<-list(x,y,z)

#Time series plots of solutions
par(mfrow=c(3,1))
plot(x,type='l',lwd=3,cex.axis=2,cex.lab=2,font.axis=2,font.lab=2)
plot(y,type='l',lwd=3,cex.axis=2,cex.lab=2,font.axis=2,font.lab=2)
plot(z,type='l',lwd=3,cex.axis=2,cex.lab=2,font.axis=2,font.lab=2)

#Automated phase space
setwd("C:\Results")  #export video to "Results" directory
animat3d.plot(Mx,20,0.5)
```

We employ the user-defined function `animat3d_udf` (Medina, 2016) built around the R package `animation` (Xie, 2017) to animate the three-dimensional attractors so that we can watch them evolve in real time (Code 6.3). The function requires input of a matrix M whose columns are three distinct time series or the lagged copies of a single time series in an embedded data matrix, a `start.angle` giving the initial viewing angle for the plot in degrees, and a `step.angle` setting the rate at which the plot rotates as it evolves. The executable file `ffmpeg` was downloaded from http://ffmpeg.org/download.html and copied into the `"C:/User Defined Functions"` directory. The video of the animated phase trajectory is directed to `"C:/Results"`.

```
#Code 6.3  User-defined function for automated phase diagram

#Author:  Miles Medina (2016), Unpublished PhD dissertation, Appendix 2,
#  University of Florida, Department of Agricultural and Biological Engineering.

#Inputs
#M:            Matrix whose columns are at least three time series.
#start.angle: Initial viewing angle for plot (degrees).
#step.angle:  Rotation rate (degrees of rotation per frame). Fractional values between
#             0 and 1 provide slow rotation. Set to 0 for no rotation.

animat3d.plot <- function(M,start.angle,step.angle){
    library(animation); library(scatterplot3d)
    ff.loc<-"C:/User Defined Functions/ffmpeg"
    #Determine axes' limits for plot
    x.min<-min(M[,1]); x.max<-max(M[,1])
    y.min<-min(M[,2]); y.max<-max(M[,2])
    z.min<-min(M[,3]); z.max<-max(M[,3])

    saveVideo(
    expr={
      ani.record(reset=TRUE) # discard previous plots
      ani.options(interval=0.05) # set interval between frames (seconds)
      for(i in 1:(dim(M)[1]-1)){
        # generate a new plot including each next row from matrix
        scatterplot3d(M[1:(i+1),1],M[1:(i+1),2],M[1:(i+1),3],box=F,
                      angle=(start.angle+step.angle*i),
                      xlim=c(x.min,x.max), ylim=c(y.min,y.max), zlim=c(z.min,z.max),
                      xlab="x",ylab="y",zlab="z",
                      type='l',lwd=2,color='blue')
        ani.record() # record the current plot as an animation frame
      }
    },
    video.name="animation.mp4",
    ffmpeg=ff.loc
    )
}

#Command to run user-defined function
setwd("C:/Results")  #export video to "Results" directory
#animat3d.plot(Mx,20,0.5)
```

The three-dimensional spiral sink is characterized by a negative real eigenvalue and a complex conjugate pair of eigenvalues with negative real parts. As a result, the time series solutions for each variable follow dampened oscillations towards the steady state at the origin (Figure 6.5a). The corresponding phase space trajectory starting from

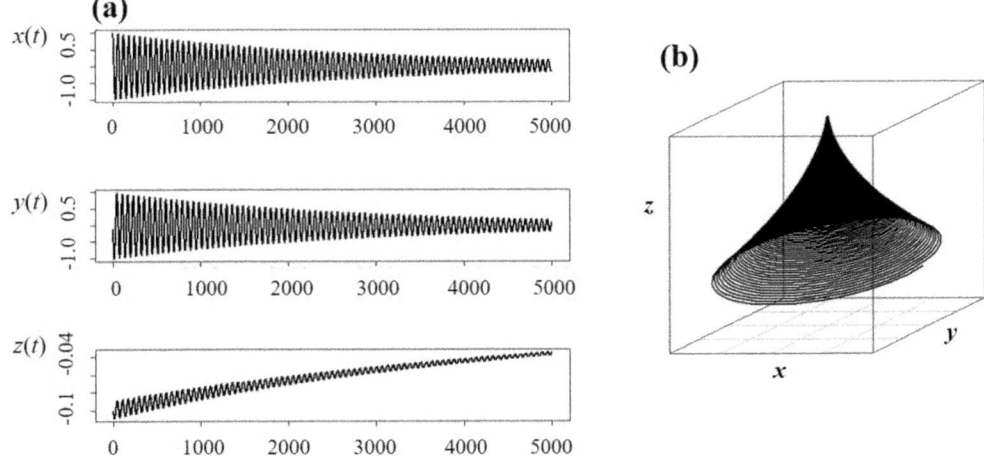

Fig. 6.5 Three-dimensional linear spiral sink dynamics. The real eigenvalue is negative and the complex conjugate pair has negative real parts. (a) The time series solutions for x, y, and z evolve towards the steady state at the origin along dampened oscillations. (b) A phase space trajectory starting from below the xy plane (for example) spirals in along the stable line given by the z axis towards a steady state.

initial conditions below the xy plane (for example) spirals unidirectionally towards a steady state along the stable line given by the z axis (Figure 6.5b). The three-dimensional trajectory has a cone-like appearance.

The spiral centre is associated with a negative real eigenvalue and a complex conjugate pair with zero real parts. Consequently, the time series solutions for x and y cycle periodically around the steady state at the origin, and z approaches the steady state asymptotically (Figure 6.6a). The phase space trajectory starting from initial conditions below the xy plane spirals unidirectionally towards the centre point orbit centred around the origin in the xy plane and, once on the orbit, never leaves it (Figure 6.6b). The three-dimensional trajectory has a cylindrical appearance.

We can distill the above linear stability results into the following lesson: If we immediately observe time series variables that exponentially dampen or explode, or evolve along periodic trajectories, we do not need NLTS methods to tell us that linear modelling suffices to reconstruct representative dynamics. However, in practice, temporal data behave in a more complex manner. Observed complexity may be attributed to linear dynamics agitated by the introduction of noise and/or nonstationarity or to deterministic nonlinear structure.

6.3 Noisy Linear Dynamics

We now illustrate how the addition of noise can make periodic linear behaviour appear more complex . We first calibrate the system (6.1) for a two-dimensional centre point characterized by a periodic orbit, and introduce noise by adding a random (uniform)

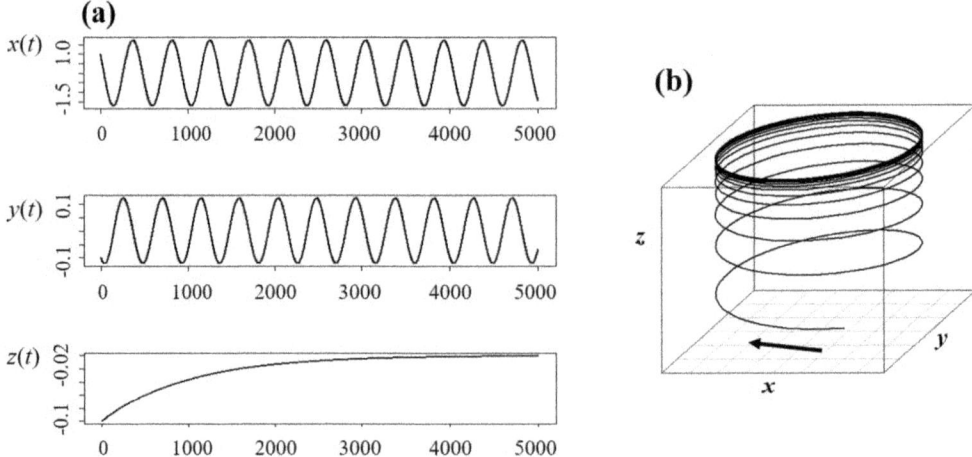

Fig. 6.6 Three-dimensional linear spiral centre dynamics. The real eigenvalue is negative and the complex conjugate pair has zero real parts. (a) The time series solutions for x and y cycle periodically, and z evolves towards the steady state at the origin. (b) The phase space trajectory starting from below the xy plane spirals in towards the centre point orbit centred at the origin.

error term ϵ_i to each equation:

$$\begin{bmatrix} \dot{x}_t \\ \dot{y}_t \end{bmatrix} = \underbrace{\begin{bmatrix} 0 & 3 \\ -3 & 0 \end{bmatrix}}_{A} \begin{bmatrix} x_t \\ y_t \end{bmatrix} + \begin{bmatrix} \epsilon_1(t) \\ \epsilon_2(t) \end{bmatrix} \tag{6.5}$$

Code 6.4 computes random error each period as a draw between -0.1 and 1 from a uniform probability distribution with the command `runif(1,-0.1,1)`. The command `set.seed(k)` (where `k` is an integer) could be added to obtain a reproducible random series. However, in this illustration, we do not need this option, so we leave it out. We use the R package **deSolve** (Soetaert *et al.*, 2016) to solve the system, and organize the code as **deSolve** requires. We specify model equations and parameters and then include instructions to plot time series solutions and the phase diagram.

```
#Code 6.4 Induce irregular dynamics with noisy linear dynamics
rm(list=ls(all=TRUE))  #clear values between runs

library(deSolve) #ODE solver package

#Set Parameters
delta<-0.1  #integration time step
state<-c(x=0,y=1) #initial conditions for variables
parameters<-c(a=0,b=3,c=-3,d=0) #parameters giving centre point

#Define ODE system
model<-function(t,state,parameters){
```

```
  with(as.list(c(state,parameters)),{
  #add uniform random error to linear ODE system
  #set.seed(3) #if want random sequence to be repeatable
  dx <- a*x+b*y+runif(1,-0.1,1)  #shocked centre point
  dy <- c*x+d*y+runif(1,-0.1,1)
  list(c(dx,dy))
  }) #end with (as.list...
} #end model function

#Integrate ODE system
times<-seq(0,100,by=delta)  #integration step
out<-ode(y=state,times=times,func=model,parms=parameters,method="rk4")
head(out)

#Plots of simulated values
par(mfrow=c(3,1))
plot(out[,"x"],type='l',lwd=3,cex.axis=1.1,font.axis=2) #time series plots
plot(out[,"y"],type='l',lwd=3,cex.axis=1.1,font.axis=2)
plot(out[,"x"],out[,"y"],type='l',lwd=3,cex.axis=1.1,font.axis=2) #phase space
```

The time series solutions exhibit the type of irregular nonperiodic behaviour that comes closer to what we typically observe in real-world data (Figure 6.7a). The phase diagram trajectory undergoes irregular oscillations randomly shifting through time (Figure 6.7b).

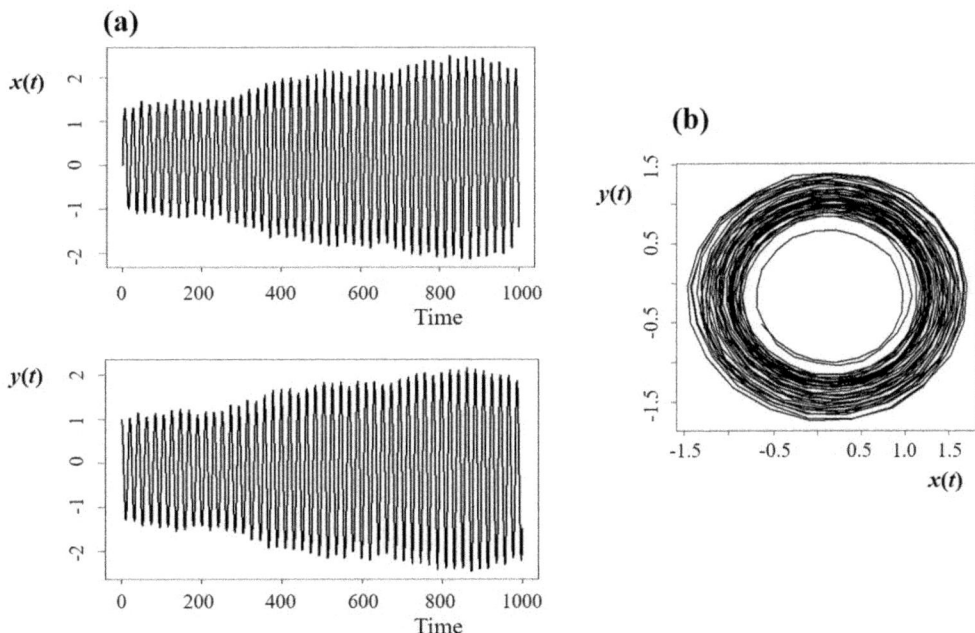

Fig. 6.7 Randomly shocked centre point. (a) The time series solutions exhibit the complexity commonly observed in real-world data. (b) The phase space trajectory undergoes irregular randomly shifting cycles.

We use Code 6.3 to visualize the impact of noise on a three-dimensional spiral centre, following the same procedure as for the two-dimensional centre point. In the model statement, we comment-in the equations without noise, comment-out the equations with randomly generated uniform noise and parameterize the model for a spiral centre. We observe that random shocks change the behaviour of the x and y solutions so that they now cycle with increasing amplitude, while the z solution fluctuates randomly (Figure 6.8a). This behaviour exhibits the complexity commonly observed in real-world data. The phase space trajectory oscillates irregularly, and consequently mimics the complex behaviour of deterministic nonlinear attractors (Figure 6.8b).

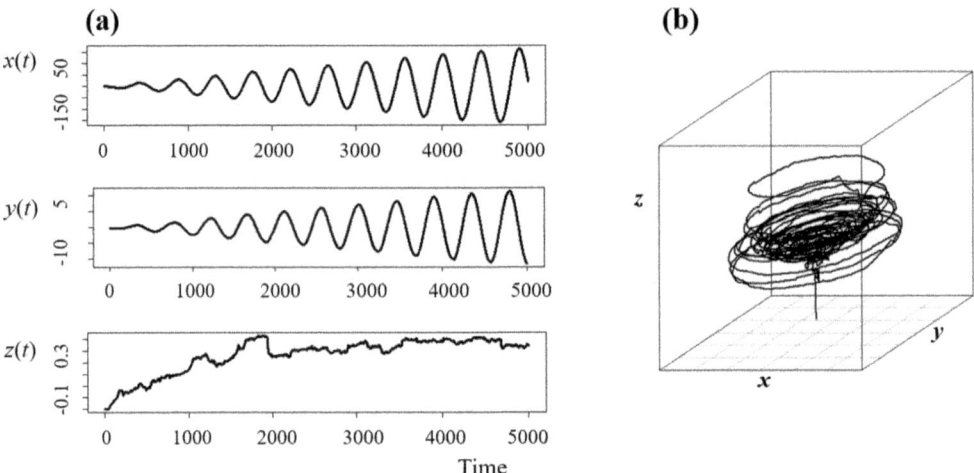

Fig. 6.8 Randomly shocked spiral centre. (a) Random shocks cause the time series solutions for x and y to oscillate with increasing amplitude, while z fluctuates randomly. (b) The corresponding phase space trajectory oscillates irregularly, similar in appearance to noisy nonlinear attractors reconstructed from real-world time series data.

6.4 Singular Spectrum Analysis

We use signal processing methods to reduce the impact of noisy data on masking other potential drivers of complex behaviour. We focus on *singular spectrum analysis (SSA)* as a data-driven method that detects structural variation in highly irregular time series data without imposing *ex ante* theoretical presumptions regarding the source of irregularity (Elsner and Tsonsis, 2010; Golyandina *et al.*, 2001). SSA separates time series data into structured variation (*signal*), including trend and oscillatory components, and residual unstructured variation (*noise*) . It does so in three steps: *matrix decomposition, grouping* and *time series reconstruction*. There are two methods of performing these steps: *singular value decomposition (SVD)-SSA* and *Toeplitz-SSA*. Each method takes a different approach to matrix decomposition, resulting in alternative mathematical measures of signal strength.

SSA provides a number of benefits beyond filtering noise. It may ease nonstationarity by allowing us to remove slow-moving trend components that may be at the root of the problem. It measures signal strength so that we can make an informed decision regarding whether to continue searching for deterministic structure in observed time series. Finally, SSA can be applied to fill in intermittent missing observations in observed data (Golyandina *et al.*, 2001). It does so by using the structured variation in the entire time series rather than relying solely on immediately surrounding observations as do moving average approaches.

In practice, we run SSA using the R package **Rssa** (Korobeynikov *et al.*, 2015). **Rssa** provides a number of visual diagnostics that the user must learn to interpret to run the package. To aid interpretation, we illustrate how SSA operates inside **Rssa** with the simple time series $x_t = (1, 3, 0, -3, -2, -1)^{\mathrm{T}}$ (superscript T denotes the transpose). We formulate code to run the illustration using SVD-SSA and then Toeplitz-SSA.

6.4.1 SVD-SSA

Code 6.5 illustrates the mathematical mechanics of SVD-SSA:

```
#Code 6.5  Illustration of SVD-SSA mechanics
rm(list=ls(all=TRUE))

x<-c(1,3,0,-3,-2,-1)  #sample time series

#Step 1: Matrix decomposition

  #Step 1a:  Trajectory matrix
library(tseriesChaos)
X<-embedd(x,3,1)  #embedded trajectory matrix
nrow<-nrow(X);ncol<-ncol(X)

  #Step 1b:  Lagged covariance matrix
S<-t(X)%*%X

  #Step 1c:  Eigensystem of S
eigensys<-eigen(S,symmetric=TRUE)
eigenvals<-eigensys$values  #eigenvalues
eigenvecs<-eigensys$vectors  #eigenvectors
#Left eigenvectors
left.1<-X%*%eigenvecs[,1]/sqrt(eigenvals[1])
left.2<-X%*%eigenvecs[,2]/sqrt(eigenvals[2])
left.3<-X%*%eigenvecs[,3]/sqrt(eigenvals[3])

  #Step 1d: Matrices in decomposition
X1<-sqrt(eigenvals[1])*left.1%*%t(eigenvecs[,1])
X2<-sqrt(eigenvals[2])*left.2%*%t(eigenvecs[,2])
X3<-sqrt(eigenvals[3])*left.3%*%t(eigenvecs[,3])

  #Step 1e:  Characteristic contributions of Xi
X1.cc<-eigenvals[1]/sum(eigenvals)
X2.cc<-eigenvals[2]/sum(eigenvals)
X3.cc<-eigenvals[3]/sum(eigenvals)

#Step 2:  Matrix grouping
X12<-X1+X2
X12.cc<-(eigenvals[1]+eigenvals[2])/sum(eigenvals)
```

```
#Step 3:  Time series reconstruction (diagonal averaging)

  #Step 3a: User-defined function for averaging of minor diagonals
diag.ave<-function(mat) {
hold<-matrix(0,(nrow+(ncol-1)))
for(i in 1:(nrow+(ncol-1))) {

  if(i==1) {d<-mat[1,1]}

  if(i>1 & i<=ncol) {d<-diag(mat[i:1,1:i])}

  if(i>ncol & i<=nrow) {d<-diag(mat[i:(i-(ncol-1)),1:ncol])}

  if(i>nrow & i<(nrow+(ncol-1))) {
    d<-diag(mat[nrow:(i-(ncol-1)),(i-(nrow-1)):ncol])}

  if(i==(nrow+(ncol-1))) {d<-mat[nrow,ncol]}

  d.ave<-mean(d)   #average minor diagonals
  hold[i,]<-d.ave
  } #end loop
return(hold)
}  #end function

  #Step 3b:  Reconstructed time series
x1<-diag.ave(X1)
x2<-diag.ave(X2)
x3<-diag.ave(X3)
x12<-diag.ave(X12)
```

Step 1 performs *matrix decomposition* . Step 1a computes a *trajectory matrix X* by *embedding* the time series x_t. For SVD-SSA, we select the embedding dimension K, but the embedding delay is fixed at $d = 1$, so that X is an $L \times K$ matrix. The number of rows is determined by $L = N - K + 1$ (where N is the number of observations in the time series) and is referred to as the *window length*. The first column of X contains the first L observations of x_t and the subsequent columns contain one-period forward-lagged copies of x_t. For our example, we select $K = 3$, so that $L = 6 - 3 + 1 = 4$ and X is the following 4×3 matrix:

$$\underset{L \times K}{X} = \begin{bmatrix} 1 & 3 & 0 \\ 3 & 0 & -3 \\ 0 & -3 & -2 \\ -3 & -2 & -1 \end{bmatrix} \tag{6.6}$$

Step 1b computes the $K \times K$ *lagged covariance matrix* $S = X^{\mathrm{T}}X$. For this example, S is

$$\underset{3 \times 3}{S} = \begin{bmatrix} 19 & 9 & -6 \\ 9 & 22 & 8 \\ -6 & 8 & 14 \end{bmatrix} \tag{6.7}$$

Step 1c calculates the *eigensystem* of S. In general, the eigensystem includes K *eigenvalues* $\lambda_i, i = 1, \ldots, K$. Each eigenvalue has a corresponding $K \times 1$ *eigenvector* EV_i and an $L \times 1$ *left eigenvector* V_i. We compute left eigenvectors using the formula

$V_i = X * EV_i / \sqrt{\lambda_i}$. For our example, the eigenvalues are $\lambda_1 = 30.07$, $\lambda_2 = 22.15$ and $\lambda_3 = 2.77$. The 3×1 eigenvectors are

$$
\underbrace{\begin{bmatrix} 0.5516144 \\ 0.8103992 \\ 0.1974201 \end{bmatrix}}_{EV_1}, \quad
\underbrace{\begin{bmatrix} -0.6392595 \\ 0.2587119 \\ 0.7241653 \end{bmatrix}}_{EV_2}, \quad
\underbrace{\begin{bmatrix} 0.5357881 \\ -0.5256627 \\ 0.6607646 \end{bmatrix}}_{EV_3}
\tag{6.8}
$$

The 4×1 left eigenvectors are

$$
\underbrace{\begin{bmatrix} 0.5439060 \\ 0.1937585 \\ -0.5153187 \\ -0.6333014 \end{bmatrix}}_{V_1}, \quad
\underbrace{\begin{bmatrix} 0.02908016 \\ -0.86900216 \\ -0.47260060 \\ 0.14366026 \end{bmatrix}}_{V_2}, \quad
\underbrace{\begin{bmatrix} -0.6255355 \\ -0.2252514 \\ 0.1534754 \\ -0.7310352 \end{bmatrix}}_{V_3}
\tag{6.9}
$$

Step 1d decomposes the trajectory matrix X into a sum of new $L \times K$ matrices: $X = X_1 + X_2 + \ldots + X_K$. Each new matrix in this decomposition is computed as the matrix product of the three components from the eigensystem of S (an *eigentriplet*):

$$
X_i = \sqrt{\lambda_i} * V_i * EV_i^{\mathrm{T}}
\tag{6.10}
$$

(Golyandina *et al.*, 2001). For our example, the $K = 3$ new matrices X_i are

$$
\underbrace{\begin{bmatrix} 1.65 & 2.42 & 0.59 \\ 0.59 & 0.86 & 0.21 \\ -1.56 & -2.29 & -0.56 \\ -1.92 & -2.81 & -0.69 \end{bmatrix}}_{X_1}, \quad
\underbrace{\begin{bmatrix} -0.09 & 0.04 & 0.01 \\ 2.61 & -1.06 & -2.96 \\ 1.42 & -0.58 & -1.61 \\ -0.43 & 0.17 & 0.49 \end{bmatrix}}_{X_2}, \quad
\underbrace{\begin{bmatrix} -0.56 & 0.55 & -0.69 \\ -0.20 & 0.20 & -0.25 \\ 0.14 & -0.13 & 0.17 \\ -0.65 & 0.64 & -0.80 \end{bmatrix}}_{X_3}
$$

$$
\tag{6.11}
$$

Step 1e computes the strength of the individual contribution of one of the new matrices X_i relative to the overall decomposition of X. This contribution is measured in terms of the *Frobenius norm* of X:

$$
\|X\|_F = \sqrt{\sum_{i=1}^{K} \lambda_i} \quad \text{or} \quad \|X\|_F^2 = \sum_{i=1}^{K} \lambda_i
\tag{6.12}
$$

That is, the square of the Frobenius norm is the sum of the eigenvalues in the decomposition. Then, the *characteristic contribution* of X_i is measured as

$$
\frac{\lambda_i}{\|X\|_F^2} \quad \text{or} \quad \frac{\lambda_i}{\sum_{i=1}^{K} \lambda_i}
\tag{6.13}
$$

that is, as the fractional share of its eigenvalue λ_i in the sum of all the eigenvalues in the SVD decomposition (Golyandina *et al.*, 2001). For our example, the characteristic contributions of X_1, X_2 and X_3 are $0.55 = 30.07/55$, $0.40 = 22.15/55$ and $0.05 = 2.77/55$.

Step 2 (*matrix grouping*) partitions the matrices in the SVD decomposition, X_i, into groups. In practice, we use a series of diagnostics to strategically group matrices so that groupings capture various signal and noise components of the data. However, in our example, we simply group the first two matrices, $X_{12} = X_1 + X_2$:

$$
\underbrace{\begin{bmatrix} 1.56 & 2.45 & 0.69 \\ 3.20 & -0.20 & -2.75 \\ -0.14 & -2.90 & -2.17 \\ -2.35 & -2.64 & -0.20 \end{bmatrix}}_{X_{12}} = X_1 + X_2 \tag{6.14}
$$

The characteristic contribution of the grouped matrix X_{12} is

$$
\frac{\lambda_1 + \lambda_2}{\sum_{i=1}^{K} \lambda_i} = 0.95 \tag{6.15}
$$

Step 3 (*time series reconstruction*) converts the grouped matrices into corresponding vector time series. For our example, X_{12} and X_3 are converted into corresponding time series x_{12} and x_3. This is accomplished by taking the *minor diagonal averages* of X_{12} and X_3. To illustrate how this diagonal averaging works in ideal circumstances, we apply it to convert the trajectory matrix X back into the original vector time series x_t; in other words, to reverse the embedding process. The minor diagonals of X are denoted by the rectangular boxes in Figure 6.9. Starting from the upper left corner of X, the average of the first minor diagonal is 1, that of the second is $3 = (3+3)/2$, and so on. Combining the averages in a single array perfectly reproduces the original time series: $x_t = 1, 3, 0, -3, -2, -1$. Diagonal averaging works ideally for *Hankel matrices* (like X), which have constant elements along the minor diagonals.

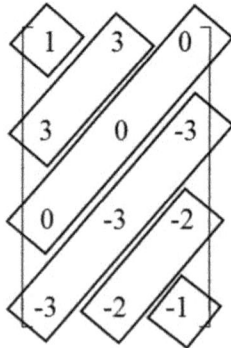

Fig. 6.9 Diagonal averaging of the trajectory matrix X.

In practice, the matrices in the SVD decomposition need not be Hankel matrices. However, the *Hankelization operator* can convert an arbitrary matrix into a Hankel matrix before diagonal averaging (Golyandina *et al.*, 2001).

Step 3a formulates a user-defined function `diag.ave` to compute minor diagonal averages of a matrix, and Step 3b applies it to the grouped matrices in the SVD decomposition. Although X_{12} and X_3 are not Hankel matrices, diagonal averaging delivers corresponding time series that when summed reproduce the original time series x_t for our simple example (Table 6.1).

Table 6.1 SVD-SSA decomposition of time series.

$x(t)$	x_{12}	x_3
1	1.56	−0.56
3	2.83	0.17
0	0.12	−0.12
−3	−2.66	−0.34
−2	−2.40	0.40
−1	−0.20	−0.80

6.4.2 Toeplitz-SSA

Code 6.6 illustrates the mechanics of Toeplitz-SSA:

```
#Code 6.6  Illustration of Toeplitz-SSA mechanics
rm(list=ls(all=TRUE))

x<-c(1,3,0,-3,-2,-1)  #sample time series

#Step 1: Matrix decomposition

  #Step 1a:  Trajectory matrix
X.traj<-matrix(c(1,3,0,-3,-2,-1,3,0,-3,-2,-1,0,
  0,-3,-2,-1,0,0),6,byrow=FALSE)
nrow<-nrow(X.traj);ncol<-ncol(X.traj)  #number of rows and columns

  #Step 1b:  Lagged covariance matrix
c11<-sum(x^2)/length(x)  #variance
#co-variance between x and lagged vector one period removed
c12<-sum((x*X.traj[,2])[1:(length(x)-1)])/(length(x)-1)
#co-variance between x and lagged vector two periods removed
c13<-sum((x*X.traj[,3])[1:(length(x)-2)])/(length(x)-2)
#lagged covariance matrix
cx<-matrix(c(c11,c12,c13,c12,c11,c12,c13,c12,c11),3,byrow=T)

  #Step 1c:  Eigensystem of cx
eigensys<-eigen(cx,symmetric=TRUE)
eigenvals<-eigensys$values  #eigenvalues
eigenvecs<-eigensys$vectors  #eigenvectors
#Left eigenvectors
left.1<-X.traj%*%eigenvecs[,1]/sqrt(eigenvals[1])
```

```
left.2<-X.traj%*%eigenvecs[,2]/sqrt(eigenvals[2])
left.3<-X.traj%*%eigenvecs[,3]/sqrt(eigenvals[3])

#Step 1d: Matrices in decomposition
X1<-sqrt(eigenvals[1])*left.1%*%t(eigenvecs[,1])
X2<-sqrt(eigenvals[2])*left.2%*%t(eigenvecs[,2])
X3<-sqrt(eigenvals[3])*left.3%*%t(eigenvecs[,3])

#Step 1e:  Contributions of Xi
X1.c<-eigenvals[1]/sum(eigenvals)
X2.c<-eigenvals[2]/sum(eigenvals)
X3.c<-eigenvals[3]/sum(eigenvals)

#Step 2:  Time series reconstruction (diagonal averaging)

#Step 2a: User-defined function for averaging of minor diagonals
diag.ave<-function(mat) {
hold<-matrix(0,(nrow+(ncol-1)))
for(i in 1:(nrow+(ncol-1))) {
  if(i==1) {d<-mat[1,1]}

  if(i>1 & i<=ncol) {d<-diag(mat[i:1,1:i])}

  if(i>ncol & i<=nrow) {d<-diag(mat[i:(i-(ncol-1)),1:ncol])}

  if(i>nrow & i<(nrow+(ncol-1))) {
    d<-diag(mat[nrow:(i-(ncol-1)),(i-(nrow-1)):ncol])}

  if(i==(nrow+(ncol-1))) {d<-mat[nrow,ncol]}

  d.ave<-mean(d)   #average minor diagonals
  hold[i,]<-d.ave
  } #end loop
return(hold)
}   #end function

#Step 2b:  Reconstructed time series
x1<-diag.ave(X1)[1:length(x)]
x2<-diag.ave(X2)[1:length(x)]
x3<-diag.ave(X3)[1:length(x)]
```

Step 1 performs *matrix decomposition*. Step 1a constructs a variation of the *trajectory matrix* X used in SVD-SSA. The first column of X contains the entire length of time series x_t. Zeros are added to the bottom of the remaining $K - 1$ one-period forward-lagged copies so that they have equal row dimension with x_t. The result is that X is an $N \times K$ matrix. For our example, we set $K = 3$ to obtain the following 6×3 matrix X:

$$
\begin{bmatrix}
1 & 3 & 0 \\
3 & 0 & -3 \\
0 & -3 & -2 \\
-3 & -2 & -1 \\
-2 & -1 & 0 \\
-1 & 0 & 0
\end{bmatrix}
\tag{6.16}
$$

Step 1b computes the $K \times K$ *lagged covariance matrix* C_X, whose elements are

$$c_{ij} = \frac{1}{N - |i - j|} \sum_{t=1}^{N - |i - j|} x(t)x(t + |i - j|) \tag{6.17}$$

where N is the length of the time series (Ghil *et al.*, 2001; Golyandina *et al.*, 2001). The diagonal elements are the variance (from zero) of the time series (c_{11}). The first diagonal removed contains the covariance between the time series and its lagged copy one period apart (c_{12}). The second diagonal removed contains the covariance between the time series and its lagged copy two periods apart (c_{13}). For this example, the lagged covariance matrix is

$$C_X = \begin{bmatrix} 4 & 2.2 & -1.5 \\ 2.2 & 4 & 2.2 \\ -1.5 & 2.2 & 4 \end{bmatrix} \tag{6.18}$$

The subsequent steps proceed as in SVD-SSA (Golyandina *et al.*, 2001). Step 1c computes the eigensystem of C_X. For our example, the eigenvalues are $\lambda_1 = 6.45$, $\lambda_2 = 5.5$ and $\lambda_3 = 0.05$. The 3×1 eigenvectors are

$$\underbrace{\begin{bmatrix} 0.44 \\ 0.79 \\ 0.44 \end{bmatrix}}_{EV_1}, \quad \underbrace{\begin{bmatrix} 0.71 \\ 0.00 \\ -0.71 \end{bmatrix}}_{EV_2}, \quad \underbrace{\begin{bmatrix} 0.56 \\ -0.62 \\ 0.56 \end{bmatrix}}_{EV_3} \tag{6.19}$$

The 6×1 left eigenvectors are

$$\underbrace{\begin{bmatrix} 1.10 \\ 0.00 \\ -1.27 \\ -1.31 \\ -0.65 \\ -0.17 \end{bmatrix}}_{V_1}, \quad \underbrace{\begin{bmatrix} 0.30 \\ 1.81 \\ 0.60 \\ -0.60 \\ -0.60 \\ -0.30 \end{bmatrix}}_{V_2}, \quad \underbrace{\begin{bmatrix} -5.84 \\ 0.00 \\ 3.35 \\ -4.42 \\ -2.21 \\ -2.48 \end{bmatrix}}_{V_3} \tag{6.20}$$

Step 1d uses this eigensystem of C_X for matrix decomposition of the Toeplitz trajectory matrix X. The $K = 3$ matrices in the decomposition are

$$
\begin{bmatrix}
1.22 & 2.20 & 1.22 \\
0.00 & 0.00 & 0.00 \\
-1.41 & -2.54 & -1.41 \\
-1.45 & -2.61 & -1.45 \\
-0.73 & -1.30 & -0.73 \\
-0.19 & -0.34 & -0.19
\end{bmatrix}
\underbrace{}_{X_1},
\quad
\begin{bmatrix}
0.5 & 0 & -0.5 \\
3.0 & 0 & -3.0 \\
1.0 & 0 & -1.0 \\
1.0 & 0 & 1.0 \\
-1.0 & 0 & 1.0 \\
-0.5 & 0 & 0.5
\end{bmatrix}
\underbrace{}_{X_2},
\quad
\begin{bmatrix}
-0.72 & 0.80 & -0.72 \\
0.00 & 0.00 & 0.00 \\
0.41 & -0.46 & 0.41 \\
-0.55 & 0.61 & -0.55 \\
-0.27 & 0.30 & -0.27 \\
-0.31 & 0.34 & -0.31
\end{bmatrix}
\underbrace{}_{X_3}
$$

$$(6.21)$$

In this illustration, we keep the three matrices grouped separately.

In Step 1e, Toeplitz-SSA measures the contribution of a matrix to the overall decomposition in relation to the total variance (from zero) in the time series. This provides an intuitive measure of signal strength for non-mathematicians. For our example, we note from the diagonal of C_X that the total variance of x_t is $c_{11} = 4$ (eqn 6.18). The trace of C_X is $\mathrm{Tr}(C_X) = 4 + 4 + 4 = 3c_{11}$, so that $c_{11} = \mathrm{Tr}(C_X)/3$. Since the trace of a square matrix equals the sum of its eigenvalues, we can compute the total variance of the time series (from zero) as the average of the eigenvalues of C_X:

$$
c_{11} = \frac{1}{K(=3)} \sum_{i=1}^{K(=3)} \lambda_i = 4 \tag{6.22}
$$

The partial variance attributed to the ith matrix in the Toeplitz decomposition is

$$
\frac{\frac{1}{3}\lambda_i}{\frac{1}{3}\lambda_1 + \frac{1}{3}\lambda_2 + \frac{1}{3}\lambda_3} = \frac{\lambda_i}{\sum_{i=1}^{K} \lambda_i} \tag{6.23}
$$

In our example, the matrices in the decomposition, X_1, X_2 and X_3, account for 54%, 46%, and 0.004% of the variance (from zero) in the time series. If we want to measure the contribution in terms of total variance in the time series from its mean, then we initially transform the data by subtracting the series mean from each observation.

In Step 2, X_1, X_2, and X_3 are converted into corresponding time series with diagonal averaging. For our example, the decomposed time series are reported in Table 6.2.

6.4.3 SSA in R

In practice, we run SSA with the R package Rssa (Korobeynikov *et al.*, 2015). We present two codes that split Rssa into the matrix decomposition and grouping steps (Code 6.9) and the time series reconstruction step (Code 6.11). The reason is that Code 6.9 provides visual diagnostics that must be consulted before running Code 6.11. We illustrate the use of Rssa to run signal processing on monthly sea levels (1928–2000) collected at the Mayport station on the Atlantic coast east of Jacksonville, Florida (PSMSL, 2016). The plotted time series exhibits substantial variability (see Figure 6.10b, left-hand plot).

Table 6.2 Toeplitz-SSA decomposition of time series.

$x(t)$	x_{12}	x_3
1	1.72	−0.72
3	2.60	0.40
0	0.10	−0.10
−3	−2.66	−0.34
−2	−2.25	0.25
−1	−0.82	−0.18

Rssa requires specification of only the SSA *window* parameter to run the matrix decomposition. In SVD-SSA, the window length L sets the number of rows in the trajectory matrix. In Toeplitz-SSA, the window length K sets the number of columns in the trajectory matrix. Since there are no precise rules for setting the window length, guidance comes from *rules of thumb* found in the literature. Ghil *et al.* (2001) present the choice as a trade-off between competing objectives. On the one hand, the window should be sufficiently wide to capture the range of dynamic behaviour exhibited by the time series. Golyandina *et al.* (2001) judge that that matrix decomposition is most precise when the window is set to approximately half the length of the time series, $N/2$. On the other hand, the window should be sufficiently narrow to generate statistical confidence in the results by allowing for as many repetitions of the features of interest as possible, i.e. as large a ratio of N/K as possible (Ghil *et al.*, 2001). Hassani (2007) suggests that window length should be proportional to a periodic component in the data, but should not exceed $N/2$. For example, if monthly data were characterized by a two-year cycle, and the time series length were $N = 1000$, then the window could be set at integer multiples (i.e. repetitions) of 24 months not exceeding $N/2 = 500$, such as $L = 24$ months × 20 repetitions = 480.

We can identify periodic components in time series data with the *Fourier power spectrum* (Kaplan and Glass, 1995). Initially, the *Fourier transform* converts the time series from the *time domain* to the *frequency domain* by decomposing it into a sum of sinusoidal functions of different frequencies. The Fourier power spectrum then presents the frequency content of the time series in a graph plotting the importance (i.e. *power*) of particular frequencies in the data over the spectrum of frequencies. The power that a particular frequency contributes to the time series is identified by the height of a spike above that frequency. Time series with strong periodic components show sharp peaks at constituent frequencies forming the harmonics of the data. A continuous *floor* appearing in the spectrum may be due to noise or deterministic nonlinear dynamics resembling noise. We cannot distinguish between the two without additional information (Kantz and Schreiber, 1997).

Code 6.7 formulates a user-defined function, spectral_udf, to run spectral analysis with the R routine spectrum. We initially run a simple illustration using a single sine wave

$$y(t) = A\sin\left[\left(a\frac{2\pi}{T}\right)t + \phi\right] \tag{6.24}$$

where T is the number of time periods t, a is the number of cycles completed over T periods, A is the amplitude of the cycles and ϕ is the phase determining where the cycle is at $t{=}0$. There are T/a periods per cycle, so that the cycle frequency is the inverse, a/T cycles per period.

```
#Code 6.7 Example of Fourier power spectrum for single sine wave
rm(list=ls(all=TRUE))

#Parameters for sine wave
phi<-0 #phase
A<-1 #amplitude
T<-1000 #time periods
a<-10 #number of cycles completed over T
frequency<-a/T;frequency

#Sine wave
t<-1:T #number of time steps
x<-A*sin(((a*2*pi/T)*t)+phi)
plot(x,type='l')

#Define user-defined function to run spectral analysis
spectral<-function(x,method_spec) {
if(method_spec) {spec.out<-spectrum(x,method="ar")} else
 {spec.out<-spectrum(x,method="pgram")}

#Power spectrum plots
#Spectral values (vertical axis of plot)
power<-spec.out$spec
#Frequencies on  horizontal axis of plot
frequency<-spec.out$freq
#Against cycle lengths (on horizontal of plot)
cycle<-1/frequency

#Sort cycles in order of magnitude of power spikes
hold<-matrix(0,(length(power)-2),1)
for(i in 1:(length(power)-2)){
  max1<-if(power[i+1]>power[i]&&power[i+1]>power[i+2])1 else (0)
  #print(max1)
  hold[i,]<-max1
 } #end loop i
max<-which(hold==1)+1
power.max<-power[max]
cycle.max<-cycle[max]
o<-order(power.max,decreasing=TRUE)
cycle.max.o<-cycle.max[o]  #cycles
results<-list(cycle.max.o)
return(results)
} #end user-defined function

results<-spectral(x,method_spec=TRUE)
cycles<-results[[1]];cycles #cycles sorted in decreasing magnitude of their power spikes
```

This code computes a sine wave that completes $a = 10$ cycles over $T = 1000$ periods with $A = 1$ and $\phi = 0$. Each cycle repeats every 100 periods with frequency 0.01 cycles per period (Figure 6.10a, left-hand plot). After computing the wave, we

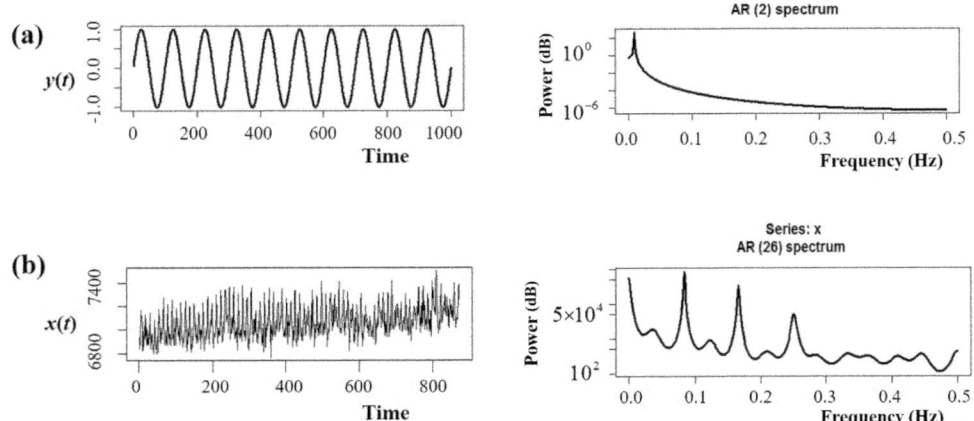

Fig. 6.10 (a) Example of a Fourier power spectrum for a single sine wave. The sine wave (left-hand plot) completes 10 cycles in 1000 periods. The Fourier power spectrum (right-hand plot) identifies the frequency of the wave with a spike over $10/1000 = 0.01$ cycles per period. (b) The left-hand plot is a time series of Mayport sea levels. The Fourier power spectrum (right-hand plot) shows dominant frequencies at 0.083, 0.167 and 0.25 Hz.

formulate the user-defined function `spectral_udf` for spectral analysis, which requires input of the time series x and the spectral method *method_spec*. If we set *method_spec = TRUE*, then the R routine `spectrum` runs the method *ar*; alternatively, if we set *method_spec = FALSE*, then `spectrum` runs the other available method, *pgram*. We retrieve output from `spectrum`, with `out$spec` giving powers and `out$freq` giving frequencies, and compute the cycle lengths as reciprocal frequencies. Finally, we identify the cycle lengths corresponding to peak power spikes. We first identify the spikes with a conditional statement assigning 1 when a spike is greater than its immediate neighbours and 0 otherwise. We then order the cycle lengths in decreasing order of magnitude of their spikes in the output vector, `cycles`. The Fourier power spectrum identifies the frequency of the sine wave with a spike over 0.01 cycles per period (Figure 6.10a, right-hand plot), and the corresponding cycle length ≈ 100 periods is reported in `cycles`.

Code 6.8 applies the Fourier power spectrum to identify periodic components in the Mayport sea level series with the mean removed:

```
#Code 6.8  Fourier power spectrum for time series
rm(list=ls(all=TRUE))

#Load user-defined functions
setwd("C:/User Defined Functions")
dump("spectral_udf", file="spectral_udf.R")
source("spectral_udf.R")

#Read in data
setwd("C:/Users/Ray/Documents/R_files/Data")
```

```
ts<-read.csv("Mayport.csv");x<-ts$observed

#Run user-defined function for spectral analysis
results<-spectral(x,method_spec=TRUE)
cycles<-results[[1]];cycles
```

This code reads in a user-defined function `spectral_udf` and the Mayport time series. The Fourier power spectrum for the Mayport data shows dominant frequencies
at 0.083, 0.167 and 0.25 Hz (Figure 6.10b, left-hand plot). The lengths of the corresponding dominant oscillations in the data are the reciprocals of the frequencies: 12,
6 and 4 months. Given that the length of the Mayport time series is 864 observations,
a reasonable SSA window length is 432 = 12-month oscillation × 36 repetitions.

Code 6.9 runs the SSA decomposition and grouping steps:

```
#Code 6.9 SSA: matrix decomposition and grouping diagnostics
rm(list=ls(all=TRUE))

#Read in data
setwd("C:/Data")
ts<-read.csv("Mayport.csv");sea<-ts$observed

#SSA Decomposition
library(Rssa)
L=432 #input SSA window length L=(12-month cycle)(30 repetitions)
s<-ssa(x,L,kind="toeplitz-ssa")   #run Rssa

#Run grouping diagnostics to group eigentriplets
   #First visual diagnostic: Eigenspectrum
eigenvalues<-plot(s,numvalues=20,col="black",lwd=2) #plot 1st 20 largest

   #Second visual diagnostic: Eigenvector plots
plot(s,type="vectors",idx=1:10,xlim=c(1,300),col="black",lwd=2)  #plot 1st 10 for 300 periods

   #Third visual diagositic: Pairwise scatterplots of eigenvectors
plot(s,type="paired",idx=1:9,plot.contrib=FALSE,col="black",lwd=2)  #plot 1st 9 pairs

   #Weighted correlation matrix
plot(w<-wcor(s,groups=c(1:20)))  #1st 20 eigentriplets
w.corr.res<-wcor(s,groups=c(1:10))  #table for 1st 10 eigentriplets
```

This code reads in the Mayport time series with the mean removed. It then sets the
window length at $L = 432$ and runs `Rssa`, selecting the `toeplitz-ssa` method so that
the contribution of a matrix to the overall decomposition is measured in relation to the
total variance (from the mean) in the time series. `Rssa` output generates three visual
diagnostics used to group the decomposed matrices X_i: the *eigenspectrum, eigenvector
plots* and *pairwise scatter plots of eigenvectors*. These diagnostics are derived from
each component of the eigentriplets defining each decomposed matrix in eqn (6.10).

The first visual diagnostic is the eigenspectrum, which plots the square roots of the
eigenvalues (*singular values*) in the eigentriplets in descending magnitude. The eigenspectrum typically has a hockey-stick appearance. The rule of thumb is that singular
values along the steep portion of the eigenspectrum are associated with eigentriplets
forming the basis of the deterministic signal. Successive singular values that are equal
create steps in the steep portion. Each step represents eigentriplet pairs forming the

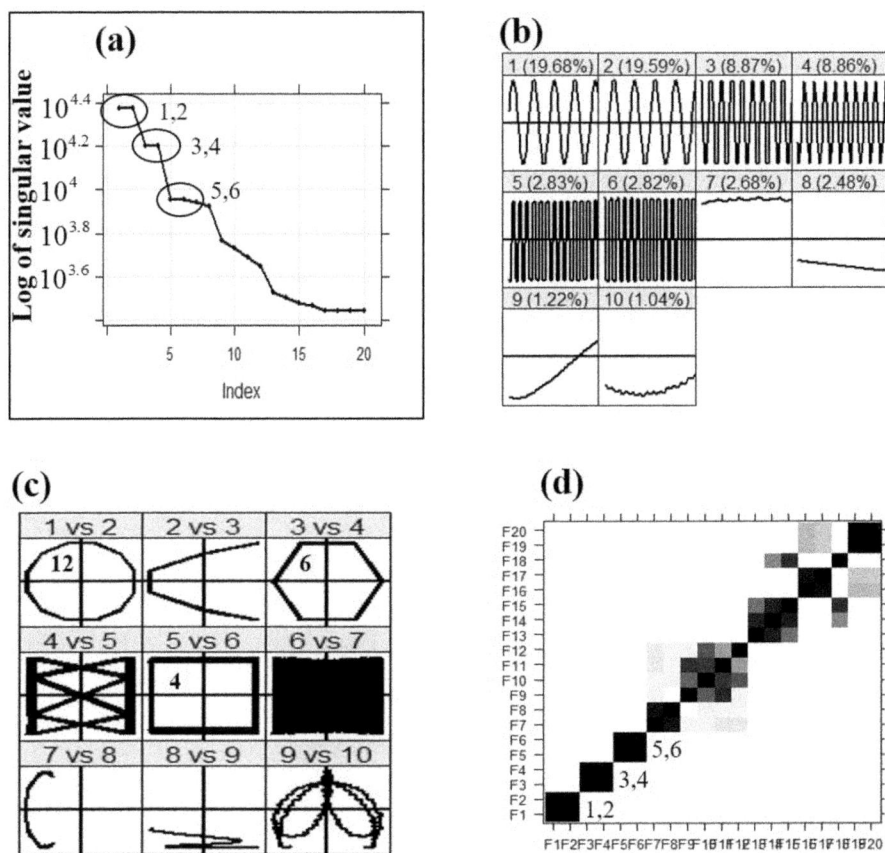

Fig. 6.11 SSA grouping diagnostics for Mayport sea level data: (a) eigenspectrum; (b) eigenvector plots (first 10); (c) scatter plots of paired eigenvectors (first 10); (d) weighted correlation matrix (first 20 paired eigentriplets).

basis of potential harmonic oscillations in the signal (Ghil *et al.*, 2001). Singular values along the flat portion of the eigenspectrum (*noise floor*) are associated with the noise component. The eigenspectrum for the Mayport data has the expected hockey-stick appearance, with steps at paired singular values $(1,2)$, $(3,4)$ and $(5,6)$ along the steep portion of the plot (Figure 6.11a).

The second visual diagnostic plots eigenvectors corresponding to singular values along the eigenspectrum. Each eigenvector is the time series representation of the dynamics captured in the corresponding eigentriplet. Paired eigenvectors associated with potential harmonic oscillations oscillate with identical frequency in phase quadrature (i.e. the oscillations are offset by a quarter cycle, or $\pi/2$ radians). The paired eigenvectors associated with the potential harmonic oscillations in the Mayport eigenspectrum, $(1,2)$, $(3,4)$ and $(5,6)$, each oscillate at an identical frequency in phase quadrature, providing further evidence for the harmonics (Figure 6.11b). The plots of eigenvectors

7 and 8 indicate a low-frequency trend in the data. The percentages reported in the upper margins of the plots measure the portion of total variance (from the mean) in the Mayport time series explained by the particular eigenvector. The partial variances explained by the relevant eigentriplet pairs are $39.27\% = 19.68\% + 19.59\%$ for pair $(1, 2)$, $17.73\% = 8.87\% + 8.86\%$ for pair $(3, 4)$ and $5.65\% = 2.83\% + 2.82\%$ for pair $(5, 6)$. The partial variance explained by eigentriplets 7 and 8 (representing trend) is $5.16 = 2.68\% + 2.48\%$. Collectively, these groups account for 67.81%, or well over half, of the total variation in the observed data.

The third visual diagnostic shows scatter plots of successive paired eigenvectors corresponding to singular values along the eigenspectrum. Scatter plots of successive eigenvectors associated with potential harmonic oscillations result in polygons whose numbers of sides indicate the oscillation period (Hassani, 2007). For example, a four-sided polygon represents a four-period cycle. The explanation is that pairwise scatter plots of pure sine and cosine functions oscillating at the same frequency generate polygons. Code 6.10 demonstrates this by generating such scatter plots for cycle lengths of 12, 6 and 4 periods. The code sets the parameters needed to compute the fundamental frequencies corresponding to these cycle lengths, calculates the corresponding sine and cosine functions, and then generates the corresponding pairwise scatter plots.

```
#Code 6.10 Scatter plot cosine and sine functions oscillating at identical frequency

#Parameters defining fundamental frequencies
a=1  #number of cycles each T
T=c(12,6,4)  #cycle length
t<-1:20  #time

w1=2*pi/T[1]  #fundamental frequency for 12-period cycle
sin1<-sin(a*w1*t)
cos1<-cos(a*w1*t)

w2=2*pi/T[2]  #6-period cycle
sin2=sin(a*w2*t)
cos2=cos(a*w2*t)

w3=2*pi/T[3]  #4-period cycle
sin3=sin(a*w3*t)
cos3=cos(a*w3*t)

plot(cos1,sin1,main="Scatterplot",type="l",lwd=2,cex.axis=1.1,font.axis=2)
plot(cos2,sin2,main="Scatterplot",type="l",lwd=2,cex.axis=1.1,font.axis=2)
plot(cos3,sin3,main="Scatterplot",type="l",lwd=2,cex.axis=1.1,font.axis=2)
```

Figure 6.12 shows the expected result: cosine and sine functions with the same cycle length produce scatter-plotted polygons whose number of sides equals the cycle length. In practice, the closer that paired eigenvectors oscillating at identical frequency in phase quadrature are to ideal sine and cosine functions, the more closely will their scatter plots resemble distinct polygons. For the Mayport data, the scatter plots of paired eigenvectors $(1, 2)$, $(3, 4)$ and $(5, 6)$ are distinct polygons whose numbers of sides indicate the dominant 12-, 6- and 4-month oscillations identified by the Fourier power spectrum (Figure 6.11c).

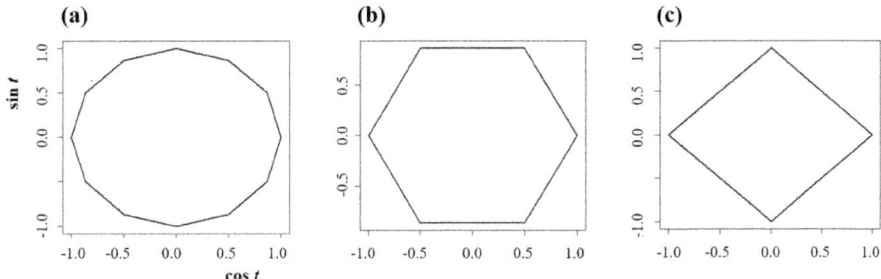

Fig. 6.12 Scatter plots of paired cosine and sine functions for (a) 12-, (b) 6- and (c) 4-period cycles.

A final diagnostic is the *w-correlation plot*, which shows weighted correlations between decomposed matrices (i.e. eigentriplets) after they have been converted to time series with diagonal averaging. If the w-correlation between two eigentriplets is low, they are deemed to be statistically independent and *separable* into different groupings. Alternatively, if the w-correlation is high, the eigentriplets are nonseparable and should not be grouped apart (Golyandina *et al.*, 2001). The w-correlation plot has the eigentriplets along each axis in rank order of their corresponding singular values. Conventionally, darker shading is used to reflect higher w-correlations. The diagonal of the plot is black because eigentriplets are perfectly correlated with themselves.

The w-correlation plot for the Mayport data is limited to the first 20 matrices in the decomposition (Figure 6.11d). The plot indicates that the paired groupings $(1, 2)$, $(3, 4)$ and $(5, 6)$ forming the basis of the oscillatory components in the data are well grouped. The eigentriplets grouped in each pair are highly correlated to each other, as indicated by the black shading. In addition, each pair is statistically independent from other eigentriplets in the decomposition as indicated by the white shading.

We conclude from the SSA diagnostics that the signal component of the Mayport data includes the first eight matrices in the decomposition grouped into pairs. The first three pairs, $(1, 2)$, $(3, 4)$ and $(5, 6)$, form the basis of 12, 6, and 4 month oscillations, respectively. The last pair in the signal, $(7, 8)$ forms the basis of a slow-moving trend. The signal is strong since it accounts for 67.8% of the total variation in the data (as has already been shown). The residual noise component groups the remaining matrices in the decomposition.

Code 6.11 uses the groupings identified in Code 6.9 to run the final time-series reconstruction step:

```
#Code 6.11 SSA: reconstruction of time series
rm(list=ls(all=TRUE)) #Remove objects in memory

#Define working directories
dir.data<-as.character("C:/Data")
dir.udf<-as.character("C:/User Defined Functions")
dir.results<-as.character("C:/Results")

#Load observed time series 'x'
```

```
setwd(dir.data)
ts<-read.csv("Mayport.csv")
x.ts<-ts$observed
x<-na.omit(x)

#Eigentriplet groups
n.groups<-4  #number of eigentriplet pairs
group.1<-c(1,2)
group.2<-c(3,4)
group.3<-c(5,6)
group.4<-c(7,8)
#group.5<-c()
#group.6<-c()

#Run singular spectrum analysis (ssa)
library(Rssa)
L<-432  #ssa window
s<-ssa(x,L,kind="toeplitz")

setwd(dir.results)
#Results for 1 eigentriplet group
if (n.groups==1) {
#Compute periods of eigentriplet groups
if(length(group.1)==2) {
  cycle.1<-parestimate(s,groups=list(group.1))[2]
  cycles<-cycle.1}

#Reconstruct time series from eigentriplet groups
r.1<-reconstruct(s,groups=group.1)  #group 1
recon.1<-r.1$F1

#Signal/Noise
#recon<-reconstruct(s,groups=list(1:5));recon<-r$F1;res<-residuals(r)
signal<-recon.1
noise<-x-signal  #residuals to composite reconstruction

#Export reconstructed series
data<-cbind(x,signal,noise,recon.1)
write.table(data,sep=",",col.names=c("data","signal","noise","recon.1"),
  "data.csv",row.names=FALSE)
} #end if n.group=1

#Results for 2 eigentriplet groups
if (n.groups==2) {
#Compute periods of eigentriplet groups
if(length(group.1)==2) {
  cycle.1<-parestimate(s,groups=list(group.1))[2]}
if(length(group.2)==2){
  cycle.2<-parestimate(s,groups=list(group.2))[2]}
if(length(group.1)==2) {
  cycles<-cbind(cycle.1,cycle.2)} else {
  cycles<-cbind(cycle.2)}

#Reconstruct time series from eigentriplet groups
r.1<-reconstruct(s,groups=group.1)  #group 1
recon.1<-r.1$F1
r.2<-reconstruct(s,groups=group.2)  #group 2
recon.2<-r.2$F1

#Signal/Noise
```

```
#recon<-reconstruct(s,groups=list(1:5));recon<-r$F1;res<-residuals(r)
signal<-recon.1+recon.2
noise<-x-signal  #residuals to composite reconstruction


#Export reconstructed series
data<-cbind(x,signal,noise,recon.1,recon.2)
write.table(data,sep=",",col.names=c("data","signal","noise","recon.1",
  "recon.2"),"data.csv",row.names=FALSE)
} #end if n.group=2


#Results for 3 eigentriplet groups
if (n.groups==3) {
#Compute periods of eigentriplet groups
if(length(group.1)==2) {
  cycle.1<-parestimate(s,groups=list(group.1))[2]}
if(length(group.2)==2){
  cycle.2<-parestimate(s,groups=list(group.2))[2]}
if(length(group.3)==2) {
  cycle.3<-parestimate(s,groups=list(group.3))[2]}
if(length(group.1)==2) {
  cycles<-cbind(cycle.1,cycle.2,cycle.3)} else {
  cycles<-cbind(cycle.2,cycle.3)}


#Reconstruct time series from eigentriplet groups
 r.1<-reconstruct(s,groups=group.1)  #group 1
 recon.1<-r.1$F1
 r.2<-reconstruct(s,groups=group.2)  #group 2
 recon.2<-r.2$F1
 r.3<-reconstruct(s,groups=group.3)  #group 3
 recon.3<-r.3$F1


#Signal/Noise
#recon<-reconstruct(s,groups=list(1:5));recon<-r$F1;res<-residuals(r)
signal<-recon.1+recon.2+recon.3
noise<-x-signal  #residuals to composite reconstruction


#Export reconstructed series
data<-cbind(x,signal,noise,recon.1,recon.2,recon.3)
write.table(data,sep=",",col.names=c("data","signal","noise","recon.1",
  "recon.2","recon.3"),"data.csv",row.names=FALSE)
} #end if n.group=3


#Results for 4 eigentriplet groups
if (n.groups==4) {
#Compute periods of eigentriplet groups
if(length(group.1)==2) {
  cycle.1<-parestimate(s,groups=list(group.1))[2]}
if(length(group.2)==2){
  cycle.2<-parestimate(s,groups=list(group.2))[2]}
if(length(group.3)==2) {
  cycle.3<-parestimate(s,groups=list(group.3))[2]}
if(length(group.4)==2){
  cycle.4<-parestimate(s,groups=list(group.4))[2]}
if(length(group.1)==2) {
  cycles<-cbind(cycle.1,cycle.2,cycle.3,cycle.4)} else {
  cycles<-cbind(cycle.2,cycle.3,cycle.4)}


#Reconstruct time series from eigentriplet groups
r.1<-reconstruct(s,groups=group.1)  #group 1
recon.1<-r.1$F1
```

```
r.2<-reconstruct(s,groups=group.2)  #group 2
recon.2<-r.2$F1
r.3<-reconstruct(s,groups=group.3)  #group 3
recon.3<-r.3$F1
r.4<-reconstruct(s,groups=group.4)  #group 4
recon.4<-r.4$F1

#Signal/Noise
#recon<-reconstruct(s,groups=list(1:5));recon<-r$F1;res<-residuals(r)
signal<-recon.1+recon.2+recon.3+recon.4
noise<-x-signal  #residuals to composite reconstruction

#Export reconstructed series
data<-cbind(x,signal,noise,recon.1,recon.2,recon.3,recon.4)
write.table(data,sep=",",col.names=c("data","signal","noise","recon.1",
  "recon.2","recon.3","recon.4"),"data.csv",row.names=FALSE)
} #end if n.group=4

#Results for 5 eigentriplet groups
if (n.groups==5) {
#Compute periods of eigentriplet groups
if(length(group.1)==2) {
  cycle.1<-parestimate(s,groups=list(group.1))[2]}
if(length(group.2)==2){
  cycle.2<-parestimate(s,groups=list(group.2))[2]}
if(length(group.3)==2) {
  cycle.3<-parestimate(s,groups=list(group.3))[2]}
if(length(group.4)==2){
  cycle.4<-parestimate(s,groups=list(group.4))[2]}
if(length(group.5)==2){
  cycle.5<-parestimate(s,groups=list(group.5))[2]}
if(length(group.1)==2) {
  cycles<-cbind(cycle.1,cycle.2,cycle.3,cycle.4,cycle.5)} else {
  cycles<-cbind(cycle.2,cycle.3,cycle.4,cycle.5)}

#Reconstruct time series from eigentriplet groups
r.1<-reconstruct(s,groups=group.1)  #group 1
recon.1<-r.1$F1
r.2<-reconstruct(s,groups=group.2)  #group 2
recon.2<-r.2$F1
r.3<-reconstruct(s,groups=group.3)  #group 3
recon.3<-r.3$F1
r.4<-reconstruct(s,groups=group.4)  #group 4
recon.4<-r.4$F1
r.5<-reconstruct(s,groups=group.5)  #group 5
recon.5<-r.5$F1

#Signal/Noise
#recon<-reconstruct(s,groups=list(1:5));recon<-r$F1;res<-residuals(r)
signal<-recon.1+recon.2+recon.3+recon.4+recon.5
noise<-x-signal  #residuals to composite reconstruction

#Export reconstructed series
data<-cbind(x,signal,noise,recon.1,recon.2,recon.3,recon.4,recon.5)
write.table(data,sep=",",col.names=c("data","signal","noise","recon.1",
  "recon.2","recon.3","recon.4","recon.5"),"data.csv",row.names=FALSE)
} #end if n.group=5

#Results for 6 eigentriplet groups
if (n.groups==6) {
```

```
#Compute periods of eigentriplet groups
if(length(group.1)==2) {
  cycle.1<-parestimate(s,groups=list(group.1))[2]}
if(length(group.2)==2){
  cycle.2<-parestimate(s,groups=list(group.2))[2]}
if(length(group.3)==2) {
  cycle.3<-parestimate(s,groups=list(group.3))[2]}
if(length(group.4)==2){
  cycle.4<-parestimate(s,groups=list(group.4))[2]}
if(length(group.5)==2){
  cycle.5<-parestimate(s,groups=list(group.5))[2]}
if(length(group.6)==2){
  cycle.6<-parestimate(s,groups=list(group.6))[2]}

if(length(group.1)==2) {
  cycles<-cbind(cycle.1,cycle.2,cycle.3,cycle.4,cycle.5,cycle.6)} else {
  cycles<-cbind(cycle.2,cycle.3,cycle.4,cycle.5,cycle.6)}

#Reconstruct time series from eigentriplet groups
r.1<-reconstruct(s,groups=group.1)  #group 1
recon.1<-r.1$F1
r.2<-reconstruct(s,groups=group.2)  #group 2
recon.2<-r.2$F1
r.3<-reconstruct(s,groups=group.3)  #group 3
recon.3<-r.3$F1
r.4<-reconstruct(s,groups=group.4)  #group 4
recon.4<-r.4$F1
r.5<-reconstruct(s,groups=group.5)  #group 5
recon.5<-r.5$F1
r.6<-reconstruct(s,groups=group.6)  #group 6
recon.6<-r.6$F1

#Signal/Noise
#recon<-reconstruct(s,groups=list(1:5));recon<-r$F1;res<-residuals(r)
signal<-recon.1+recon.2+recon.3+recon.4+recon.5+recon.6
noise<-x-signal  #residuals to composite reconstruction

#Export reconstructed series
data<-cbind(x,signal,noise,recon.1,recon.2,recon.3,recon.4,recon.5,
  recon.6)
write.table(data,sep=",",col.names=c("data","signal","noise","recon.1",
  "recon.2","recon.3","recon.4","recon.5","recon.6"),"data.csv",
  row.names=FALSE)
} #end if n.group=6

#Cycle lengths
cycles
```

We initially enter the eigentriplet groups comprising the signal. Code 6.11 accommodates up to six eigentriplet groups. We enter the number of groups, n.groups, and then specify each paired group. For this illustration, we enter n.groups<-4 and the four groups. For example, the first group is composed of eigentriplets 1 and 2: group.1<-c(1,2). We next call Rssa (using the same window $L = 432$) to convert the eigentriplet groups into corresponding time series with diagonal averaging. The reconstruction results are organized in a number of conditional statements based on the number of eigentriplet groups entered: n.groups. In particular, the **parestimate** command confirms the periods of the oscillatory pairings, and the **reconstruct** command reconstructs the time series for each group. The signal is calculated by adding together the corresponding reconstructed time series, and the noise by calculating the residual between the mean-adjusted time series and the signal. A matrix is exported that reports the original (mean-adjusted) time series in the first column followed by time series reconstructions of the signal, noise, and oscillatory and trend components. Finally, the periods of the oscillatory pairings are reported.

The plots (generated in Excel) for the SSA reconstructions of the Mayport data are shown in Figure 6.13. The original data (grey line) exhibit a slight trend (black line) (Figure 6.13a). The signal exhibits the slight trend and dominant oscillations in the data (Figure 6.13b). The noise displays the unstructured variation in the data (Figure 6.13c). The isolated 12-, 6- and 4-month oscillations are shown in Figures 6.13d, 6.13e and 6.13f.

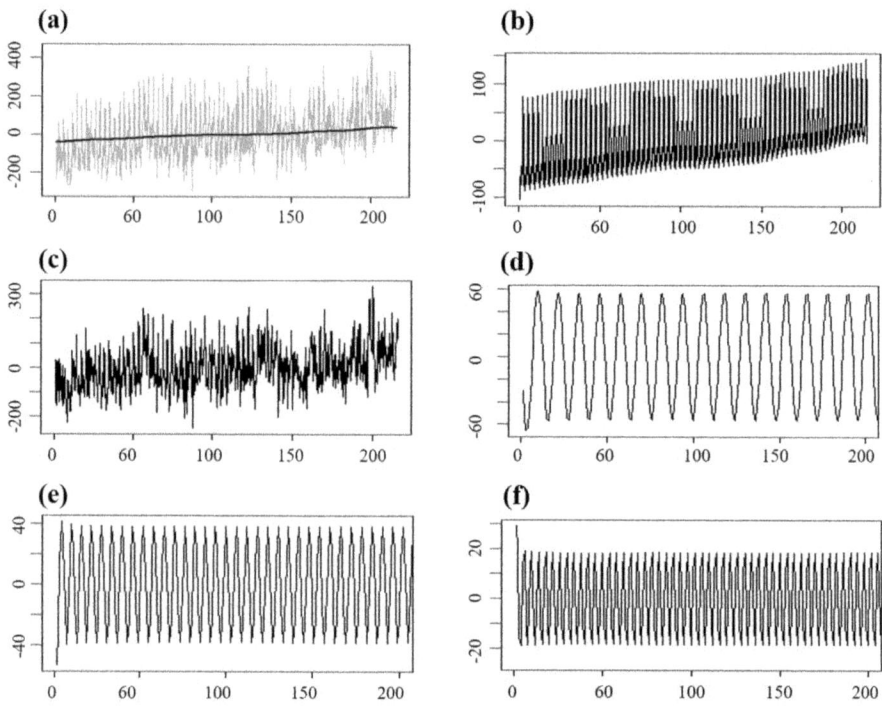

Fig. 6.13 SSA reconstruction step: plots (generated in Excel) of (a) time series and trend, (b) signal, (c) noise, (d) 12-month oscillation, (e) 6-month oscillation and (f) 4-month oscillation.

6.5 Nonstationary Dynamics

We now investigate how a nonstationary linear model can generate irregular behaviour, and present a test to detect nonstationary dynamics in observed data. Broadly defined, a nonstationary time series exhibits changes in the dynamics during the measurement period (Schreiber, 1997). Code 6.12 introduces nonstationarity into the linear system (6.1) by modifying Code 6.1 so that one of the previously fixed parameters varies slowly over time (see e.g. Schreiber (1997)). In particular, the parameter a becomes a linear function of time: $a(t) = -1.2 + 0.2t$. The other parameters remain at levels generating the centre point dynamics in Figures 6.4c and 6.4d.

```
#Code 6.12 Induce complex dynamics with nonstationary linear dynamics
rm(list=ls(all=TRUE))  #clear values between runs

library(deSolve) #ODE solver package

#Set Parameters
delta<-0.1  #integration time step
state<-c(x=0,y=1) #initial conditions for variables
parameters<-c(a=0,b=3,c=-3,d=0) #parameters giving centre point

#Define ODE system
model<-function(t,state,parameters){
  with(as.list(c(state,parameters)),{
  dx <-(-1.2+0.2*t)*x+b*y  #make parameter 'a' time-dependent
  dy <- c*x+d*y
  list(c(dx,dy))
  }) #end with (as.list...
}  #end model function

#Integrate ODE system
times<-seq(0,12,by=delta)  #integration step
out<-ode(y=state,times=times,func=model,parms=parameters,method="rk4")
head(out)

#Plots of simulated values
#par(mfrow=c(3,1))
#plot(out[,"x"],type='l',lwd=3,cex.axis=1.1,font.axis=2)
#plot(out[,"y"],type='l',lwd=3,cex.axis=1.1,font.axis=2)
plot(out[,"x"],out[,"y"],type='l',lwd=3,cex.axis=1.1,font.axis=2)
```

The time-dependent parameter $a(t)$ is designed to drag the linear system through three abrupt changes (i.e. *bifurcations*) in system dynamics through time. During the initial time interval $t = (0, 5)$, a is negative, and the system dynamics are characterized by a stable focus (Figure 6.14a). At $t = 6$, $a = 0$, and the system abruptly shifts for an instant to a centre point dynamic (Figure 6.14b). During the remaining time interval $t = (7, 12)$, a is positive, and the system dynamics convert to an unstable focus (Figure 6.14c). A trajectory over the entire time horizon mimics irregular nonlinear dynamics by seamlessly blending the three linear dynamical structures (Figure 6.14d). Initially, the trajectory approaches equilibrium at $(0, 0)$, orbits around it, and then escapes from it in explosive oscillations.

6.6 Testing for Nonstationarity in Time Series Data

There are multiple tests for nonstationarity in time series data that depend on how one defines *changes* in dynamical behaviour (Schreiber, 1997). Linear time series analysis requires only *weak* stationarity in which *second-order moments*, such as the mean and variance, do not fluctuate over time. Testing for weak stationarity generally divides

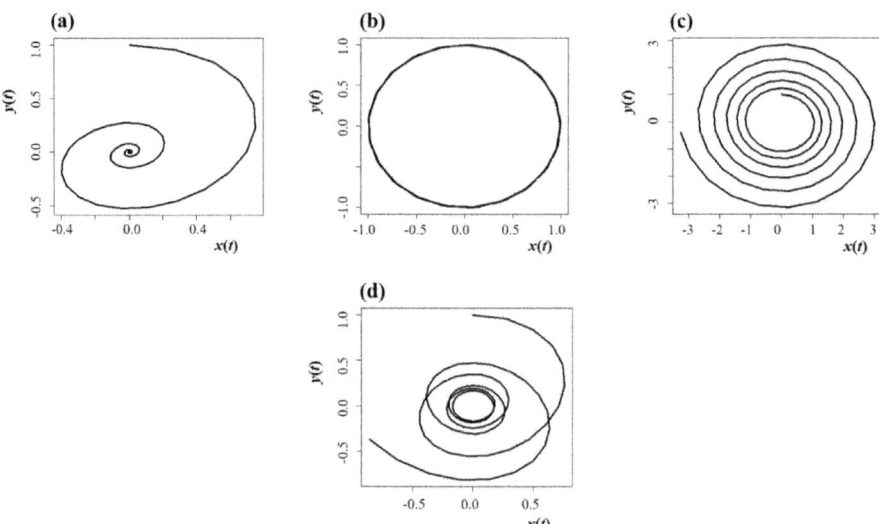

Fig. 6.14 Numerically generated phase space dynamics for a nonstationary linear model. We make the parameter a in the linear system (6.1) time-dependent, and the plots here show how the dynamics undergoes three abrupt changes, from a stable focus (a), to a centre point orbit (b) and finally to an unstable focus (c). A trajectory over the entire time horizon blends the three linear dynamical structures (d). The trajectory mimics irregular nonlinear dynamics by initially approaching equilibrium, orbiting around it, and then escaping from it in explosive oscillations.

the time series into several segments and computes mean and variance in each segment. The time series is considered stationary if the computed moments do not vary significantly across segments.

Weak stationarity does not suffice in nonlinear time series analysis. Nor can we simply replace the mean and variance with nonlinear measures of complexity (e.g. the correlation dimension and Lyapunov exponents) in weak stationarity tests. This would be unreliable, because an attractor can change shape while these measures remain largely unchanged (Schreiber, 1997). Consequently, Schreiber (1997) proposed a nonlinear *cross-prediction stationarity test* for use when reconstructing nonlinear system dynamics from time series data (see also (Kantz and Schreiber, 1997; Heathcote and Elliott, 2011)). The test looks for similar nonlinear behaviour across the segments themselves rather than similar statistical parameters computed from each segment. In particular, it divides the time series into non-overlapping segments, and applies nonlinear prediction methods to measure the skill with which each segment cross-predicts the others. The time series is stationary if each segment cross-predicts the others with equal skill. In other words, cross-predictive skill does not decrease significantly when segments are further apart in time.

In this section, we build R code to run the nonlinear cross-prediction stationarity test. The test is based on *nonlinear prediction* methods that require *time-delay embed-*

ding methods to reconstruct a phase space attractor from the time series. Consequently, we first formulate user-defined functions for time-delay embedding and nonlinear prediction. For illustration, we test whether the signal isolated from the Mayport sea level data is nonstationary.

6.6.1 Time-Delay Embedding

Selecting time-delay embedding parameters is not an exact science. Our hope is to unearth values that result in a faithful reconstruction of the real-world phase space. We could follow a brute force approach by reconstructing phase spaces over a grid of embedding values to identify ranges providing attractors with the most visual regularity. However, we can run a more directed search with the aid of statistical tests designed to home in on the most promising values. Code 6.13 formulates a user-defined function `embed_udf` to run these tests for a given time series by automating routines in the R package `tseriesChaos` (Di Narzo, 2015). The user-defined function requires input of only the signal separated from the time series, x, and is organized into four steps. The first three steps estimate the embedding parameters in the order required by successive `tseriesChaos` routines. The fourth step uses the parameters to embed the time series.

```
#Code 6.13 Compute embedded data matrix for observed time series using statistical tests
rm(list=ls(all=TRUE))  #clear objects between runs

#Read in data
setwd("C:/Data")
ts<-read.csv("Mayport.csv");x<-ts$signal

#User defined function to embed x(t)
embed_udf <-function(x){

#Step 1: Calculate embedding delay
  #Step 1a. Compute average mutual information (AMI) function
  library(tseriesChaos)
  mutual.out<-mutual(x)  #mutual(tseriesChaos)

  #Step 1b: Use embedding approach to calculate delay at which AMI hits
    #its first minimum
  mutual.em<-embedd(mutual.out,2,1)  #embedd(tseriesChaos)
  mutual.adj.length<-length(mutual.out)-1    #lose 1 observation to delay

  mutual.hold<-matrix(0,mutual.adj.length,1)
  for (i in 1:mutual.adj.length){  #loop to compute successfive value differences
    mutual.test<-if(mutual.em[i,1]>mutual.em[i,2])TRUE else break
    mutual.hold[i,1]<-mutual.test  #"TRUE' = 1 in R, so min delay occurs at sum(TRUE)
  }  #end iloop
  mutual.hold.sum<-sum(mutual.hold)

  #Step 1c: Estimate embeddeding delay (d).  If the mutual information function is
    #decreasing across all 20 delays, set delay at max = 20
  d<-if(mutual.hold.sum<20)mutual.hold.sum else 20  #Embedding delay

#Step 2: Compute Theiler window parameter (tw) required for false nearest
#         neighbours test
  #Step 2a: Autocorrelation function
  lag.max=100
```

```
acf.run<-acf(x,lag.max)
acf.out<-acf.run$acf #array of acf values

#Step 2b: Use embedding approach to calculate delay at which AMI hits
  #its first minimum.
acf.em<-embedd(acf.out,2,1)  #embedd(tseriesChaos)
acf.adj.length<-length(acf.out)-1   #lose 1 observation to lag

acf.hold<-matrix(0,acf.adj.length,1)
for (i in 1:acf.adj.length){ #loop to compute successfive value differences
  acf.test<-if(acf.em[i,1]>acf.em[i,2])TRUE else break
  acf.hold[i,1]<-acf.test  #"TRUE' = 1 in R, so min delay occurs at sum(TRUE)
} #end iloop

#Step 2c: Estimate Theiler window (tw)
tw<-sum(acf.hold)

#Step 3: Embedding dimension (m)
  #Step 3a: False nearest neighbours function
  m.max<-6  #maximum number of embedding dimensions to consider
  fn.out<-false.nearest(x,m.max,d,tw)  #false.nearest(tseriesChaos)
  fn.out[is.na(fn.out)] <- 0  #set NA in fn.out to zero
  #plot(fn.out)

  #Step 3b: Find delay at which false nearest neighbours decrease below set tolerance
  #Output vector of fnn percentages from fn.out
  fnp<-c(fn.out[1],fn.out[3],fn.out[5],fn.out[7],fn.out[9],fn.out[11])
  fnp.tol<-fnp>0.15 #If fnp greater than tolerance of 15%, T entered into fnp.tol
  fnp.tol.sum<-sum(fnp.tol)  #sum up number of T's
  m<-if(fnp.tol.sum<m.max)fnp.tol.sum+1 else m.max #Embedding dimension

#Step 4: Embed time series (Mx)
  #If m=1, embedd routine crashes due to 'subscript out of bounds' error--need to
    #guarantee an embedding dimension of at least two:
  if(m<=1){m<-2} else {m}
  Mx<-embedd(x,m,d) #embedd(tseriesChaos)

#Results
  results.embed_udf<-list(d,m,tw,Mx)
  return(results.embed_udf)
} #end user-defined function

embed<-embed_udf(x)  #run user-defined function
d<-embed[[1]]  #embedding delay
m<-embed[[2]]  #embedding dimension
tw<-embed[[3]] #Theiler window
Mx<-embed[[4]] #embedded data matrix

#Plot reconstructed attractor
if(m==2){plot(Mx[,1],Mx[,2],type='l')}
if(m==3) {library(scatterplot3d);scatterplot3d(Mx,type='l')}
```

Step 1 of **embed_udf** estimates the embedding delay *d*. We search for an embedding delay that introduces the required statistical independence between the time series and successive delayed copies of it through time. A delay that is too short does not give the system dynamics time to evolve, while one that is too long skips over important dynamical structure (Williams, 1997). The embedding delay is conventionally selected with the *average mutual information function (AMI)*. The AMI is a probabilistic measure

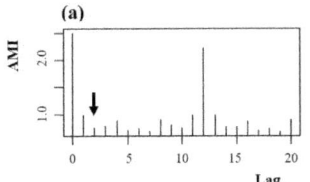 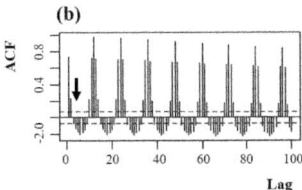 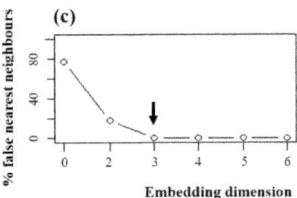

Fig. 6.15 Plotted output from Code 6.13: (a) AMI function; (b) autocorrelation function; (c) percentage of false nearest neighbours. The first minima of these functions are indicated by the arrows, and provide estimates of the embedding delay d, the Theiler window tw and the embedding dimension m, respectively.

of the extent to which a time series is related to delayed copies of itself at different points in time.

Step 1a computes the AMI function with the routine `mutual(tseriesChaos)`. The output file, `mutual.out`, plots the AMI function (Figure 6.15a) and provides the numerical AMI values for each value of d as follows:

```
> mutual.out
        0         1         2         3         4         5         6         7
2.5014699 0.9801214 0.7518974 0.7877724 0.8877943 0.6954625 0.7392803 0.6847265
        8         9        10        11        12        13        14        15
0.8917962 0.8005872 0.7442711 0.9804990 2.2150976 0.9818154 0.7644367 0.7667776
       16        17        18        19        20
0.8690444 0.7003054 0.7279985 0.6793871 0.8902865
```

The AMI function and numerical values indicate that the first minimum occurs at an embedding delay $d = 2$.

Step 1b provides a routine to compute the first minimum so that we do not need to visually identify it in the output. We use the routine `embedd(tseriesChaos)` to embed the AMI values with embedding parameters $d = 1$ and $m = 2$. The first column of the embedded matrix, *mutual.em*, contains the AMI values and the second column contains the immediately successive values. We use the **head** command to report only the first six rows:

```
> mutual.em
          V1/0      V1/1
[1,] 2.5014699 0.9801214
[2,] 0.9801214 0.7518974
[3,] 0.7518974 0.7877724
[4,] 0.7877724 0.8877943
[5,] 0.8877943 0.6954625
[6,] 0.6954625 0.7392803
```

We then loop through each row of the embedded AMI matrix to run a test returning 1 if the value in the first column is greater than the value in the second. The loop breaks when the test reaches a row for which the opposite is true, and a 0 is entered from that row forward. For this example, the loop breaks at row 2:

```
< mutual.hold
[,1] [,2] [,3] [,4] [,5]
   1    1    0    0    0
```

The sum of the test vector computes the first minimum estimate $d = 2$.

Step 2 of `embed_udf` computes the *Theiler window* parameter tw, required by a subsequent test. The Theiler window corrects for the following problem: neighbouring points in phase space may be determined to be close simply because they are proximate in time. We might mistake temporal correlation for the attractor's geometric structure. Consequently, points within the Theiler window are excluded in reconstructing the attractor (Kantz and Schreiber, 1997). Step 2a follows a conventional method for selecting the Theiler window as the lag giving the first minimum of the *autocorrelation function (ACF)*. The ACF measures the linear correlation of a time series with delayed copies of itself at different points in time. We compute the ACF with the routine `acf`. This routine requires input of the time series x_t and a maximum delay, which we set at `lag.max = 100`. The routine provides the ACF plot shown in Figure 6.15b. The output file `acf.out` reports numerical ACF values, with the first eight being as follows:

```
> acf.out[1:8]
[1]  1.00000000  0.73553060  0.23342248 -0.06898751 -0.13652360 -0.16947446
[7] -0.20100525 -0.17621465
```

We see that the first minimum occurs at a lag of six periods (since the first value, 1.00, is associated with a zero lag), so that the Theiler window is $tw = 6$. Step 2b uses the embedding method formulated in Step 1b to numerically identify the Theiler window. We note that space–time separation plots provide a more rigorous, and potentially more effective, estimate of the Theiler window (Provenzale *et al.*, 1992).

Step 3 of `embed_udf` estimates the embedding dimension . The embedding dimension m is conventionally estimated with the *false nearest neighbours test* (Kennel *et al.*, 1992). This test reconstructs an attractor from the observed time series for a given embedding delay d in a given embedding dimension, and computes the distances between points in phase space. The test omits points whose time indices differ by less than the Theiler window tw. It then repeats this exercise in the next higher dimension and calculates the percentage of false nearest neighbours, that is, points that do not remain close but grow apart. The embedding dimension selected is that for which the percentage of false nearest neighbours falls below a given tolerance level. An attractor reconstructed in fewer dimensions does not have sufficient space to fully express itself, while one reconstructed in more dimensions loses resolution (Williams, 1997). In Chapter 3, we illustrated false nearest neighbours with a couple of real-world examples (see Section 3.4.1).

Step 3a runs the false nearest neighbours R routine `false.nearest(tseriesChaos)`. This routine requires input of the time series x_t, a maximum embedding dimension set at $m.max = 6$, the embedding delay d, and the Theiler window tw. For our example, $d = 2$ and $tw = 6$. The percentages of false nearest neighbours from one dimension to the next, `fnp`, are recovered from the output file `fn.out`:

```
> fn.out
                 m1           m2           m3           m4           m5
fraction 7.662959e-01 1.824837e-01 3.654080e-03 4.773270e-04 5.274262e-04
total    4.890800e+04 7.376000e+03 4.926000e+03 4.190000e+03 3.792000e+03
                 m6
fraction 5.753740e-04
total    3.476000e+03
```

These values are plotted in Figure 6.15c with the command plot(fn.out), and we see that the percentage of false nearest neighbours drops to zero at $m = 3$. Step 3b computes this level numerically. If the percentage of false neighbours at each dimension exceeds the tolerance level set at 15%, then *TRUE* is entered into the output vector fnp.tol; otherwise *FALSE* is entered:

```
> fnp.tol
[1]   TRUE   TRUE FALSE FALSE FALSE FALSE
```

Summing fnp.tol counts *TRUE* as 1 and *FALSE* as 0, resulting in 2. Consequently, consistent with the plotted false nearest neighbour percentages in Figure 6.15c, the minimum embedding dimension that meets the tolerance level is $m = sum(fnp.tol) + 1 = 3$.

Finally, Step 4 of embed_udf embeds the time series x_t for the estimated embedding dimension m and delay d using the R routine embedd(tseriesChaos). The embedded data matrix M_x is an $N \times m$ matrix whose row dimension is $N = n - (m - 1)d$, where n in the length of x_t. The N rows of M_x are m-dimensional points on a reconstructed phase space attractor . The first six rows of the embedded data matrix for the Mayport time series ($m = 4$, $d = 2$, $n = 864$) are

```
> head(Mx)
           V1/0        V1/2        V1/4
[1,]   -91.19922 -104.48767 -44.63266
[2,]  -101.68877  -78.60739 -52.35350
[3,]  -104.48767  -44.63266 -77.75044
[4,]   -78.60739  -52.35350 -42.05107
[5,]   -44.63266  -77.75044  44.90718
[6,]   -52.35350  -42.05107  78.35371
```

Scatter-plotting M_x plots the reconstructed attractor shown in Figure 6.16.

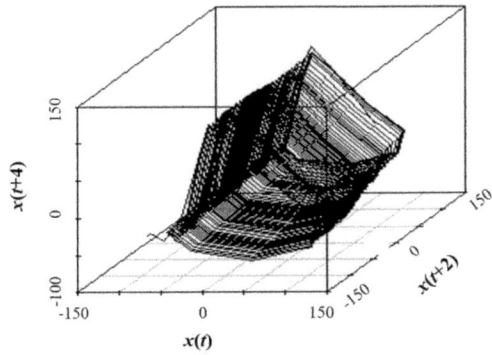

Fig. 6.16 Reconstructed attractor from Mayport data.

6.6.2 Nonlinear Prediction

Nonlinear prediction of the time series x_t proceeds by reconstructing phase space using *time-delay embedding* (Kaplan and Glass, 1995; Kantz and Schreiber, 1997). This generates an embedded data matrix M_x whose first column is x_t followed by its forward-delayed copies. The trajectory matrix shown in eqn (6.6) is an example of such an embedded data matrix. The rows of M_x are multidimensional points $D(t)$ on the reconstructed attractor in each time period t. For example, if $D(t) = (3, 5, 2)$, then 3 is the time series observation at t, and 5 and 2 are forward-delayed observations. Next, the points on the attractor (i.e. the rows of M_x) are split into an initial *learning* set used to predict the points in the remaining *test* set. In the first iteration, the nearest neighbouring points to the final point in the learning set, $D(T)$, are identified, and advanced one period (for one-step-ahead prediction). The one-period-advanced points are then averaged to predict the first point in the test set, $D(T + 1)$. At each iteration, the learning set is augmented by the next point in the test set until all test points (except for the final point) have been predicted. The corresponding time series prediction for t is the first element of the forecasted point on the attractor (the remaining elements are forward-delayed predictions). Finally, the time series predictions and the actual (test) values at each t are collected in vectors, which are compared for goodness-of-fit.

A user-defined function for nonlinear prediction, `predict_np_udf`, is formulated in Code 6.14. After reading in the signal separated from the Mayport series, x_t, Step 1 loads the user-defined function `embed_udf` formulated in Code 6.13. The command `setwd(dir.udf)` specifies the directory where `embed_udf` resides, and the commands `dump("embed_udf", file="embed_udf.R")` and `source("embed_udf.R")` retrieve it. The function computes the embedded data matrix for the Mayport time series, M_x, whose rows are points on the reconstructed attractor. The attractor reconstructed with the computed embedding parameters, m=3 and d=2, exhibits noticeable oscillatory structure (Figure 6.16).

```
#Code 6.14 Nonlinear prediction (1-step-ahead)
rm(list=ls(all=TRUE))  #remove objects between runs

#Read in data
setwd("C:/Data")
ts<-read.csv("Mayport.csv");
x<-ts$signal #ssa signal

#Step 1: Time-delay embedding
#Load user-defined functions
setwd("C:/Users/Ray/Documents/R_files/Code/User Defined Functions")
dump("embed_udf", file="embed_udf.R")
source("embed_udf.R")

#Embed time series (x)
library(tseriesChaos)
results.embed_udf<-embed_udf(x)
m<-results.embed_udf[[2]]  #embedding dimension
Mx<-results.embed_udf[[4]] #embedded data matrix

#Step 2: Define user-defined nonlinear prediction function
predict_np_udf <- function(Mx,frac.learn){

  #Step 2a: Partition Mx into learning and test sets
frac.learn<-0.5 #fraction in learning set
learn.rows<-round(frac.learn*nrow(Mx))
learn.em.0<-Mx[1:learn.rows,]   #initial learning set
test.em<-Mx[(learn.rows+1):nrow(Mx),]   #initial test set
```

```
#Step 2b: Prediction
hold.test<-matrix(0,(nrow(test.em)),1)
hold.pred<-matrix(0,(nrow(test.em)),1)
for(i in 1:(nrow(test.em))) {
#print("i");print(i)
learn.em<-Mx[1:((nrow(learn.em.0)+i)-1),]
#print("learn.em");print(learn.em)

    #Step 2b(1): Calculate nearest neighbours to last row in learning set (learn.em.0)
#Distance between points on the attractor
ref.point<-nrow(learn.em)  #index of reference point
library(fields)
dist<-rdist(learn.em)  #distance matrix
dist.ref<-dist[,ref.point] #distances from reference point to other points
sc<-max(dist.ref)  #find maximum distance from reference point
dist.ref.sc<-dist.ref/sc  #scale distances from reference point to max distance

#Order distances from closest to farthest from reference point
o<-order(dist.ref.sc)
#Remove distance of reference point to itself
remove<-which(o==nrow(learn.em));o1<-o[-remove]
#Indicies of m+1 smallest distances from reference point (m=embedding dimension)
o2<-o1[1:(m+1)];o2<-na.omit(o2)
dist.ordered<-dist.ref.sc[o2] #ordered distances

#Increment neighbouring indices by 1 period for use in prediction algorithm
o.pred<-o2+1
pred.ngh<-learn.em[o.pred,] #Nearest neighbours

    #Step 2b(2): Compute prediction as average of neighbouring points weighted by
    #            distances from reference point
#Compute weights (Sughihara et al., 2012)
u.denom<-dist.ordered[1] #distance from reference point to nearest neighbour
hold<-matrix(0,(m+1),1)
for(k in 1:(m+1)){  #summands in w.denom
  u.vector<-exp(-dist.ordered[k]/u.denom)
  hold[k,1]<-u.vector
  } #end loop k
u.vector<-hold[1:(m+1)]
w.denom<-sum(u.vector)
w.vector<-u.vector/w.denom

#Prediction of next point on attractor (row in test.em)
pred.point<-w.vector%*%pred.ngh

#Prediction of time series observation is first (unlagged) element of
#pred.point
pred.ts<-pred.point[1]
#print("pred.ts");print(pred.ts)

  #Step 2b(3):  Test point on attractor (row in test.em)
test.point<-test.em[i,]
#print("test.point");print(test.point)
#Time series observation to be validated is first (unlagged) element of test.point
test.ts<-test.point[1]

hold.test[i,]<-test.ts  #1st element is original data point
hold.pred[i,]<-pred.ts

} #end i loop through Mx
```

```
#Step 3: Calculate modified Nash-Sutcliffe model efficiency (nse)
num<-sum((hold.test-hold.pred)^2)
den<-sum((hold.test-mean(hold.test))^2)
nse<-1-(num/den)

#print("nse");print(hold.mnse)
#results<-cbind(learn.0,hold.test,hold.pred)
#print("learn,test,pred");print(results)

results<-list(nse,hold.test,hold.pred)
return(results)
}  #end function

results.np<-predict_np_udf(Mx,frac.learn=0.5)
nse<-results.np[[1]]
test<-results.np[[2]]
prediction<-results.np[[3]]
```

Step 2 of Code 6.14 formulates the user-defined function for nonlinear prediction: `predict_np_udf`. It takes as arguments the embedded data matrix M_x and the fraction of attractor points to be used for the learning set (`fract.learn`). Step 2a partitions rows of M_x into an initial learning set (`learn.em.0`) and a test set (`test.em`). Step 2b runs nonlinear prediction by iterating through the rows of M_x, beginning with the last row of `learn.em.0` and then adding a row from the test set at each iteration. The loop ends with the penultimate row of M_x, which is used to predict the final point on the attractor (i.e. the last row of M_x). The vectors `hold.test` and `hold.pred` store the test and predicted values of x_t at each iteration. Finally, Step 3 computes the goodness of fit between the two vectors.

We now discuss prediction Step 2 in more detail. In the first iteration, Step 2b(1) calculates the nearest neighbouring points on the attractor to a reference point given by the last row in the initial learning set (`learn.em.0`). We locate the reference point on the attractor with the command `ref.point<-nrow(learn.em)` and then use the command `rdist` to compute the distance matrix measuring Euclidean distances between points on the attractor (i.e. rows in the learning matrix).

The *Euclidean distance* $d(p, q)$ between two n-dimensional points is

$$d(p, q) = \sqrt{\sum_{i=1}^{n}(p_i - q_i)^2} \tag{6.25}$$

As an illustration, we compute the distance matrix for the matrix

$$M = \begin{bmatrix} 2 & 3 & 15 \\ 5 & 4 & 5 \\ 8 & 10 & 6 \end{bmatrix}$$

The distance matrix is

$$\begin{bmatrix} 0 & 10.49 & 12.88 \\ 10.49 & 0 & 6.78 \\ 12.88 & 6.78 & 0 \end{bmatrix}$$

Elements along the diagonal measure zero distances between a row and itself. The matrix is symmetric because $d(p_i, q_j) = d(p_j, q_i)$. If we identify the first row of M as the reference point, then we use the first column of the distance matrix to measure the Euclidean distances from the reference point to the second $(d(1, 2) = 10.49)$ and third $(d(1, 3) = 12.88)$ rows.

Step 2b(1) proceeds by isolating the column in the distance matrix, `dist`, giving the distances between the reference point and remaining points on the attractor, and scales these distances to the largest distance in the column. We apply the R command `order` to put the scaled distances in ascending order from the closest to the farthest removed, and delete the zero distance between the reference point to itself. Finally, we isolate the $m + 1$ smallest distances (where m is the embedding dimension of the reconstructed attractor), identify the corresponding points as nearest neighbours to the reference point, and determine their next-period values. The $m + 1$ nearest neighbours are the minimum required to run the *simplex* averaging algorithm formulated by Sugihara *et al.* (2012) to compute predicted points. For the reconstructed Mayport attractor (Figure 6.16), we use the next-period values of the $m + 1 = 4$ nearest neighbours.

Step 2b(2) codes the *simplex* prediction algorithm formulated in the supplemental materials to Sugihara *et al.* (2012). The algorithm computes predicted values from the locally weighted mean of the next period values of the $m + 1$ nearest neighbours. The weight for a given nearest neighbour is its distance from the reference point relative to the total distances of all nearest neighbours from the reference point. The output of Step 2b(2) is the predicted point on the attractor for the iteration. The corresponding predicted time series observation is the first element of the multidimensional attractor point. Step 2b(3) identifies the contemporaneous test point on the attractor, and the corresponding test time series observation (again the first element of the attractor point). When the iterations are completed (i.e. all test values have been predicted), Step 2b(3) collects test and predicted values in vectors `hold.test` and `hold.pred`, respectively.

Finally, Step 3 adopts the dimensionless *Nash–Sutcliffe coefficient of efficiency*, nse, (Nash and Sutcliffe, 1970) prevalent in hydrological modelling (Ritter and Munoz-Carpena, 2013) to measure the goodness of fit between the test and predicted values:

$$nse = 1 - \frac{\sum_{i=1}^{N}(x_i - x_{Pi})^2}{\sum_{i=1}^{N}(x_i - \overline{x})^2} = 1 - \left(\frac{RMSE}{\sigma}\right)^2 \tag{6.26}$$

where i denotes the periods in the test base, x_i is the test value in period i, x_{Pi} is the contemporaneous predicted value, $\overline{x}$ and σ are the mean and standard deviation of the time series, and $RMSE$ is the root-mean-square error of the predicted against the test values. If $nse = 1$, the predicted and test values are a perfect match. If $nse = 0$, the time series average predicts equally well. A value $nse > 0.65$ is often proposed as a model quality threshold (Ritter and Munoz-Carpena, 2013). Nonlinear prediction works very well for the Mayport data, since $nse = 0.998$.

6.6.3 Nonlinear Cross Prediction

Nonlinear cross prediction probes for nonstationarity in the signal separated from time series data by applying nonlinear prediction methods to determine the skill with which segments of the time series can predict the others. The time series is nonstationary for purposes of NLTS if the nonlinear predictive skill deteriorates substantially as segments become more remote in time. Code 6.15 runs nonlinear cross prediction. In a nutshell, the time series is divided into equal segments and then each segment is used as a learning set to nonlinearly predict the other segments serving as test sets. The segment size is a trade-off: longer segments allow the system dynamics to evolve, while shorter segments contribute to finer test resolution (Schreiber, 1997). The output is a matrix of Nash–Sutcliffe model efficiencies (**nse**). A column in the output matrix measures the skill with which a particular learning segment cross-predicts increasingly remote test segments (rows in the output matrix). Declining *nse* values from top to bottom in a column indicate nonstationarity over the time interval for the particular learning segment.

```
#Code 6.15 Nonlinear cross prediction (1-step-ahead) to probe nonstationarity
rm(list=ls(all=TRUE))  #remove objects between runs

#Directories
dir.data<-as.character("C:/Data")
dir.udf<-as.character("C:/User Defined Functions")

#Read in data
setwd(dir.data)
ts<-read.csv("Mayport.csv");
x<-ts$signal #ssa signal

#Load user-defined functions
setwd(dir.udf)
dump("embed_udf", file="embed_udf.R")
source("embed_udf.R")
dump("predict_np_udf", file="predict_np_udf.R")
source("predict_np_udf.R")

#Step 1: Segment time series x(t)
#Set number of segments
nseg<-5
seg.length<-floor(length(x)/nseg)  #length of each segment
x<-x[1:(seg.length*nseg)]  #limit x length to even multiple of seg.length
segs<-split(x,cut(seq_along(x),nseg,labels=FALSE))  #segment x(t)
#Put segments into matrix columns
seg.matrix<-matrix(0,seg.length,nseg)
for (i1 in 1:nseg) {
  seg<-unlist(segs[i1])
  seg.matrix[,i1]<-seg
}  #end i1 loop
seg.matrix<-seg.matrix  #columns are nseg segments of time series

#Step 2: Nonlinear cross prediction

#Outer loop (i2) iterates over learning sets used to predict test sets.
#The matrix 'nse.matrix' stores Nash-Sutcliffe efficiencies where the
#columns are the learning sets, and the rows are predicted
#test sets.
```

```
nse.matrix<-matrix(0,ncol(seg.matrix),ncol(seg.matrix))

for(i2 in 1:ncol(seg.matrix)) { #loop over learning sets

#Inner loop (i3) iterates over test sets to be predicted by each learning set.
#The matrix 'hold.nse' is a column vector storing Nash-Sutcliffe efficiencies
#for the predicted test sets (rows) by a given learning set.
hold.nse<-matrix(0,ncol(seg.matrix),1)

for(i3 in 1:ncol(seg.matrix)) { #loop over test sets
  #print("i3");print(i3)
  seg.matrix.2<-seg.matrix[,c(i2,i3)] #matrix isolating learning and test sets
  learn<-seg.matrix.2[,1]  #learning set for forecast
  test<-seg.matrix.2[,2]  #test set

  #Step 2a:  embedding
  #embed learning set with user-defined function embed_udf
  results.embed_udf<-embed_udf(learn)
  d<-results.embed_udf[[1]];m<-results.embed_udf[[2]]
  learn.em.0<-results.embed_udf[[4]]

  #embed test set with same embedding parameters using 'embedd' from
  #R package 'tseriesChaos'
  library(tseriesChaos)
  test.em<-embedd(test,m,d)  #initial embedded test set

  #Stack learn.em and test.em for use in prediction loop
  Mx<-rbind(learn.em.0,test.em)

  #Step 2b: Nonlinear prediction
  results.np<-predict_np_udf(Mx,frac.learn=0.5)
  nse<-results.np[[1]]
  hold.nse[i3,]<-nse
  } #end i3 loop through test sets

nse.matrix[,i2]<-hold.nse
nse.matrix[is.na(nse.matrix)] <- 1  #set any na's along diagonal = 1

} #end i2 loop through learning sets

#Plot of nse.matrix. Each line shows the Nash-Sutcliffe efficiencies for a given
  #learning set (column of nse.matrix) in 1-step prediction of each test set
  #(rows of nse.matrix) along horizontal axis.
plot(nse.matrix[,1],type='l',xlab="test sets predicted",ylab="nse",ylim=c(0,1),
  font=2,font.lab=2,cex.axis=2,cex.lab=2)
for(i4 in 2:ncol(nse.matrix)) {
  lines(nse.matrix[,i4])
} #end loop
```

Step 1 divides x_t into equal-length segments, where the number of segments (nseg) is set by the user. For our example, we divide the time series into nseg $= 5$ segments. We then formulate a loop to load each segment as a column into matrix seg.matrix. An outer loop (index $i2$) iterates over each column to select the segment to serve as the learning set. An inner loop (index $i3$) iterates over the remaining columns to select the test segment to be cross-predicted. Once the learning and test segments have been identified, Step 2 runs nonlinear cross prediction. Step 2a embeds the learning segment using the user-defined function formulated in Code 6.13 (embed_udf), which selects the embedding delay and dimension parameters. The test segment is embedded using the same parameter values with the **embedd** command from the R package **tseriesChaos** (Di Narzo, 2015). The embedded data matrices for the learning and test segments are stacked into the matrix M_x. Consequently, rows in the first half of M_x contain the initial learning set and rows in the second half contain the initial test set. Step 2b runs nonlinear prediction on M_x using the user-defined function formulated in Code 6.14 (**predict_np_udf**).

Code 6.15 concludes by computing Nash–Sutcliffe efficiencies stored in the matrix **nse.matrix**. This matrix is reproduced in Table 6.3 for the Mayport data. The **nse** values for each learning segment (the columns) do not deteriorate for more remote test segments (as we proceed down the columns of the matrix from top to bottom). Consequently, we reject nonstationarity in the Mayport data for the purposes of NLTS. We show these results in graphical form by using Excel to plot each column of the matrix (Figure 6.17). Each line in the plot shows the skill with which a selected learning segment cross predicts the test segments along the horizontal axis (Heathcote and Elliott, 2011). None of the lines slope downwards (indicating deteriorating cross-predictive skill) for the Mayport data.

Table 6.3 Code 6.15 output: **nse** matrix. The columns $[, 1], \ldots, [, 5]$ are the learning segments and the rows $[1,], \ldots, [5,]$ are the predicted test segments. The results show that the Mayport time series is stationary for the purposes of nonlinear time series analysis, since each segment cross-predicts the others with high skill.

	$[, 1]$	$[, 2]$	$[, 3]$	$[, 4]$	$[, 5]$
$[1,]$	1.000	0.970	0.916	0.940	0.874
$[2,]$	0.984	1.000	0.958	0.978	0.922
$[3,]$	0.925	0.952	1.000	0.968	0.930
$[4,]$	0.959	0.984	0.960	1.000	0.920
$[5,]$	0.918	0.961	0.949	0.965	1.000

6.7 Endogenous Complexity with Nonlinear Dynamics

We have illustrated how noisy and nonstationary data can exhibit irregular behaviour. We have also investigated how we can empirically test for these factors and begin to eliminate them as potential drivers of real-world dynamics. To conclude this chapter, we illustrate how irregular dynamics can be generated by a deterministic system of at least three nonlinear ODEs (Kantz and Schreiber, 1997). Put another way, these systems need at least three *available degrees of freedom* (Theiler, 1990). This is a necessary but not sufficient condition for deterministic irregularity. Multidimensional nonlinear systems of ODEs introduce the complication of multiple equilibria, since nonlinear nullclines can intersect more than once. However, depending on parameters, the local stability of these equilibria may be restricted to the types found in linear ODE systems (Glendinning, 1994). The presence of three or more available degrees of freedom opens the door to endogenous irregular dynamics, but does not guarantee them.

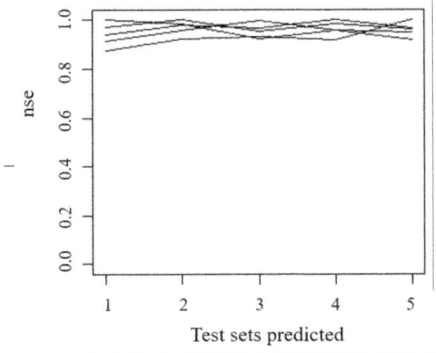

Fig. 6.17 Plots of Nash–Sutcliffe model efficiencies *nse*. Each line represents the skill at which a particular learning segment cross-predicts the other segments (horizontal axis).

Chaos is the quintessential example of irregular dynamics emerging from an ODE system with at least three available degrees of freedom. Chaotic systems are *self-organizing* in that trajectories in phase space are attracted to a lower-dimensional subspace, where they remain trapped, oscillating irregularly along a *strange attractor*. The dimension of this subspace reduces the total available degrees of freedom in the system to the *active degrees of freedom* required to represent long-term system dynamics along the attractor (Theiler, 1990). In chaotic systems, and *dissipative* dynamical systems in general, not all of the system's available degrees of freedom are active in portraying lower-dimensional attractor dynamics.

The classic Lorenz equations can be calibrated to generate a strange attractor (Lorenz, 1963). The Lorenz equations, the attractor and solutions were described in Section 3.4.

Code 6.16 formulates a user-defined function `lorenz_udf` to numerically solve the Lorenz system using the R package `deSolve`. This function takes as inputs the initial values of the system variables (x, y and z), the system parameters (`sigma`, `beta` and `rho`), and integration parameters including the end period (`t.end`) and the integration step (`delta`).

```
#Code 6.16 User-defined function to solve Lorenz system
rm(list=ls(all=TRUE))  #clear values between runs

lorenz_udf<-function(x0,y0,z0,sigma,beta,rho,t.end,delta) {

#Initial conditions and parameters
state<-c(x=x0,y=y0,z=z0) #initial conditions
parameters<-c(sigma,beta,rho) #parameters giving chaotic dynamics

#ODE model
model<-function(t,state,parameters){
  with(as.list(c(state,parameters)),{
  dx<- sigma*(y-x) #Lorenz equations
  dy<- x*(rho-z)-y
  dz<- x*y-beta*z
  list(c(dx,dy,dz))
```

```
  }) #end with (as.list...
} #end model function

#Solution
times<-seq(0,t.end,by=delta)  #integration step
library(deSolve)  #ODE solver package
out<-ode(y=state,times=times,func=model,parms=parameters,method="lsoda")

#Solution variables
x<-out[,"x"];y<-out[,"y"];z<-out[,"z"]

results<-list(x,y,z)
return(results)

} #end user-defined function

#User-defined function run statement
lorenz_results<-lorenz_udf(x0=1,y0=1,z0=1,sigma=10,beta=8/3,rho=28,
  t.end=30,delta=0.01)

x<-lorenz_results[[1]]
y<-lorenz_results[[2]]
z<-lorenz_results[[3]]

#Time series plots of solutions
par(mfrow=c(3,1))
plot(x,type='l',lwd=3,cex.axis=1.1,font.axis=2)
plot(y,type='l',lwd=3,cex.axis=1.1,font.axis=2)
plot(z,type='l',lwd=3,cex.axis=1.1,font.axis=2)

#Phase space
library(scatterplot3d)
scatterplot3d(x,y,z,type="l",lwd=3,cex.axis=1.1,font.axis=2)
```

The organization of `lorenz_udf` follows the way that information is entered into the deSolve routine. We set the parameters, specify the model, and then call up the lsoda integration routine. The output of `lorenz_udf` comprises the numerical solutions x, y and z. After formulating `lorenz_udf`, Code 6.16 plots each solution variable separately, and then scatter-plots them in phase space with the `scatterplot3D` command found in the R package `scatterplot3d`. Figure 3.7 shows how apparent randomness in the time series solutions for the three variables conceals the dynamical order made visible in the phase space *butterfly* attractor. We note that the Lorenz equations produce linear behaviour in the form of a stable node if $\rho = 0.5$, or a stable focus if $\rho = 10$, with the other two parameters held at their existing values. Chaos provides a benchmark of nonlinear low-dimensional deterministic behaviour that we will use in Chapter 7 to distinguish between linear stochastic and deterministic explanations for complexity in observed time series.

6.8 Summary

We seek to reconstruct real-world system dynamics from time series data on a single variable. This is challenging because real-world data often exhibit a highly volatile and irregular appearance potentially driven by several diverse factors. NLTS methods help eliminate less likely drivers of dynamic irregularity. We set a benchmark for *regular* behaviour by investigating how linear systems of ODEs are restricted to exponential and periodic dynamics, and illustrating how irregular behaviour can arise if regular linear dynamics are corrupted with noise or shift over time (i.e. nonstationarity).

We have investigated how data can be preprocessed to control for the noise and nonstationarity potentially camouflaging nonlinear deterministic drivers of observed complexity. We can apply signal-detection methods, such as *singular spectrum analysis (SSA)* to separate signal from noise in the data, and test the signal for nonstationarity potentially corrected with SSA. For example, if the root of nonstationarity is a slow-moving cycle whose period extends beyond the length of the time series, then SSA may be able to isolate and remove it from the signal. In addition, SSA measures signal strength, which provides a useful initial indicator of whether we should continue searching for endogenous nonlinear drivers of complexity. Finally, we have used the classic Lorenz equations to illustrate how a deterministic nonlinear system of ODEs with at least three equations can generate observed irregular dynamics endogenously without aid of exogenous shocks or nonstationary dynamics.

Where does this leave us? Ideally, data preprocessing has eliminated nonstationarity in the data and mitigated the impact of noise. We say *mitigated*, because we cannot reasonably expect that SSA will filter all the noise from complex real-world data. The signal might still be corrupted by lingering noise (i.e. unstructured variation). Consequently, our remaining task in diagnosing deterministic nonlinear structure in the potentially noisy signal is to eliminate linear stochastic dynamics as a likely driver. We do this with `surrogate data testing`, covered in the next chapter.

7
Surrogate Data Testing

'It is more than possible, it is probable.'
        Arthur Conan Doyle, Silver Blaze

7.1 Introduction

The dynamics of real-world biophysical systems are conventionally investigated by relying on theory to formulate an abstract model, calibrating the model to available data and characterizing simulated model dynamics (Kaplan and Glass, 1995). The conventional approach raises a number of concerns, including the following:

- Even if everyone agrees on theory, there are multiple ways to model processes that can generate widely disparate behaviour. How can we tell which model specification corresponds best to the real-world behaviour that we want to understand?
- Even if everyone agrees on theory and a particular model, the same model often can be parameterized to display vastly different behaviour (Hornberger and Spear, 1981). How can we tell which set of parameters is the most reasonable?
- If there is disagreement over theory, and flexible models can be calibrated to produce behaviour consistent with the expectations of multiple theories, how can we reliably use models to distinguish among competing theories?

These concerns cut across any attempt to capture real-world behaviour with abstract models, and boil down to one question: How can we demonstrate correspondence between our model and the real world? We must abstract from an open-ended reality to which there is only limited access. Moreover, we cannot logically verify that a particular abstraction is an accurate representation of reality (Oreskes *et al.*, 1994). Similar to the courtroom, we are lawyers trying to convince a jury of peers that our model reasonably corresponds to the reality that we attempt to explain. In the European Union, the jury of peers includes a government agency that formally audits models used in public policy. What type of evidence makes a compelling case for real-world correspondence in a model audit?

In this book, we join the chorus of voices recommending 'getting to know your data' as a preliminary evidentiary step in modelling (Crawley, 2015). The first thing we notice about time series data is that they are often highly fluctuating, with a random appearance. Observed volatility is commonly attributed to external random shocks to real-world systems that are otherwise stable. If we believe this, we can model the

Nonlinear Time Series Analysis with R. Ray Huffaker, Marco Bittelli and Rodolfo Rosa, Oxford University Press (2017).
© Ray Huffaker, Marco Bittelli and Rodolfo Rosa. DOI: 10.1093/oso/9780198782933.001.0001

stable portion of the system with linear dynamics and excite volatility by adding random shocks. However, breakthroughs in nonlinear dynamics raise another possibility: Highly complex dynamics can emerge from astoundingly parsimonious deterministic nonlinear models (Glendinning, 1994). The implication is that we can no longer make a compelling case for our model by presuming that observed complexity is due to exogenous random forces. We must find a way to replace presumption with hard empirical evidence regarding the driving dynamics.

We can do this by following an empirical process of elimination. First, we know that noisy and nonstationary data can distort deterministic structure, and make linear dynamical structure appear nonlinear. Consequently, we can begin with signal processing to separate signal (structured variation) from noise (i.e. unstructured variation) in the data , and a nonlinear stationarity test to rule out shifting dynamical structure .

Next, we attempt to reconstruct the dynamics of the entire system from the single isolated stationary signal and test the dynamics for deterministic structure. If the dynamics of the underlying real-world dynamical system are *dissipative*, then trajectories converge to a subset of phase space where they oscillate aperiodically along an attractor forever after . We cannot realistically construct a real-world attractor as direct evidence of deterministic dynamics, because we would need to know, and have histories of, all system variables. However, we can work in reverse to reconstruct a *shadow* version of a real-world attractor from an available time series without prior knowledge of system equations (Kaplan and Glass, 1995; Theiler *et al.*, 1992) . If we apply conventional *time-delay embedding*, the coordinates of shadow phase space are the signal and delayed-coordinate copies that serve as surrogates for unobserved system variables . Takens (1981) proved a sufficient condition guaranteeing that shadow phase space preserves essential mathematical properties of the original phase space, since time-delay embedding provides a one-to-one mapping of points from the shadow attractor to points on the original attractor (Figure 7.1). The sufficient condition is $m > 2D + 1$, where m is the embedding dimension of the shadow phase space and D is the fractal dimension of the real-world attractor. Since the dimension of the real-world attractor is unknown, we cannot use the sufficient condition to directly estimate the required embedding dimension. Statistical tests are available to choose time-delay embedding parameters approximating the ideal conditions.

Successful reconstruction of a shadow attractor provides preliminary empirical evidence that the signal may be generated by deterministic dynamics. However, because we cannot reasonably expect signal processing to purge the signal of all noise in practice, and because noisy linear behaviour can be visually indistinguishable from nonlinear behaviour, the possibility remains that noticeable regularity detected in a shadow attractor may be fortuitously reconstructed from data generated by a linear stochastic process (Schreiber and Schmitz, 2000; Kantz and Schreiber, 1997; Theiler *et al.*, 1992). This chapter investigates how we can test this null hypothesis using ***surrogate data testing***. The combination of a noticeably regular shadow attractor, along with strong statistical rejection of fortuitous regularity, increases the probability that observed data are generated by deterministic real-world dynamics.

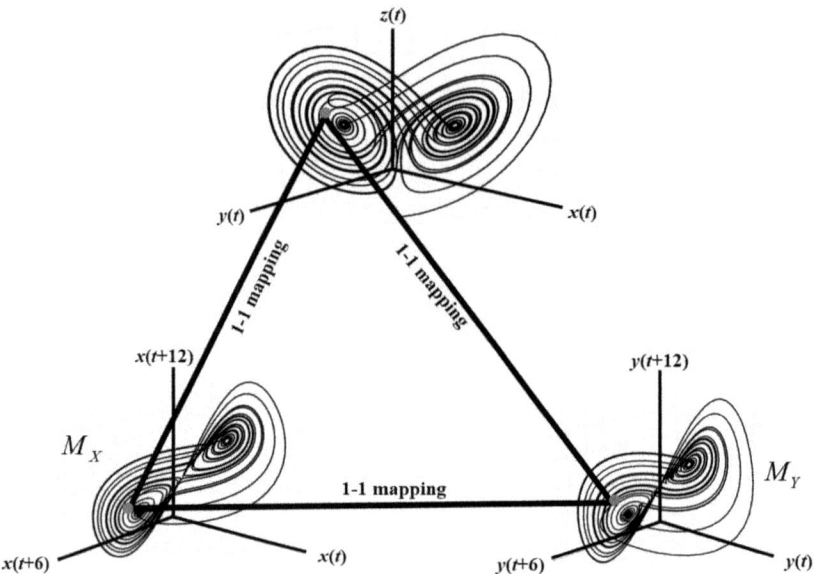

Fig. 7.1 One-to-one mapping of points on shadow Lorenz attractors to each other and to points on the original attractor.

7.2 Surrogate Data Testing in a Nutshell

Surrogate data testing proceeds in three steps . First, we frame null hypotheses testing for the simplest explanations for observed dynamics in the time series and generate surrogate data vectors compatible with these hypotheses (Schreiber and Schmitz, 2000). For example, the simplest hypothesis is that the signal observations are generated as *independent and identically distributed (IID)* random variables (Theiler *et al.*, 1992), so that the reconstructed shadow attractor is simply IID noise . To generate surrogate data vectors compatible with this hypothesis, we shuffle the time order of the signal multiple times. Shuffling destroys serial structure in the signal, and the surrogates are effectively random draws (without replacement) from the same probability distribution as the signal (Small and Tse, 2003). In sum, IID noise surrogates test the null hypothesis that all structure in the signal is given by the probability distribution. In general, surrogates are developed to destroy temporal structure in the signal while preserving statistical properties hypothesized to account for dynamic behaviour.

Second, we reconstruct the shadow phase space from the preprocessed signal and then from each surrogate vector. We estimate *discriminating statistics* measuring the extent to which shadow attractors satisfy particular characteristics of nonlinear deterministic behaviour .

Finally, we test the null hypothesis that preserved statistical properties in the surrogates account for dynamical behaviour in the signal. Applying nonparametric rank-order statistics, we accept the null hypothesis if discriminating statistics taken from the signal attractor do not fall within the extreme ranges of values taken from surrogate attractors. Alternatively, rejecting the null hypothesis leaves untested dynamical structures (such as nonlinear deterministic dynamics) in play.

Surrogate data testing is an application of the *bootstrapping* resampling method from statistics (Efron, 1979). Appendix B further acquaints readers with this important method.

7.3 Surrogate Types

We discuss three algorithms for generating surrogates that test data for various types of linear stochastic dynamics, as well as the associated null hypotheses tested (Small and Tse, 2003).

7.3.1 Algorithm 1 Surrogates

If the data are Gaussian (i.e. observations are drawn from a normal distribution), Algorithm 1 tests the null hypothesis that the time series is driven by linear dynamics subjected to Gaussian white noise inputs:

$$x_t = \sum_{i=1}^{M} a_i x_{t-i} + \sum_{i=0}^{N} b_i \varepsilon_{t-i} \tag{7.1}$$

The first sum is an *autoregressive (AR)* model of time-delayed data vectors, the second sum is a *moving average (MA)* model of time-delayed noise terms composed of Gaussian (i.e. normally distributed) random errors, and a_i and b_i are fixed parameters (Schreiber and Schmitz, 2000). Equation (7.1) is also referred to as *linearly filtered noise* (Small and Tse, 2003). Moreover, we must test the *composite* null hypothesis that the data are generated by a broad class of linear Gaussian process, that is, not just by the process defined by a particular set of parameter values. We do this by testing for broad statistical properties. Algorithm 1 surrogates are designed to have the same Fourier power spectrum as the time series. This is accomplished by taking the Fourier transform of the time series, randomizing the phases, and generating the surrogates by inverting the transform (Theiler, 1990). Consequently, testing the null hypothesis of linear stochastic dynamics with Algorithm 1 boils down to whether structure in the data is given by the Fourier power spectrum (Theiler *et al.*, 1992).

7.3.2 Algorithm 2 (AAFT) Surrogates

Alternatively, if the data are non-Gaussian, then we generate Algorithm 2 *amplitude-adjusted Fourier transform (AAFT)* surrogates to test for stochastic linear dynamics. These are the most common surrogates in use. The data are transformed into a Gaussian series that is randomized in a manner that produces surrogates preserving both the probability distribution and the Fourier power spectrum of the data. This is accomplished with a *static monotonic nonlinear transformation* of linearly filtered noise (Theiler *et al.*, 1992; Small and Tse, 2003). Consequently, testing the null hypothesis of linear stochastic dynamics for non-Gaussian data boils down to whether all structure in the data is given by the probability distribution and the Fourier power spectrum. The mathematical details of how data are shuffled in Algorithm 2 can be found in

Theiler *et al.* (1992), Small and Tse (2003) and Kaplan and Glass (1995). We will rely on the R package `fractal` (Constantine and Percival, 2015) to generate AAFT surrogates.

7.3.3   PPS Surrogates

If the data are characterized by aperiodic (nonrepeated) oscillations, we can generate *PPS surrogates* with an algorithm formulated by Small and Tse (2003). PPS surrogates test the null hypothesis that aperiodic oscillations are generated by a randomly shifting periodic orbit characteristic of noisy linear dynamics. Rejecting the null hypothesis leaves the door open to the possibility that such oscillations are generated by chaotic dynamics characterized by 'nonlinear nonperiodic determinism' (Small and Tse, 2003). The PPS algorithm implements a random walk over a reconstructed shadow attractor and takes as a surrogate data vector the first coordinates of the randomly selected points on the attractor. Consequently, PPS surrogates preserve coarse deterministic features (e.g. periodic trends) but destroy fine structure (e.g. deterministic chaos) (Small and Tse, 2003).

Since there is no R routine available to run the PPS algorithm, Code 7.1 formulates a user-defined function `PPS_udf` based on Small and Tse (2003). The function takes as inputs x (the time series), `nsurr` (the number of surrogates to be generated), and `lb` and `ub` (determining respectively the smallest and largest distances from a reference point for which another point qualifies as a close neighbour). We import the signal separated from the Mayport sea level time series introduced in Chapter 6. After computing the embedded data matrix M_x from the time series (Step 1), we compute the distance matrix whose columns give distances from a particular point on the attractor to the other points (Step 2) . We then take a random walk on the attractor (Step 3). We first randomly select a reference row in M_x (Step 3a), and turn to the relevant column in the distance matrix to select neighbouring rows within the specified lower (`lb`) and upper (`ub`) bounds (Step 3b). Finally, we pick one of these neighbouring rows at random (Step 3c). The element of a surrogate data vector is given by the initial coordinate of this point, i.e. $M_x[,1]$. The inner loop i fills the elements of a single surrogate vector, and the outer loop j collects the surrogate vectors in the output matrix `surr.PPS`.

```
#Code 7.1  User-defined function to generate PPS surrogates
rm(list=ls(all=TRUE))

#Directories
dir.data<-as.character("C:/Data")
dir.udf<-as.character("C:/User Defined Functions")
dir.results<-as.character("C:/Results")

#Load user-defined functions
setwd(dir.udf)
dump("embed_udf", file="embed_udf.R")
source("embed_udf.R")

#Load Data from an external .csv file
setwd(dir.data)
ts<-read.csv("Mayport.csv")
```

```
x<-ts$signal

PPS_udf<-function(x,nsurr,lb,ub){

#Step 1: Embed time series
results.embed_udf<-embed_udf(x)
Mx<-results.embed_udf[[4]] #embedded data matrix

#Step 2: Distances between points on attractor (i.e. row of Mx)
   #Step 2a: Calculate distance matrix from embedded time series matrix 'Mx'
   library(fields)
   dist<-rdist(Mx)

   #Step 2b: Scale each column element to maximum distance within that column in 'dist'
   sf<-apply(dist,2,max) #apply(base) [find maximum distance in each column]
   dist.tr<-t(dist)
   dist.sc<-t(dist.tr/sf)

#Step 3: Random walk on attractor (i.e. over rows of embedded time series matrix 'x.em'.
   surr.PPS<-matrix(0,length(x),nsurr)   #matrix to hold PPS surrogate vectors
   for(j in 1:nsurr){
      surr.vector<-matrix(0,length(x),1)   #matrix to hold PPS vector as it is created
                                           #element by element

      for(i in 1:length(x)){

         #Step 3a: Pick reference point on attractor at random
         rp.1<-sample(nrow(Mx),1)

         #Step 3b: Identify neighbouring points
         ngh.lb<-which(lb < dist.sc[,rp.1]) #indices satisfying lower bound 'lb' of nghbhd
         ngh.ub<-which(ub > dist.sc[,rp.1]) #indices satisfying upper bound 'ub' of nghbhd
         int<-intersect(ngh.lb,ngh.ub)    #indices satisfying both bounds in neighbourhood

         #Step 3c: Pick point in neighbourhood at random
         rp.2<-sample(int,1)
         surr.vector[i,1]<-Mx[,1][rp.2] #1st column of Mx gives element of surrogate vector
      }   #end iloop

      surr.PPS[,j]<-surr.vector
   }  #end jloop

   return(surr.PPS)
}  #end user-defined function

surr.PPS<-PPS_udf(x,nsurr=4,lb=0.01,ub=0.4)
surr.PPS
```

7.3.4   A Strategy for Selecting Surrogate Types

The AAFT algorithm has been criticized on the basis that mathematical assumptions required to run the static monotonic nonlinear transformation may not be met by real-world data (Kugiumtzis, 2002). If so, AAFT surrogates may not incorporate linear correlations in the data (Kugiumtzis, 2002) and may produce a biased (flatter) Fourier power spectrum (Schreiber and Schmitz, 2000). Moreover, AAFT surrogates do not test the hypothesis of a noisy orbit in aperiodic data. Our strategy is to follow conventional practice in first testing the null hypothesis that structure in the data is given by the probability distribution and the Fourier power spectrum with AAFT surrogates. We then follow the recommendation of Kugiumtzis (1999) and Kugiumtzis (2002) to re-test a more general null hypothesis with other surrogates. Since volatile data commonly exhibit aperiodicity, we re-test with PPS surrogates.

7.4 Discriminating Statistics

Measures developed to quantify characteristic features of deterministic chaos are also used to discriminate between linear stochastic and nonlinear deterministic dynamics in surrogate data testing. Conventional statistics used to detect hallmarks of deterministic structure in time series data include the *correlation dimension* (measuring the fractional dimension of a reconstructed attractor), the *maximum Lyapunov exponent* (measuring the sensitivity to initial conditions and the consequential spreading of trajectories over time) and the *nonlinear prediction error* (measuring the skill with which a reconstructed attractor can forecast in the short term) (Kaplan and Glass, 1995; Kantz and Schreiber, 1997). An extensive description of these discriminating statistics has been provided in Chapter 4. In this chapter, we revisit these measures to further motivate them and introduce other ways to estimate them.

The literature strongly advises against using these measures as positive evidence that real-world systems are chaotic, because they are based on asymptotic properties requiring vast amounts of high-quality data while practitioners typically are faced with finite noisy and possibly nonstationary records. However, there is general agreement that the measures can be used reliably to discriminate between deterministic and random structure in surrogate data testing (Kantz and Schreiber, 1997; Schreiber and Schmitz, 2000; McSharry, 2011).

7.4.1 Correlation Dimension

Phase space trajectories in dissipative dynamic systems converge to a subspace where they remain oscillating aperiodically on an attractor . Unlike a zero-dimensional point, a one-dimensional line, a two-dimensional surface or a three-dimensional volume, an attractor can be a fractal geometric object with a fractional (noninteger) dimension. Fractal geometric objects are characterized by *self-similarity*, i.e. by irregular patterns that repeat themselves at different scales (Theiler, 1990) . An attractor's fractal dimension reduces the total *available degrees of freedom* in the system to the *active degrees of freedom* required to portray long-term system dynamics, and thus provides an informative measure of system complexity (Theiler, 1990). The available degrees of freedom correspond to the number of variables in the system (i.e. system dimensionality).

In this chapter, we focus on numerical estimation of the correlation dimension using the Grassberger–Procaccia correlation algorithm (Grassberger and Procaccia, 1983*b*) and the Takens estimator (Takens, 1985).

We begin with Code 7.2 to provide a simple example of how to compute the *correlation integral* at the core of the *Grassberger–Procaccia algorithm*:

```
#Code 7.2  A simple example of how to calculate C(epsilon;N)
rm(list=ls(all=TRUE))  #remove values between runs

x<-c(5,8,2,0,1,3) #time series
```

```
#Step 1: Embed time series
d<-1;m<-3 #embedding delay and dimension
library(tseriesChaos)
Mx<-embedd(x,m,d) #embedded data matrix
n<-nrow(Mx) #number of Mx rows

#Step 2: Distance matrix
library(fields)
dist<-rdist(Mx)   #Distance matrix for Mx

#Step 3: Identify distances below threshold level
eps<-6.5  #distance threshold
  #Step 3a: Exclude diagonal distances
diag(dist)<-eps+1 #puts diagonal distances above threshold
  #Step 3b: Count distances under threshold in entire matrix
dist.vec<-as.vector(dist)
distances.thresh<-dist.vec[-c(which(dist.vec>eps))]

#Step 4: Compute fraction of distances within threshold
frac<-length(distances.thresh)/(n*(n-1))
```

In Step 1, we compute an embedded data matrix M_x, whose rows represent multidimensional points on a reconstructed attractor. In Step 2, we compute the distance matrix containing *Euclidean distances* between the rows of M_x:

$$
\begin{array}{cccc}
0.0 & 7.0 & 8.6 & 8.7 \\
7.0 & 0.0 & 6.4 & 8.6 \\
8.6 & 6.4 & 0.0 & 3.0 \\
8.7 & 8.6 & 3.0 & 0.0
\end{array}
$$

Step 3 specifies threshold distance $\epsilon = 6.5$. We observe from the distance matrix that there are four distances below the threshold (excluding the diagonal distances), so that the fraction of distances below the threshold is $4/(4)(3) = 0.33$. We numerically calculate this fraction by first excluding the diagonal distances in the distance matrix (Step 3a). This is done by putting the zero distance between a point and itself above the threshold so that it is not counted. We then count the nondiagonal distances under the threshold (Step 3b). If we relax the distance threshold to $\epsilon = 7.1$, the fraction of distances below the threshold increases to 0.5. Finally, Step 4 computes a correlation integral as the fraction of distances below the threshold.

As explained in Chapter 4, the estimated *correlation dimension* of an attractor, D_2, is the exponent in a *power law* scaling the correlation integral to distance thresholds:

$$
C(\epsilon; N) \sim k\epsilon^{D_2} \tag{7.2}
$$

We can solve for D_2 by first taking the logarithm of both sides and then taking a derivative requiring $\epsilon \to 0$ and $N \to \infty$. Using this derivative to directly compute D_2 is generally infeasible in practice because the smallest ϵ is bounded below by the shortest distance between two points on the reconstructed attractor, and the largest N is bounded above by the length of the time series and the magnitude of estimated embedding parameters.

Code 7.3 estimates the correlation dimension from a time series by applying procedures formulated in the literature to work around these mathematical problems. In this illustration, we use the x solution to the Lorenz equations generated with the user-defined function `lorenz_udf` (see Code 6.16), so that we can check the performance of Code 7.3 against the estimated dimension of the Lorenz attractor in the literature: $v \approx 2.06$ (McGuinness, 1983).

```
#Code 7.3 Compute correlation dimension for time series using Takens estimator
rm(list=ls(all=TRUE))   #remove values between runs

#Directories
dir.udf<-as.character("C:/User Defined Functions")

#Load user-defined functions
setwd(dir.udf)
dump("lorenz_udf", file="lorenz_udf.R")
source("lorenz_udf.R")
dump("embed_udf", file="embed_udf.R")
source("embed_udf.R")
dump("cd_takens_udf", file="cd_takens_udf.R")
source("cd_takens_udf.R")

#Generate Lorenz data with user-defined function 'lorenz_udf'
lorenz_results<-lorenz_udf(x0=1,y0=1,z0=1,sigma=10,beta=8/3,rho=28,
  t.end=50,delta=0.01)
x<-lorenz_results[[1]]

#Step 1:  Use 'd2(tseriesChaos) to plot correlation integrals
          #to view "scaling range"

  #Step 1a: Embed time series
embed<-embed_udf(x)  #run user-defined function
d<-embed[[1]]   #embedding delay
tw<-embed[[3]] #Theiler window
Mx<-embed[[4]] #embedded data matrix

  #Step 1b: Set radius parameter required by 'd2(tseriesChaos)'
library(fields)
dist<-rdist(Mx)   #Distance matrix
r<-0.05
eps.min<-r*max(dist) #set eps.min for d2()

  #Step 1c: Plot correlation integrals for increasing
           #embedding dimension (m)
library(tseriesChaos)
m_max<-10  #maximum embedding dimension
d2.out<-d2(x,m_max,d,tw,eps.min)  #run d2 routine

#Step 2:  Compute Takens estimator of correlation dimension for
          #multiple embedding dimensions (cd_takens_udf)
cd.m<-cd_takens(x,m_max)

#Step 3: Plots
par(mfrow=c(2,1))
  #Step 3a: Plot correlation integrals over m (for scaling range)
plot(d2.out)
  #Step 3b: Plot correlation dimension against embedding dimension.
           #Cancel default x-axis (xaxt="n") to define custom axis
           #with only integers
```

```
m<-1:m_max  #revised x-axis
cd.m.plot<-plot(m,cd.m,xaxt="n",xlab="",ylab="correlation dimension",
  pch=15,type="b")
#Custom x-axis, side=1 is x-axis, "at" defines 'tick-mark' numbers
axis(side=1,at=1:length(cd.m))
#Label custom x-axis, "line=2.5" puts label below tick-mark numbers
mtext(side=1,"embedding dimension",line=2,cex=0.8)
#Label points in plot
cd.m<-round(cd.m,digits=2)
text(m,cd.m,cd.m,cex=0.7,pos=1,offset=0.4)
```

Step 1a embeds the time series, and then the R routine d2(tseriesChaos) is used to compute several correlation integrals, each of which varies over the distance threshold ϵ for a fixed embedding dimension (Steps 1b, c). The d2(tseriesChaos) routine requires inputs including the time series x, the maximum number of embedding dimensions for which to plot correlation integrals $m_{\max}$, the embedding delay d, the Theiler window tw omitting pairs of points in the distance matrix with time indices differing by less than tw so that temporally correlated points are not mistaken for the attractor's geometry (Kantz and Schreiber, 1997), and a minimum starting threshold distance *eps.min*. There is no fixed rule for setting *eps.min*. In Step 1b, we set *eps.min* as a proportion r of the maximum distance in the distance matrix. Step 1c plots the correlation integral for each embedding dimension up to $m_{\max}$ (Figure 7.2a).

The correlation integrals are plotted against distance thresholds ϵ on a double logarithmic scale so that we can easily detect whether the data exhibit the power law relationship (7.2) used to estimate the correlation dimension v. The power law holds for a *scaling range* of ϵ for which the plots drawn for increasing embedding dimensions m are reasonably straight parallel lines . The correlation dimension D_2 is estimated as the slope of these lines. This provides evidence of *self-similar* geometry in a reconstructed attractor, because the scaling exponent D_2 is independent of scale (i.e. of m). Put another way, the attractor's correlation dimension is independent of the dimensionality of phase space in which it is embedded. We can easily identify a scaling range for ϵ in the correlation integrals computed from the Lorenz data (Figure 7.2). After the uppermost plot $m = 2$, the descending correlation integrals show parallel straight lines for approximately $2 < \epsilon < 10$.

There are several problems with attempting to estimate the correlation dimension D_2 directly from the slope of the correlation integral (Theiler, 1990). Consequently, Step 2 computes D_2 by formulating a user-defined function cd_takens to compute the *Takens estimator*. The Takens estimator is a maximum likelihood procedure that avoids the need to compute the slope (Takens, 1985):

$$D_2(\varepsilon_0) = \frac{-1}{\left\langle \log\left(\dfrac{\varepsilon_{i,j}}{\varepsilon_0}\right) \right\rangle} \tag{7.3}$$

Here, ε_0 is an upper bound on the distance threshold and $\langle \cdot \rangle$ is an arithmetic average of the logarithms of all distances less than the upper bound (Theiler, 1990).

Step 3 of Code 7.3 plots the correlation dimension D_2 against the embedding dimension m (Figure 7.2b). The estimated level is where the plot saturates at $v \approx 2.07$,

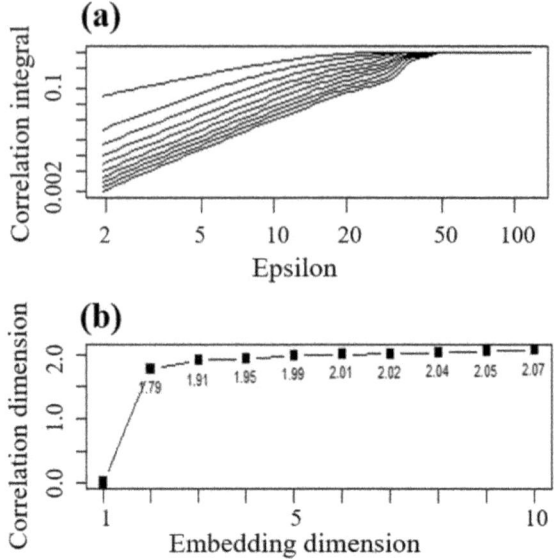

Fig. 7.2 (a) Plots of correlation integrals against distance thresholds computed from Lorenz data. The plots are on a double logarithmic scale to test whether the data satisfy the power law relationship in eqn (7.2). Each plot is associated with a different embedding dimension, starting with $m=2$ (uppermost curve) and ending with $m=10$ (lowermost curve). Within the scaling range $2 < \epsilon < 10$, the plots are reasonably straight parallel lines showing self-similar behaviour since the power law holds independently of m. The estimated correlation dimension is the slope of these parallel lines. (b) Plots of correlation dimension against embedding dimension. The estimated correlation dimension is where the plot saturates at $v \approx 2.07$.

which is close to the correlation dimension reported for the Lorenz attractor in the literature ($v \approx 2.06$).

Code 7.4 expands the user-defined function `cd_takens_udf` used in Step 2 of Code 7.3 to estimate the correlation dimension with eqn (7.3):

```
#Code 7.4 Formulate user-defined function to compute correlation dimension for
         #time series using Takens estimator (cd_takens_udf)

rm(list=ls(all=TRUE))   #remove values between runs

#Directories
dir.udf<-as.character("C:/User Defined Functions")

#Load user-defined functions
setwd(dir.udf)
dump("lorenz_udf", file="lorenz_udf.R")
source("lorenz_udf.R")
dump("embed_udf", file="embed_udf.R")
source("embed_udf.R")

#Generate Lorenz data with user-defined function 'lorenz_udf'
```

```
lorenz_results<-lorenz_udf(x0=1,y0=1,z0=1,sigma=10,beta=8/3,rho=28,
 t.end=50,delta=0.01)
x<-lorenz_results[[1]]


cd_takens<-function(x,m_max) {

cd.m<-matrix(0,m_max,1) #store estimated correlation dimensions
for(j in 2:m_max) { #iterate over embedding dimensions (m)

#Step 1: Compute embedding delay and Theiler window
embed<-embed_udf(x)
d<-embed[[1]]  #embedding delay
tw<-embed[[3]] #Theiler window

#Step 2: Compute embedded data matrix
library(tseriesChaos)
Mx<-embedd(x,j,d)  #embedded data matrices
n<-nrow(Mx)

#Step 3: Compute distance matrix
library(fields)
dist<-rdist(Mx)  #Distance matrix
dist.vec<-as.vector(dist)

#Step 4: Adjust distance matrix to Theiler window

  #Step 4a: reduce to relevant rows and columns
dist.1<-dist[(tw+1):n,1:(n-tw)]

  #Step 4b: reduce to relevant distances in each column
dist.2<-matrix(0,(n-tw),(n-tw)) #storage matrix for loop
for(i in 1:(n-tw)) { #iterate over columns of dist.1
 column<-dist.1[,i][i:(n-tw)]
 length(column)<-n-tw #make reduced columns same length for storage
 dist.2[,i]<-column
} #end loop i

#Step 5: Put distances in matrix into vector
dist.vec1<-as.vector(dist.2)  #distances not satisfying Theiler window
                              #have NAs
dist.vec2<-na.omit(dist.vec1) #remove NAs, only distances satisfying
                              #Theiler window remain

if(tw==0) {distances<-dist.vec} else {distances<-dist.vec2}

#Step 6: Set upper threshold on distance
x.thresh<-0.5*sqrt(sum(x^2)/length(x)) #1/2 root mean squared error (rms)
distances.thresh<-distances[-c(which(distances>x.thresh))]  #omit distances over threshold

#Step 7: Takens estimation of correlation dimension
cd<--1/mean(log(distances.thresh/x.thresh))

cd.m[j,]<-cd  #vector storing correlation dimensions over range of
              #embedding dimensions
}  #end loop j

return(cd.m)
} #end user-defined function


cd_m<-cd_takens(x,m_max=10)
```

The user-defined function (`cd_takens_udf`) requires input of the time series x and the maximum embedding dimension m_{max}. The function estimates the correlation dimension for several embedding dimensions with a loop running from $m = 2$ through m_{max}. Each iteration through the loop initially computes the embedded data matrix and the associated distance matrix (Steps 1–3).

Step 4 then adjusts the distance matrix to the Theiler window tw, to omit pairs of points in the distance matrix with time indices differing by less than tw. Figure 7.3 illustrates this for a 4×4 distance matrix, where $d(i,j)$ is the distance between points i and j. If $tw=2$, we are left with three distances: $d(1,3)$, $d(1,4)$ and $d(2,4)$. Since the distance matrix is symmetric, we omit the redundant distances $d(3,1)$, $d(4,1)$ and $d(4,2)$. We also omit the diagonal distances. Step 4 performs this reduction in two stages. In Step 4a, the first $tw = 2$ rows and last $tw = 2$ columns are removed from the matrix, leaving the submatrix in the lower left quadrant. In Step 4b, loop i is used to eliminate the remaining distances in the lower left quadrant whose time indices differ by less than tw (i.e. $d(2,3)$). During the first pass through the loop ($i = 1$), the command `column<-dist.1[,i][i:(n-tw)]` picks out both elements of the first column of the submatrix, i.e. $d(1,3)$ and $d(1,4)$, both of which satisfy the Theiler window. During the second pass ($i = 2$), the command picks out the second element of the second column satisfying the Theiler window, and replaces the first element not satisfying the Theiler window with `NA`.

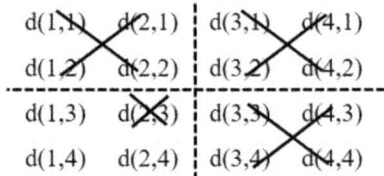

Fig. 7.3 Illustration of adjustment of the distance matrix to the Theiler window.

Step 5 stacks the columns of the submatrix, removes `NA` elements and leaves us with a vector of distances satisfying the Theiler window.

Step 6 follows convention in setting the upper threshold bound ε_0 at half of the root-mean-square (rms) amplitude of the time series (Theiler *et al.*, 1992; Kantz and Schreiber, 1997) and omits distances in the vector exceeding this bound. Finally, Step 7 calculates the Takens estimate of the correlation dimension for a particular embedding dimension using eqn (7.3).

7.4.2 Maximum Lyapunov Exponent

The *maximum Lyapunov exponent* measures the average rate at which initially neighbouring points on an attractor exponentially diverge or converge, and consequently the extent to which an attractor exhibits the sensitivity to initial conditions characteristic of deterministic chaos (Kaplan and Glass, 1995; Kantz and Schreiber, 1997). We illustrate this sensitivity in the case of the chaotic Lorenz attractor by applying the user-defined function `lorenz_udf` to plot the x coordinate of a solution trajectory

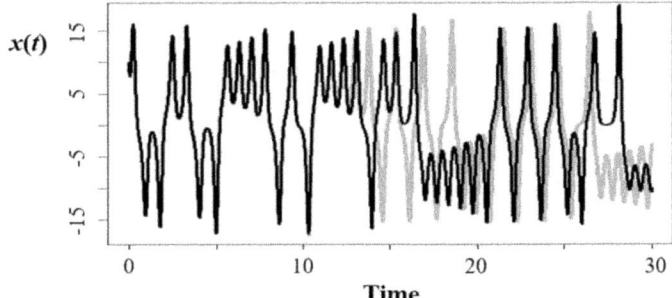

Fig. 7.4 Sensitivity to initial conditions. The solution for the x variable from the Lorenz equations is plotted for initial conditions $x(t = 0) = 10$, $y(t = 0) = 0.1$ and $z(t = 0) = 10$ (black line) and for initial conditions with $x(t = 0)$ increased slightly to 10.0001 with the others held constant (grey line). The trajectories begin to separate substantially after about 14 periods.

for 30 periods with an integration step of 0.01 (Figure 7.4). First, we solve the system from initial conditions $x(t = 0) = 10$, $y(t = 0) = 0.1$ and $z(t = 0) = 10$ (black line). Then we slightly increase $x(t = 0)$ to 10.0001 while leaving the other initial levels constant (grey line). Initially, the neighbouring x trajectories are very close. After about 14 periods, they begin to separate, and finally become uncorrelated, although they are generated by the same dynamical system. The maximum Lyapunov exponent measures the separation rate averaged over the entire attractor, i.e. over all initial conditions (Kaplan and Glass, 1995). A key signature of deterministic chaos is that separation occurs at an exponential rate, as measured by a positive maximum Lyapunov exponent. A surprising implication of sensitivity to initial conditions is that, given the numerical imprecision of measuring initial conditions, long-term prediction in chaotic systems is impossible even though the governing equations are deterministic.

We can numerically estimate the maximum Lyapunov exponent with the R routine `lyap_k(tseriesChaos)` based on an algorithm in Kantz and Schreiber (1997). Code 7.5 illustrates how this algorithm works in the simpler setting of numerically characterizing sensitivity to initial conditions from a single initial point on the Lorenz attractor:

```
#Code 7.5   Illustration of Kantz-Schreiber algorithm for a
            #single initial point on the Lorenz attractor.

rm(list=ls(all=TRUE))   #clear values between runs

#Directories
dir.udf<-as.character("C:/User Defined Functions")

#Load user-defined functions
setwd(dir.udf)
dump("lorenz_udf", file="lorenz_udf.R")
source("lorenz_udf.R")

library(deSolve) #ODE solver package
```

```
library(fields) #distances between rows of same matrix
library(pdist)  #distances between two matrices
library(scatterplot3d) #plot phase diagram


#Step 1: Solve Lorenz equations from initial point
t.end<-30
delta<-0.01
#Generate Lorenz data with user-defined function 'lorenz_udf'
lorenz_results<-lorenz_udf(x0=10,y0=0.1,z0=10,sigma=10,beta=8/3,rho=28,
 t.end,delta)
x<-lorenz_results[[1]];y<-lorenz_results[[2]];z<-lorenz_results[[3]]
M<-cbind(x,y,z)


#Step 2: Calculate nearest neighbours to reference point (row in M)
ref<-1  #Set reference point as 1st row of M (for example)
dist<-rdist(M)  #distance matrix
dist.ref<-dist[,ref]  #distances from reference row to other rows in M


#Step 3: Identify indices of points whose distances from ref are
          #under threshold
#Set threshold (eps) as fraction r of maximum distance in dist.  Set r
#to achieve at least 10 near neighbours
r<-0.202
eps<-r*max(dist.ref)
indices<-which(dist.ref<eps) #points whose distances under eps
indices<-indices[-1] #omit reference point from nearest neighbours
neigh.number<-length(indices) #number of nearest neighbours


#Use indices to isolate nearest neighbouring rows of M to ref point
neigh<-matrix(0,length(indices),3) #matrix whose rows are nearest points
for(i in 1:length(indices)) {
  rows<-M[indices[i],]
  neigh[i,]<-rows
  } #end loop i


#Step 4: Compute trajectories starting from nearest neighbouring points

#Loop j iterates over each row of neigh. Each column of dist.neigh.mat
#stores the distances between the trajectory starting from the reference
#row to the trajectory starting from a nearest neighbour for each iteration.

dist.neigh.mat<-matrix(0,((t.end/delta)+1),nrow(neigh))
for(j in 1:nrow(neigh)) {

  #use lorenz_udf to solve for Lorenz trajectories from neighbouring points
  lorenz_results<-lorenz_udf(x0=neigh[j,1],y0=neigh[j,2],z0=neigh[j,3],
  sigma=10,beta=8/3,rho=28,t.end,delta)
  x<-lorenz_results[[1]];y<-lorenz_results[[2]];z<-lorenz_results[[3]]
  M.neigh<-cbind(x,y,z)

  #Compute distances between M and M.neigh at each iteration
  dist.all<-as.matrix(pdist(M,M.neigh)) #includes distances across rows
  dist.neigh<-log(abs(diag(dist.all))) #isolate distances between same rows
  dist.neigh.mat[,j]<-dist.neigh
  } #end loop j
dist.neigh.mat


#Step 5:  Average distances across rows of dist.neigh.mat
dist.ave<-apply(dist.neigh.mat,1,mean)
```

```
#Step 6: Semilog plot of log(dist.ave) against time periods.

#The slope of the linear portion of the curve represents an average
#exponential growth in distance between neighbouring trajectories. When
#averaged over several initial reference points, this serves as an estimate
#of the maximum Lyapunov exponent.

dist.log<-log(abs(dist.ave))
time1<-1:length(dist.log)  #iteration steps
time2<-time1*delta  #convert to time period
plot(time2,dist.log,type='l',cex.axis=1.4,lwd=3,font=2)
```

Step 1 applies the user-defined function `lorenz_udf` (see Code 6.16) to solve the Lorenz equations for a single reference trajectory on the attractor from initial conditions $x(t = 0) = 10$, $y(t = 0) = 0.1$ and $z(t = 0) = 10$ over 30 time periods, with an integration step $delta = 0.01$. The output is a matrix M whose columns are solutions for x, y and z, respectively, and whose rows are three-dimensional points on the Lorenz attractor.

Step 2 sets a reference point on the reference trajectory as the first row of M and then computes the distance matrix measuring Euclidean distances to all other points on the attractor. Step 3 identifies the indices of the nearest neighbouring points to the reference point (i.e. their row number in M). A point qualifies as a nearest neighbour if its distance from the reference point falls below a given threshold. The most important consideration in setting a threshold is that it results in a sufficient number of neighbouring points to allow the analysis to be performed. Kantz and Schreiber (1997) advise that at least ten nearest neighbours are needed. We set the threshold `eps` as a fraction of the maximum distance of a point from the reference point and compute the resulting number of nearest neighbours, `neigh.number`. We use the indices of neighbouring points to identify the actual points and collect these points as rows in the matrix `neigh` with `loop i`.

Step 4 formulates `loop j` to iterate over rows of `neigh`, each of which serves as the initial condition for a trajectory initially close to the reference trajectory before separation. After re-applying `lorenz_udf` to solve for initially neighbouring trajectories, we compute the distance between the reference and neighbouring trajectories at each iteration. This results in the matrix `dist.neigh.mat`, whose columns contain the distances between the reference point and the associated neighbouring trajectory.

Step 5 averages the distances between the reference point and all of the neighbouring trajectories at each iteration by taking the average of each row of `dist.neigh.mat`. Step 6 plots the logarithm of average distance against time (Figure 7.5), and, given the semilogarithmic scale, the average exponential separation rate is the slope of a linear portion of this curve. This slope gives an estimate of the maximum Lyapunov exponent when it is derived from a curve averaged over multiple reference trajectories computed from many initial points on the attractor. The routine `lyap_k(tseriesChaos)` generalizes the illustration in Code 7.5 in this and other important ways, including the incorporation of a Theiler window to distinguish between temporal correlation and attractor geometry.

Code 7.6 runs the routine `lyap_k(tseriesChaos)` after generating the required inputs:

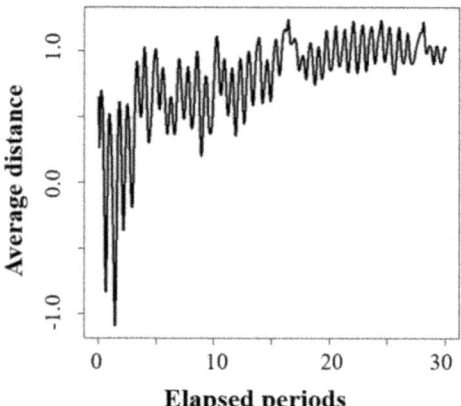

Fig. 7.5 Plot of the separation of initially neighbouring trajectories from a single initial condition on a semilogarithmic scale. The slope of a linear portion of the curve is the average exponential separation rate from this single point. This slope provides an estimate of the maximum Lyapunov exponent when it is computed from a curve averaged over trajectories originating from many initial points on the attractor.

```
#Code 7.6  Numerical estimation of maximum Lyapunov exponent
          #from time series using lyap_k(tseriesChaos)

rm(list=ls(all=TRUE))

#Directories
dir.udf<-as.character("C:/User Defined Functions")

#Load user-defined functions
setwd(dir.udf)
dump("lorenz_udf", file="lorenz_udf.R")
source("lorenz_udf.R")
dump("embed_udf", file="embed_udf.R")
source("embed_udf.R")

#Step 1: Generate Lorenz data with user-defined function 'lorenz_udf'
lorenz_results<-lorenz_udf(x0=5,y0=5,z0=5,sigma=10,beta=8/3,rho=28,
 t.end=500,delta=0.1)
x<-lorenz_results[[1]]

#Step 2: Embed time series (x)
results.embed_udf<-embed_udf(x)
d<-results.embed_udf[[1]]
m<-results.embed_udf[[2]]
tw<-results.embed_udf[[3]]   #embedding dimension
Mx<-results.embed_udf[[4]]   #embedded data matrix

#Step 3: Estimate maximum Lyapunov exponent
  #Step 3a: Convert imported time series to R time series object
n<-length(x)-1  #length of time series
frequency<-10 #iterations per period (e.g. months/year)
periods<-n/frequency   #periods in data set x (e.g. years)
x.ts<-ts(x,start=c(0,1),end=c(periods,1),frequency)
```

```
#Step 3b: Set radius parameter (eps_lyap) for lyap_k(tseriesChaos)
library(fields)
dist<-rdist(Mx)  #Distance matrix
r_lyap<-0.1
eps_lyap<-r_lyap*max(dist)  #set eps required by 'lyap_k'

#Step 3c: Run lyap_k(tseriesChaos)
library(tseriesChaos)
lyap_out<-lyap_k(x.ts, m, d=2,s=10*frequency, tw, ref=2000, k=2, eps=eps_lyap)
plot(lyap_out,lwd=1) #Semilog plot of log(dist.ave) against time periods.

#Step 3d: Fit regression line to linear region
LE.max<-lyap(lyap_out,1,3.4);LE.max
```

We use the x solution to the Lorenz equations for our time series (Step 1). We then embed the solution to provide inputs to `lyap_k(tseriesChaos)`, including the embedding dimension m, the embedding delay d and the Theiler window tw (Step 2). The routine `lyap_k(tseriesChaos)` also requires that the data x be converted to an R *time series object* x.ts (Step 3a). The command takes the form x.ts<-ts(x,start=c(0,1), end=c(periods,1),frequency). The time series object runs from `start=1` to `end=periods`. In this example, `periods` equals the `t.end=500` periods set in solving for the Lorenz time series (Step 2), and `frequency` is set to 10 since the integration step is set at `delta=0.1` in the Lorenz solution. This means that each of the 500 periods is split into 10 iterates, resulting in 5000 iterates (5001 including the initial condition). In another example, if we were converting 72 years of monthly data (864 observations) to an R time series object, the conversion command would read x.ts<-ts(x,start=c(0,1),end=c(72,1),frequency=12). The routine `lyap_k(tseriesChaos)` further requires that we set the threshold distance parameter `eps`, which we again do with the distance matrix from embedding (Step 3b).

In Step 3c, we run `lyap_k(tseriesChaos)`, which generates a semilogarithmic plot of the logarithm of average separation against time periods with the command `lyap_out`. The s parameter in the command line sets an upper limit on periods in the plot (x axis) and the `ref` parameter sets the number of reference points to consider. We observe a strong linear section to the semilogarithmic plot (dashed line) running from about 1 to 3.4 periods along the x axis (see Figure 4.3). The routine `lyap_k(tseriesChaos)` estimates the slope of the linear portion with the command `LE.max<-lyap(lyap_out,1,3.4)` (Step 3d). This slope gives an estimate of the maximum Lyapunov exponent `LE.max=0.9`, which is consistent with numerical estimates in the literature (Sprott, 2016).

7.4.3 Nonlinear Prediction Skill

A hallmark of deterministic behaviour is the skill with which an empirically reconstructed attractor can make short-term forecasts (Kaplan and Glass, 1995). Nonlinear prediction skill is strongly recommended for use as a discriminating statistic in surrogate data testing, given the numerical challenges of estimating reliable measures of correlation dimension and maximum Lyapunov exponents with short and noisy time series data (Kantz and Schreiber, 1997). Code 6.14 in Chapter 6 formulates a user-defined function to run nonlinear prediction that we shall use in the surrogate data testing code in the present chapter.

7.4.4 Entropy Complexity Measure

Permutation entropy (Brandt and Pompe, 2002) is a popular probability-based complexity measure that is used as a discriminating statistic in surrogate data tests (Gotoda and Kobayashi, 2017). It modifies the classic *Shannon H* measure for application to finite noisy time series data. We briefly summarize some important properties of H from Chapter 5: (1) H measures the information content in the time series; (2) if $H = 0$, then the time series is perfectly predictable from past values; and (3) H achieves a maximum value when time series values are independent and uniformly distributed. In sum, relatively large values of H indicate more highly complex random behaviour as opposed to low-dimensional deterministic behaviour.

We have also learned from Chapter 5 that H is a function of a discrete random variable X. In the context of time series analysis, X represents a particular realization of a time series $\{x_t\}$, with support $S(X)$ providing the set of all possible values that X can take. When X can take an infinite number of values, conventional practice is to partition these into a finite sequence (*library*) of symbols. Permutation entropy pursues a data-driven approach by taking partitions based on comparison of neighbouring time series values. Following the notation of Brandt and Pompe (2002), the measure is based on n permutations π of order n, which enumerates the possible orderings of n different numbers. The relative frequency of each permutation is given by

$$p(\pi) = \frac{NUM\{t \leq T - n, \ (x_{t+1}, \ldots, x_{t+n}) \text{ has type } \pi\}}{T - n + 1} \tag{7.4}$$

where NUM means 'the number of' and T is the time series length. Equation (7.4) is a probability density function (PDF) extracted from time series data (Kowalski *et al.*, 2012). It can be computed with the R package `pdc` (Brandmaier, 2015).

The *permutation entropy* measure is a function of this PDF:

$$H(n) = -\sum p(\pi) \log p(\pi) \tag{7.5}$$

where the sum runs over all n permutations π of order n $(n \geq 2)$. In summary, $H(n)$ measures the information in the time series gained from comparing n consecutive values. It lies within the interval $0 \leq H(n) \leq \log n!$, where the upper bound indicates a completely random system and an intermediate value indicates the potential presence of structural dynamics. We shall compute $H(n)$ with the R package `pdc` (Brandmaier, 2015) in our surrogate data testing code.

7.5 Rank Order Statistics

We use nonparametric rank order statistics to test the null hypothesis of linear stochastic dynamics in surrogate data testing (Theiler *et al.*, 1992; Schreiber and Schmitz, 2000). We need to set two parameters. One is α, which sets the probability of false rejection and establishes the level of significance for the test: $(1 - \alpha) * 100$. The other is k, which determines the number of surrogates required for given values of α. A two-tailed hypothesis test requires $S = (2k/\alpha) - 1$ surrogates. We reject the null hypothesis if a discriminating statistic measured from the empirically reconstructed attractor is among the k largest or smallest values in the ensemble of statistics measured from surrogate attractors. Single-tailed tests require $S = (k/\alpha) - 1$ surrogates. We reject the null hypothesis if a discriminating statistic from the empirically reconstructed attractor is among the k largest (for an upper-tailed test) or k smallest (for a lower-tailed test) of those measured from the surrogates. Reducing the probability of false rejection (α) or increasing k enlarges the number of surrogates required, and produces more sensitive tests.

We conduct an upper-tailed test for nonlinear prediction skill so that the null hypothesis is rejected only when an attractor reconstructed from the time series predicts with more skill than its surrogate counterparts. We conduct a lower-tailed test for the permutation entropy measure to reject the null hypothesis only when the measure taken from the time series is significantly less than those taken from the collection of surrogate data vectors. We conduct two-tailed tests for the correlation dimension and maximum Lyapunov exponent discriminating statistics.

7.6 R Code for Surrogate Data Testing

We now consolidate Codes 7.1–7.6 into two 'master' codes. Code 7.7 computes discriminating statistics from an observed time series. Code 7.8 uses surrogate data to test null hypotheses that the data are generated by various types of linear stochastic dynamics.

Kantz and Schreiber (1997) stress the importance of visually inspecting the logarithmic plot of correlation integrals for a *scaling range* in estimating the correlation dimension and the semilogarithmic plot of average separation against time for a robust linear increase in computing the maximum Lyapunov exponent. Without these visual diagnostics, the estimates provided by the algorithms may not be meaningful. Accordingly, we first run Code 7.7 to generate these diagnostic plots. Identification of a scaling range legitimizes the Takens estimate of the correlation dimension. Identification of a robust linear increase provides the information needed to re-run the code to estimate the maximum Lyapunov exponent.

```
#Code 7.7 Discriminating statistics for time series
rm(list=ls(all=TRUE))

#Directories
dir.udf<-as.character("C:/User Defined Functions")
dir.data<-as.character("C:/Data")

#Load user-defined functions
setwd(dir.udf)
dump("embed_udf", file="embed_udf.R")
source("embed_udf.R")
dump("cd_takens_udf", file="cd_takens_udf.R")
source("cd_takens_udf.R")
dump("predict_np_udf", file="predict_np_udf.R")
source("predict_np_udf.R")

library(fields) #distance matrix
library(tseriesChaos) #correlation dimension, Lyapunov exponent
library(pdc)  #permutation entropy
library(scatterplot3d)  #3-D phase space plot
```

```
#Step 1: Load data from an external .csv file
setwd(dir.data)
ts<-read.csv("Mayport.csv")
x<-ts$signal

#Plot time series
x.plot<-plot(x,type="l",main="Time Series",xlab="",ylab="")

#Step 2: Embed time series (x)
results.embed_udf<-embed_udf(x)
d<-results.embed_udf[[1]]
m<-results.embed_udf[[2]]
tw<-results.embed_udf[[3]]  #embedding dimension
Mx<-results.embed_udf[[4]] #embedded data matrix
dist<-rdist(Mx) #Distance matrix, rdist(fields)

#Step 3: Correlation dimension
  #Step 3a: Compute correlation integrals
r_cd<-0.05  #parameters needed for d2(tseriesChaos)
eps.min<-r_cd*max(dist)
m_max<-10
d2.out<-d2(x,m_max,d,tw,eps.min) #correlation integrals, (d2(tseriesChaos)

  #Step 3b:  Compute Takens estimator of correlation dimension for
            #mulitiple embedding dimensions
cd.m<-cd_takens(x,m_max) #user-defined function cd_takens_udf

#Step 4: Maximum Lyapunov exponent
   #Step 4a: Convert data to time series object for lyap_k(tseriesChaos)
n<-length(x)  #length of time series
frequency<-12 #iterations per period (e.g. months/year)
periods<-n/frequency   #periods in data set x (e.g., years)
x.ts<-ts(x,start=c(0,1),end=c(periods,1),frequency)

  #Step 4b: Set radius parameter for lyap_k(tseriesChaos)
r_lyap<-0.2
eps_lyap<-r_lyap*max(dist)  #set eps required by 'lyap_k'

  #Step 4c: Run lyap_k(tseriesChaos)
#s<-20*12  #20 years in stretching factor plot
lyap_out<-lyap_k(x.ts, m, d, s=(20*frequency), tw, ref=10, k=2, eps=eps_lyap)

  #Step 4d: Fit regression line to linear region over time periods 10-20
#LE.max<-lyap(lyap_out,1,3)

#Step 5: Nonlinear prediction skill
results.np<-predict_np_udf(Mx,frac.learn=0.5)
nse.x<-results.np[[1]]

#Step 6: Permutation entropy
entropy<-entropyHeuristic(x,m.min=3,m.max=7,t.min=1,t.max=20)
plot(entropy)
entropy.values<-unlist(entropy$entropy.values)
entropy.value<-min(entropy.values[,3])

#Step 7: Diagnostic plots
par(mfrow=c(2,2))

  #Plot reconstructed attractor
if(m==2){
```

```
PSR<-plot(Mx[,1],Mx[,2],xlab="x(t)",ylab="x(t-d)",main="Phase Space Reconstruction",
type="l")} else {
v=cbind(Mx[,1],Mx[,2],Mx[,3]) #if m>3, do a 3D projection of Mx scatterplot
PSR<-scatterplot3d(v,main="Phase Space Reconstruction: 3D Projection",xlab="x(t)",
  ylab="x(t-d)",zlab="x(t-2d)",box=TRUE,type="l",label.tick.marks=FALSE)}

#Plot correlation integrals (review for scaling range)
plot(d2.out)

#Plot correlation dimensions (cancel default x-axis (xaxt="n") in
  #plot to define custom axis with only integers).
m1<-1:m_max  #revised x-axis
cd.m.plot<-plot(m1,cd.m,xaxt="n",xlab="",ylab="correlation dimension",
  pch=15,type="b")
#Custom x-axis, side=1 is x-axis, "at" defines 'tick-mark' numbers
axis(side=1,at=1:length(cd.m))
#Label custom x-axis, "line=2.5" puts label below tick-mark numbers
mtext(side=1,"embedding dimension",line=2,cex=0.8)
#Label points in plot
cd.m<-round(cd.m,digits=2)
text(m1,cd.m,cd.m,cex=0.7,pos=1,offset=0.4)

#Plot semilogarithmic  average separation
plot(lyap_out)

#Summary of output
cd.m    #correlation dimensions against embedding dimension
LE.max[2] #maximum Lyapunov exponent
nse.x #nonlinear prediction skill (Nash-Sutcliffe model efficiency)
entropy.value
```

After loading and embedding the signal separated from the Mayport sea level time series (Steps 1 and 2), Code 7.7 estimates the discriminating statistics. The logarithmic plot of correlation integrals is provided by the routine d2(tseriesChaos) and the correlation dimension is estimated with the user-defined function cd_takens (Step 3). The semilogarithmic plot of the average separation is provided by the routine lyap_k(tseriesChaos) after the information required by the routine has been input (Steps 4a, b). This includes the frequency parameter, which is set at 12 since we are working with a monthly time series, and the radius parameter r_lyap used to find close neighbouring points. The command LE.max<-lyap(lyap_out,1,3) is commented out on the first run because we have not yet visually identified a scaling range (Step 4d). The user-defined function predict_np generates the Nash–Sutcliffe model efficiency nse, measuring the reconstructed attractor's short-term nonlinear prediction skill (Step 5). Step 6 computes the permutation entropy measure (7.5). Finally, the diagnostic plots are generated (Step 7).

The diagnostic plots for the Mayport signal are shown in Fig. 7.6. We first observe a reconstructed attractor with striking visual regularity (Figure 7.6a). However, we do not obtain reliable estimates of the correlation integral and the maximum Lyapunov exponent (probably because of the short length of the Mayport series), and consequently we do not use them as discriminating statistics in surrogate data analysis. The estimate of the correlation dimension is unreliable because, although the Takens estimates of the correlation dimension reach a plateau at cd=1.07 (Figure 7.6c), the logarithmic plot of the correlation integral does not display an obvious scaling range

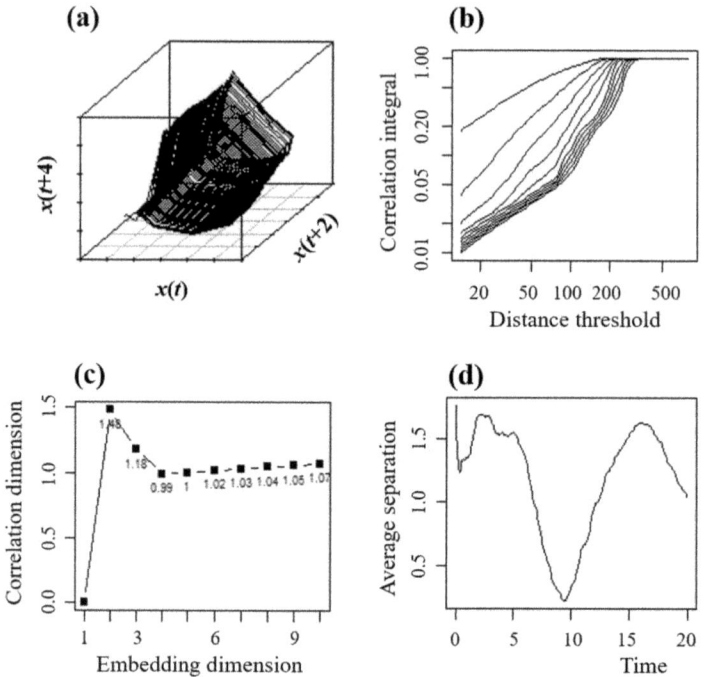

Fig. 7.6 Diagnostic plots for estimating discriminating statistics. (a) The reconstructed phase space attractor exhibits visual regularity. (b) Correlation integrals plotted for multiple embedding dimensions do not exhibit an obvious scaling region. (c) The correlation dimension appears to level off with increasing embedding dimension, but, given the absence of a scaling region, we do not use it as a discriminating statistic in surrogate data analysis; (d) The semilogarithmic plot of the average separation against time does not exhibit a robust linear portion, so we do not use the maximum Lyapunov exponent in surrogate data analysis. Consequently, we select short-term nonlinear prediction skill and permutation entropy as discriminating statistics.

(Figure 7.6b). The estimate of the maximum Lyapunov exponent is unreliable because we do not observe a robust linear scaling range in the semilogarithmic plot of the average separation (Figure 7.6d). As a result, we select short-term nonlinear prediction skill and permutation entropy as discriminating statistics.

Code 7.8 runs follow-up surrogate data testing. In Step 1, we set test parameters. We select the surrogate algorithm by entering **TRUE** for AAFT surrogates generated by the R routine **surrogate(fractal)** (Constantine and Percival, 2015) or **FALSE** for PPS surrogates generated by the user-defined function **PPS_udf**. If we select PPS surrogates, we must also set the random walk parameters (Step 1a). In Step 1b, we set the parameters for rank order testing (k=5 and **alpha**=0.05), which generates 99 surrogates vectors for one-sided tests and 199 vectors for two-sided tests. In Step 1c, we specify which discriminating statistics to use in surrogate data testing based on the visual diagnostics generated in Code 7.7. Entering TRUE uses the statistic in surrogate data testing. Finally, we enter the discriminating statistics measured from the observed time series signal in Code 7.7 (Step 1d).

```
#Code 7.8 Surrogate data testing
rm(list=ls(all=TRUE))  #remove values between runs

#Define working directories
dir.udf<-as.character("C:/User Defined Functions")
dir.data<-as.character("C:/Data")
dir.results<-as.character("C:/Results")

#Load user-defined functions
setwd(dir.udf)
dump("embed_udf", file="embed_udf.R");source("embed_udf.R")
dump("cd_takens_udf", file="cd_takens_udf.R");source("cd_takens_udf.R")
dump("predict_np_udf", file="predict_np_udf.R");source("predict_np_udf.R")
dump("PPS_udf", file="PPS_udf.R");source("PPS_udf.R")
dump("surr_table_udf", file="surr_table_udf.R");source("surr_table_udf.R")

#Load R libraries
library(fractal)  #generate linear stochastic surrogates
library(fields)    #distance matrix
library(tseriesChaos)
library(scatterplot3d)
library(pdc)

#Load Data from an external .csv file
setwd(dir.data)
ts<-read.csv("Mayport.csv")
x<-ts$signal

#Step 1:  Set parameters for surrogate testing

  #Step 1a: Select surrogate algorithm: If TRUE, use linear stochastic
  #surrogates for surrogate(fractal); if FALSE, use PPS surrogates
surr_algorithm<-TRUE
method="aaft" #type of linear stochastic surrogate if TRUE
lb=0.01;ub=0.4  #PPS random walk parameters if FALSE

  #Step 1b: Number of surrogates and significance level for testing
k=5  #used to set number of surrogates for 1- and 2-sided tests
alpha=0.05  #significance level
nsurr_1s=(k/alpha)-1    #surrogates for 1-sided hypothesis test
nsurr_2s<-(2*k/alpha)-1  #surrogates for 2-sided hypothesis test

#Step 1c: Select discriminating statistics
corr.dim<-FALSE #TRUE means using in surrogate testing
lyap<-FALSE
predict<-TRUE
entropy.measure<-TRUE

#Step 1d: Set discriminating statistics computed for time series
cd.x<-NA  #correlation dimension
LE.x<-NA     #maximum Lyapunov exponent
nse.x<-1     #prediction skill
entropy.x<-0.18 #entropy measure

#Step 2:  Generate surrogates
if(surr_algorithm){ #if TRUE generate linear stochastic surrogates
```

```
#linear stochastic surrogates needed for 1-sided test
surr_1s<-matrix(0,length(x),nsurr_1s)
for(i1 in 1:nsurr_1s){
  surr_out<-surrogate(x,method)
  surr<-surr_out[1:length(surr_out)]
  surr_1s[,i1]<-surr
}  #end loop i1

#linear stochastic surrogates needed for 2-sided test
surr_2s<-matrix(0,length(x),nsurr_2s)
for(i2 in 1:nsurr_2s){
  surr_out<-surrogate(x,method)
  surr<-surr_out[1:length(surr_out)]
  surr_2s[,i2]<-surr
}  #end loop i2

} else { #generate PPS surrogates
  surr_1s<-PPS_udf(x,nsurr_1s,lb,ub)  #1-sided test
  surr_2s<-PPS_udf(x,nsurr_2s,lb,ub)  #2-sides test
  }  #end else statement

#Step 3: Two-tailed test for correlation dimension (cd)
  #Step 3a: Compute cd for surrogates
if(corr.dim){
cd.surr<-matrix(0,nsurr_2s,1) #correlation dimensions for each surrogate
for(i3 in 1:nsurr_2s) {
cd.m<-cd_takens(surr_2s[,i3],m_max=5) #user-defined function cd_takens_udf
cd.surr[i3,]<-cd.m[length(cd.m)]
}  #end loop i3

  #Step 3b: Rank order test
cd.low<-sort(cd.surr)[k]   #upper limit on k lowest values
cd.high<-sort(cd.surr)[length(cd.surr)-(k-1)]   #lower limit on k largest values
#print("cd.surr");print(cd.surr)
}  else {cd.x=NA;cd.low=NA;cd.high=NA}

#Step 4. Two-tailed test for maximum Lyapunov exponent (LE.max)
  #Step 4a: Compute LE.max for surrogates
if(lyap){
LE.surr<-matrix(0,nsurr_2s,1) #max Lyapunov exponent for each surrogate
for(i4 in 1:nsurr_2s) {

    #Step 4a(1): Generate embedding parameters for lyap_k(tseriesChaos)
results.embed_udf<-embed_udf(surr_2s[,i4])
d1<-results.embed_udf[[1]]
m1<-results.embed_udf[[2]]
tw1<-results.embed_udf[[3]]   #embedding dimension
Mx1<-results.embed_udf[[4]] #embedded data matrix
dist<-rdist(Mx1) #Distance matrix

    #Step 4a(2): Convert data to time series object for lyap_k(tseriesChaos)
n<-length(x)  #length of time series
frequency<-12 #iterations per period (e.g. months/year)
periods<-n/frequency   #periods in data set x (e.g. years)
x.ts<-ts(surr_2s[,i4],start=c(0,1),end=c(periods,1),frequency)

    #Step 4a(3): Set radius parameter for lyap_k(tseriesChaos)
r_lyap<-0.3
eps_lyap<-r_lyap*max(dist)
```

```
    #Step 4a(4): Run lyap_k(tseriesChaos)
lyap_out<-lyap_k(x.ts, m=m1, d=d1, s=(20*12), t=tw1, ref=10, k=2, eps=eps_lyap)

    #Step 4a(5): Fit regression line to linear region over time periods 10-20
LE.max<-lyap(lyap_out,1,3) #same interval used to compute LE.max for time series
LE.surr[i4,]<-LE.max[2]
}  #end loop i4

  #Step 4b: Rank order test
LE.low<-sort(LE.surr)[k]  #upper limit on k smallest values
LE.high<-sort(LE.surr)[length(LE.surr)-(k-1)]  #lower limit on k largest values
}  else {LE.x=NA;LE.low=NA;LE.high=NA}

#Step 5: Single-tailed test for nonlinear prediction skill (nse)
  #Step 5a: Compute nse for surrogates
if(predict){
nse.surr<-matrix(0,nsurr_1s,1)  #Nash-Sutcliffe efficiency for each surrogate
for(i5 in 1:nsurr_1s) {
  embed<-embed_udf(surr_1s[,i5]) #embed surrogate data vectors
  Mx2<-embed[[4]] #embedded data matrix
  results.np<-predict_np_udf(Mx=Mx2,frac.learn=0.6)
  nse.surr[i5,]<-results.np[[1]] #nse
}  #end loop i5

  #Step 5b: Rank-order test
nse.low<-NA
nse.high<-sort(nse.surr)[length(nse.surr)-(k-1)]  #lower limit on k largest values
}  else {nse.x=NA;nse.low=NA;nse.high=NA}

#Step 6: Lower-tailed test for entropy measure
if(entropy.measure){
  #Step 6a: Compute entropy measure for surrogates
entropy.surr<-matrix(0,nsurr_1s,1)  #Nash-Sutcliffe efficiency for each surrogate
for(i6 in 1:nsurr_1s) {
  entropy<-entropyHeuristic(surr_1s[,i6],m.min=3,m.max=7,t.min=1,t.max=20)
  entropy.val<-unlist(entropy$entropy.values)
  entropy.min<-min(entropy.val[,3])
  #print("entropy.min");print(entropy.min)
  entropy.surr[i6,]<-entropy.min #entropy value
}  #end loop i6

  #Step 6b: Rank order test
entropy.low<-sort(entropy.surr)[k]  #upper limit on k lowest values
entropy.high<-NA
print("entropy.surr");print(entropy.surr)
}  else {entropy.x=NA;entropy.low=NA;entropy.high=NA}

####################################
#Export table of surrogate results#
####################################
setwd(dir.results)
surr_results(corr.dim=corr.dim,lyap=lyap,predict=predict,entropy.measure=entropy.measure,
 k=k,alpha=alpha,cd.x=cd.x,cd.low=cd.low,cd.high=cd.high,LE.x=LE.x,LE.low=LE.low,
 LE.high=LE.high,nse.x=nse.x,nse.low=nse.low,nse.high=nse.high,entropy.value=entropy.value,
 entropy.low=entropy.low,entropy.high=entropy.high)

###############################
#Error messages and corrections#
###############################
#(1) Set m_max lower if correlation dimension results have NA.
```

```
#(2) If 'r_lyap' is set too low, so that the radius for finding 'nearest neighbours' is
     #too small, R may return the following error messages: "Error in lyap_k(x.em.vec,
     #m = m.vec, d = d.vec, t = theiler(x), k = 1,  : not enough neighbours found" or
     #"Error: cannot allocate memory block of size 68719476736.0 Gb". A correction is
     #to increase 'r_lyap'.  If 'r_lyap' is set too high, the correlation dimension may
     #be underestimated.

#(3) PPS Surrogates:  If 'lb' and 'ub' are set too close together, so that the
     #neighbourhood for the random walk is too small, R may return the following error
     #message: "Error in sample(int, 1) : invalid first argument". A correction is to
     #increase the distance between 'lb' and 'ub'.

#(4) "Error: NA/NaN/Inf in foreign function call (arg 1)" This comes up if some
     #columns in data matrix are longer than others. R wants them the same length,
     #or it will fill in NAs in the shorter columns.

##########################################
#Hypothesis Result: Correlation Dimension#
##########################################
if(corr.dim){
if(cd.x<=cd.low|cd.x>=cd.high)
    {print("Reject Ho: Data not generated by linear stochastic process")} else
    {print("Accept Ho: Data generated by linear stochastic process")}
} #end if corr.dim

#############################################
#Hypothesis Result: Maximum Lyapunov Exponent#
#############################################
if(lyap){
if(LE.x<=LE.low|LE.x>=LE.high)
    {print("Reject Ho: Data not generated by linear stochastic process")} else
    {print("Accept Ho: Data generated by linear stochastic process")}
}  #end if lyap

###############################################
#Hypothesis Result: Nonlinear Prediction Skill#
###############################################
if(predict){
if(nse.x>=nse.high)
    {print("Reject Ho: Data not generated by linear stochastic process")} else
    {print("Accept Ho: Data generated by linear stochastic process")}
}  #end if predict

###################################
#Hypothesis Result: Entropy Measure#
###################################
if(entropy.measure){
if(entropy.x<=entropy.low)
    {print("Reject Ho: Data not generated by linear stochastic process")} else
    {print("Accept Ho: Data generated by linear stochastic process")}
}  #end if entropy.measure
```

Step 2 of Code 7.8 generates the number of surrogates required for one- and two-sided tests. Steps 3–6 estimate the discriminating statistics for each surrogate data vector using commands outlined in Code 7.7. If we use the maximum Lyapunov exponent, then we need to set the **frequency** and **r_lyap** parameters in Steps 4a(2) and 4a(3). Steps 3b and 4b specify two-tailed rank-order tests when correlation dimension and maximum Lyapunov exponent, respectively, are used as discriminating statistics.

The null hypothesis of stochastic dynamics is rejected if the correlation dimension and maximum Lyapunov exponent computed from the time series fall within the k smallest or k largest values. The code computes the upper limit on the k smallest and the lower limit on the k largest values. Consequently, the null hypothesis is rejected if values computed from the time series fall outside the interval given by these limits. Step 5b specifies the upper-tailed test when nonlinear prediction skill is used as a discriminating statistic. The code computes the lower limit on k largest values, and rejects the null hypothesis if the statistic computed for the time series rests above this limit. Step 6 computes the permutation entropy measure, and rejects the null hypothesis if the measure taken from the time series falls below the upper limit on the k smallest values taken from the surrogates (Step 6b). The code exports a table summarizing results (`surrogate.results.csv`). A user-defined function `surr_table_udf` conforms table labels to the particular discriminating statistics tested. Finally, the code specifies possible error messages and corrections, and posts statements at the bottom of the run summarizing the results of hypothesis testing for each discriminating statistic.

The summary tables for the Mayport signal tested with AAFT and PPS surrogates are shown in Figure 7.7a and 7.7b, respectively. We reject the null hypothesis that the signal is generated by a linear stochastic process for all discriminating statistics tested. The alternative hypothesis that the data are generated by nonlinear dynamics remains viable.

(a)

AAFT surrogates	Time Series	Surr (low)	Surr (high)	H0
Predictive Skill	1	NA	0.95	reject
Entropy	0.18	0.70	NA	reject
k	5			
Significance level	0.05			

(b)

PPS surrogates	Time Series	Surr (low)	Surr (high)	H0
Predictive Skill	1	NA	0.95	reject
Entropy	0.18	0.98	NA	reject
k	5			
Significance level	0.05			

Fig. 7.7 Results tables from Code 7.8 for (a) AAFT surrogates and (b) PPS surrogates. For both surrogate ensembles, we reject the null hypothesis that the Mayport signal is generated by linear stochastic dynamics.

7.7 Summary

We posed the following question in the introduction to this chapter: How can we demonstrate correspondence between our model and the real world? The answer is that we should 'get to know our data' before attempting to model the real-world system generating it. This is challenging because the random appearance of volatile data potentially conceals nonlinear structural variation. Nonlinear time series analysis allows us to distinguish between stochastic and deterministic drivers of observed complexity with an empirical process of elimination. We can preprocess data to begin eliminating noise and nonstationary dynamics as drivers of complexity. We can further eliminate broad categories of linear stochastic dynamics by reconstructing a low-dimensional attractor from a stationary signal and supporting this preliminary visual evidence of deterministic dynamics with surrogate data testing. If we statistically reject broad categories of linear stochastic drivers of observed complexity, then nonlinear dynamics remain a viable deterministic modelling alternative.

If we can make a compelling case that observed data are generated by deterministic nonlinear dynamics, then the door is open to further nonlinear data diagnostic techniques and to the reproduction of empirically reconstructed dynamics with phenomenological (data-driven) modelling techniques. We investigate these techniques in subsequent chapters.

8
Empirically Detecting Causality

'There should be no combination of events for which the wit of
man cannot conceive an explanation.'
                    Arthur Conan Doyle, The Valley of Fear

8.1   Introduction

Detecting causal interactions among climatic, environmental, and human forces in
complex biophysical systems is essential for understanding how these systems function
and how public policies can be devised that protect the flow of essential services to
biological diversity, agriculture and other core economic activities. Data-driven causal
detection methods infer interactions from time series data on hypothesized system
variables. Correlation (a common method of relating variables) does not imply causation. The Latin version of this proposition is *cum hoc non propter hoc* ('with this,
not because of this'). A classic example is that a negative correlation between average global temperature and number of pirates does not indicate that we can cool the
planet with more pirates.

Selecting an effective empirical causal-detection method depends on whether observed time series variables exhibit continuous or irregular-sporadic time records, and
whether continuous records are generated by deterministic or stochastic system dynamics. In this chapter, we present the convergent cross mapping (CCM) method,
which tests for causal interaction between variables that have continuous records and
are generated by deterministic, low-dimensional and nonlinear dynamics. We also
present the SST change-point detection method (based on singular spectrum analysis), which tests whether a continuous record exhibiting a strong signal undergoes
abrupt structural changes in dynamic behaviour corresponding to the irregular timing
of variables exhibiting only occasional activity. Finally, we investigate the probabilistic
tipping-point detection method (based on histograms constructed from the observed
data) for use with highly noisy continuous records. Armed with these methods, we
can detect causal interactions when confronted with a broad array of heterogeneous
historic records.

We consider first how selection of an empirical causal detection method depends on
a diagnosis of system dynamics generating observed time series data. *Granger causality*
is designed for use in stochastic systems (Granger, 1969). Variable y *Granger-causes*
x if the predictability of x decreases when y is removed from the set of possible causal

Nonlinear Time Series Analysis with R. Ray Huffaker, Marco Bittelli and Rodolfo Rosa, Oxford University Press (2017).
© Ray Huffaker, Marco Bittelli and Rodolfo Rosa. DOI: 10.1093/oso/9780198782933.001.0001

factors. However, Sugihara *et al.* (2012) demonstrated that Granger causality does not apply when system dynamics are deterministic. Deterministic interactions encode information about y into x, and this information does not disappear from x when y is removed from the system. As noted by the famous naturalist John Muir (1911), 'When we try to pick something up by itself, we find it hitched to everything else in the universe.'

Sugihara *et al.* (2012) developed *convergent cross mapping (CCM)* to detect causal networks in real-world systems diagnosed with deterministic, low-dimensional and non-linear dynamics. CCM tests whether there is any correspondence between phase spaces reconstructed from observed time series variables. If x and y interact in the same dynamical system, then attractors reconstructed with delayed x coordinates (M_x) or delayed y coordinates (M_y) map one-to-one to the original system attractor (M), and consequently map one-to-one to each other (Figure 7.1). Variable y is found to causally drive x ($y \Rightarrow x$) if M_x can be used to skillfully cross-map points on M_y (x xmap y).

We illustrate this with Lorenz attractors reconstructed from the y and x solutions in Figure 8.1. We first select a reference time t_{ref} and identify the corresponding points on each attractor, namely $M_x(t_{\text{ref}})$ and $M_y(t_{\text{ref}})$ (Figure 8.1a, b). We next identify the time indices corresponding to two (for example) nearest neighbouring points to $M_x(t_{\text{ref}})$, namely $M_x(t_a)$ and $M_x(t_b)$ (Figure 8.1a). If there is a one-to-one mapping between M_x and M_y, then these time indices also will identify the nearest neighbouring points to $M_y(t_{\text{ref}})$, namely $M_y(t_a)$ and $M_y(t_b)$, and y is determined to drive x (Figure 8.1b). CCM measures how well attractors cross-map with a nonlinear prediction algorithm. The $m + 1$ nearest neighbouring points identified on M_y are advanced one period, and averaged to predict the reference value $M_y(t_{\text{ref}})$, where $m = 3$ is the embedding dimension for the Lorenz attractor in this illustration. The future value of each nearest neighbouring point is weighted in the average by its distance from $M_y(t_{\text{ref}})$ relative to the total distances of all nearest neighbours from $M_y(t_{\text{ref}})$ (see Sugihara *et al.* (2012), supplementary materials).

The *convergence requirement* of CCM is that cross-mapped predictions approach the reference point $M_y(t_{\text{ref}})$ as the portion of the times series used to reconstruct the attractor M_x (i.e. the *library*) increases in size. The reasoning is that if M_x is indeed an attractor, then M_x will increasingly *fill out* when reconstructed from larger libraries. Trajectories on M_x, and thus its neighbouring points, come closer together, and cross-prediction should become more skillful (Figure 8.1c, d). Predictive skill can be measured with the Pearson correlation coefficient. Consequently, a plot of correlation coefficients against library size should converge to a relatively large positive fraction to indicate a strong causal relationship.

8.2   Convergent Cross Mapping with R

Code 8.1 formulates a user-defined function `ccm_udf` to do two jobs. The first job is to run pairwise CCM on a collection of time series vectors using the R package `multispatialCCM` (Clark, 2015). The second job is to prepare an *adjacency* matrix used to summarize CCM results in a *network diagram* with arrows connecting causally interactive variables. For this illustration, our data are provided by `model.1`, which

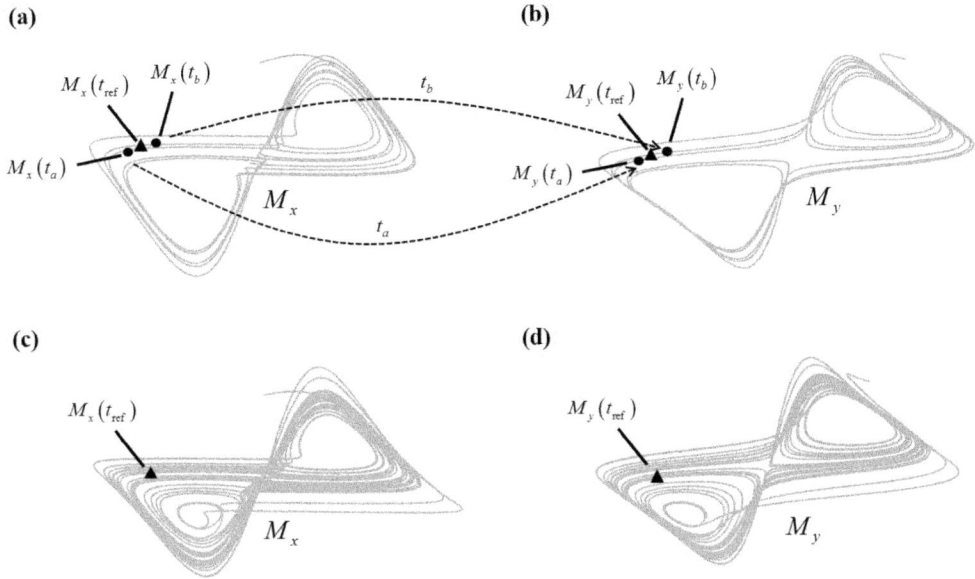

Fig. 8.1 Illustration of how CCM works when using a variable x and its lags to cross-map a variable y (x xmap y). (a) We select a reference point ($M_x(t_{\text{ref}})$) on the attractor reconstructed from x and its lags, and identify (for this illustration) two nearest neighbouring points: $M_x(t_a)$ and $M_x(t_b)$. (b) The time indices for these points (i.e. t_{ref}, t_a and t_b) are transferred from M_x to the attractor reconstructed from y and its lags, M_y. If y causally drives x, then $M_y(t_a)$ and $M_y(t_b)$ also will be nearest neighbours to $M_y(t_{\text{ref}})$. (c, d) Convergence requires that cross-mapped predictions approach the observed reference point $M_y(t_{\text{ref}})$ as the portion of the time series used to reconstruct M_x increases. The reason is that cross prediction should be more skillful as trajectories on the attractor, and thus neighbouring points, come closer together as the attractor fills out.

generates two time series vectors v and w, with coupled logistic difference equations found in Ye *et al.* (2015):

$$v(t+1) = v(t)[3.78 - 3.78v(t) - 0.07w(t)]$$
$$w(t+1) = w(t)[3.77 - 3.77w(t) - 0.08v(t)]$$

(8.1)

We run this model with the commands `model.1<-TRUE` and `model.2<-FALSE`, and formulate a user-defined function `system` to iteratively solve it for initial values v_0 and w_0, number of iterations $nstep = 1000$, and parameters a, b, c, d.

```
#Code 8.1 User-defined function for convergent cross mapping (ccm_udf) (logistic model example)
rm(list=ls(all=TRUE))

#Define working directories
```

```
dir.udf<-as.character("C:/User Defined Functions")
dir.results<-as.character("C:/Results")

#Load user-defined functions
setwd(dir.udf)
dump("embed_udf", file="embed_udf.R")
source("embed_udf.R")

#Load libraries
library(tseriesChaos)
library(multispatialCCM)

#Select model to run
model.1<-TRUE     #TRUE means run model.1
model.2<-FALSE    #TRUE means run model.2

#Solve ODE system to generate data
system<-function(v0,w0,nstep,a,b,c,d){
vt<-numeric();wt<-numeric()
v<-v0;w<-w0
vt[1]<-v;wt[1]<-w
for(i in 2:nstep){
 if(model.1){v1<-v*(a-(a*v)-(b*w)) #System (1)
   w1<-w*(c-(c*w)-(d*v))}
 if(model.2){v1<-v*(a-(a*v)) #System (2)
   w1<-w*(c-(c*w)-(d*v))}
 v<-v1;w<-w1
 vt[i]<-v;wt[i]<-w
 } #end loop i
solutions<-cbind(vt,wt)
return(solutions)
} #end user-defined function

if(model.1){solutions<-system(0.2,0.4,1000,3.78,0.07,3.77,0.08)} #run System (1)
if(model.2){results<-solutions(0.2,0.4,1000,3.8,0,3.1,0.8)} #run System (2)
v<-solutions[,1];w<-solutions[,2]
data<-cbind(v,w)
names<-colnames(data)

#Parameter required for ccm_udf
#'frac' is a parameter used in ccm_udf to compute the 'weighted
#adjacency matrix'. (1-frac) sets the fraction of points used to calculate
#the asymptote of the ccm curve (plotting the correlation coefficient
#against the library of points on the attractor used to cross-predict
#the other variables.
frac<-0.66

#Formulate user-defined function 'ccm_udf' to run ccm

ccm_udf <-function(data,frac){

#Loop j specifies each variable in turn as the response process (x),
  #and loop k iterates over the remaining variables as forcing (driving)
  #processes (y)

wam<-matrix(0,ncol(data),ncol(data)) #weighted adjacency matrix
for(j in 1:ncol(data)){  #response process x
  print("j");print(j)
  x<-data[,j]
```

```
#Step 1: Set parameters required by CCM_boot(multispatialCCM)
#results.embed<-embed_udf(x)  #In general, we use this to embed response process x
E<-2  #E<-results.embed[[2]]
tau<-1  #tau<-results.embed[[1]]
lib<-((E*(tau-1)+(E+1)):length(x)-E+2)  #ccm library
lib.size<-length(lib) ;print("lib.size");print(lib.size)

#ccm correlations where x is the response variable and columns are driving variables (y)
corr.mat.x<-matrix(0,lib.size,ncol(data))
for(k in 1:ncol(data)){  #forcing process y, response process x
  if(j==k) next  #don't cross-map to same attractor
  print("k");print(k)
  y<-data[,k]

#Step 2: Run CCM_boot(multispatialCCM)
  ccm.yx<-CCM_boot(y,x,E,tau,DesiredL=lib,iterations=100)
  corr.yx<-ccm.yx$rho  #ccm correlation coefficients over libraries
  length.corryx<-length(corr.yx)
  print("corr.yx");print(corr.yx)

  #Step 2a: Export corr.yx plots for x xmap y
  setwd(dir.results)
  plot(corr.yx,type='l',xlab="library",ylab="rho",ylim=c(0,1),
    lwd=3,cex.axis=1.1,font=2,font.lab=2,cex.lab=1.1)
  file<-as.character("")
  names.var<-as.matrix(names)
  file<- as.character(paste(substr(file,1,nchar(file)-4),
    names.var[j],"_xmap_",names.var[k],".png", sep=""))
  dev.copy(png,filename=paste(file,".png",sep=""))
  dev.off()

#Step 3: Store corr.yx as columns in 'corr.mat.x'. Each corr.yx may
  #have a different library length, since each response series can have
  #different embedding dimensions and delays. So, we might have to store
  #vectors of different lengths.

  #Set length of corr.yx at least as long as the library length
    #associated with jth response variable (lib.size)
  length(corr.yx)<-lib.size

  #The storage matrix will contain "NA" if length(corr.yx)<lib.size
  corr.mat.x[,k]<-corr.yx  #columns are ccm coefficients
  corr.x<-corr.mat.x[1:length.corryx,]  #remove "NA"'s from corr.mat.x
  } #end loop k

  #Step 3a: Export 'corr.x' matrix for each response variable (x)
file<-as.character("")
#file<-as.character("ccm.csv")
names.var<-as.matrix(colnames(data))
file<- as.character(paste(substr(file,1,nchar(file)-4),
      names.var[j],"_as response variable",".csv", sep=""))
write.table(corr.x,file,col.names=colnames(data),
          row.names=FALSE,sep=",")

#write.table(corr.mat.x,file,col.names=colnames(data),
#           row.names=FALSE,sep=",")

#Step 4: Prepare results to be portrayed in network diagram
  #Convert corr.mat.x matrices into weighted adjacency matrix 'wam'.
n<-frac*nrow(corr.x);n<-round(n)
```

```
corr.x.1<-corr.x[(n+1):nrow(corr.x),]
corr.ave<-colMeans(corr.x.1)
wam[j,]<-corr.ave  #weighted adjacency matrix (driving variables along column)
wam<-na.omit(wam)
}  #end loop j

   #Step 4a: Export 'wam' matrix after inserting row and column labels
names.var<-as.matrix(colnames(data))
row.names <- c(names.var)
col.names<-c("",names.var)
matrix1<-cbind(row.names,wam)
matrix2<-rbind(col.names,matrix1)
write.table(matrix2,"wam.csv",col.names=FALSE,row.names=FALSE,sep=",")

results<-list(wam)  #wam is output of user-defined function
return(results)

}  #end user-defined function

#Step 5: Run ccm_udf
results_ccm<-ccm_udf(data,frac)
wam<-results_ccm[[1]] #weighted adjacency matrix

#Step 6:  Convert the 'wam' matrix into a scaled adjacency matrix 'adjacency'
  #by assigning a scale of integers to the approximated asymptotes in 'wam'. For
  #example, asymptotes greater than 0.65 are assigned 1, with 0 otherwise.
mat.1<-wam
for(row in 1:nrow(wam)){#row
  for(col in 1:nrow(wam)){#column
      if(mat.1[row,col]>=0.65){mat.1[row,col]=1}
      if(mat.1[row,col]<0.65){mat.1[row,col]=0}
  } #end col
} #end row
#Reading file: Driving variable listed in column.
adjacency<-mat.1

  #Step 6a: Export 'adjacency' matrix after inserting row and column labels
row.names <- c(names)
col.names<-c("",names)
mat1<-cbind(row.names,adjacency)
mat2<-rbind(col.names,mat1)
write.table(mat2,"adjacency.csv",col.names=FALSE,row.names=FALSE,sep=",")
```

Code 8.1 next formulates `ccm_udf`, requiring input of a matrix `data`, whose columns are the time series to be tested for pairwise causal interactions, and a parameter (`frac`). In this illustration, the first and second columns of `data` are the v and w solutions of `model.1`, which we select with the commands `model.1<-TRUE` and `model.2<-FALSE`. The function `ccm_udf` runs through two loops. The outer loop j fixes each time series in turn as the response series x, and the inner loop k then iterates over each time series taken as the forcing series y (i.e. x xmap y). Step 1 provides the input parameters required by the R package `CCM_boot` (`multispatialCCM`). These include the embedding parameters required to reconstruct an attractor from the response series x (the R package denotes these as embedding dimension E and embedding delay tau) and a sequence of feasible library lengths (`lib`). In general, `ccm_udf` uses the user-defined embedding function `embed_udf` to estimate embedding parameters. However, for this illustration, we follow Ye *et al.* (2015) in setting $E = 2$ and $tau = 1$. When we apply `ccm_udf` to other data, we must remove this restriction. We set the parameter `lib` to run from the minimum to the maximum feasible lengths given in Clark (2015). We compute the library size `lib.size` as the number of libraries of different lengths included in `lib`. For example, if `lib` runs from 1 to 5, then `lib.size` = 5.

In Step 2, loop k runs `CCM_boot(multispatialCCM)` for each forcing series, except that it jumps the loop when the response and forcing series are the same with the command `if(j==k) next`. (A response series is trivially causally interactive with itself.) The vector `corr.yx` contains the correlation coefficients for a cross-mapping x xmap y. Step 2a exports plots of each cross mapping. In this illustration, the plot for which v is the response variable, for example, is named `v_xmap_w.png`. The cross mappings in both directions, v xmap w (Figure 8.2a) and w xmap v (Figure 8.2b), converge to relatively high correlation coefficients, demonstrating that CCM successfully detects the bi-causal interactions built into `model.1`.

Step 3 stores each `corr.yx` vector as a column in the output matrix `corr.mat.x`. Each `corr.yx` vector measures the extent to which a variable selected as the forcing series, y, drives the variable fixed as the response series, x. A complicating computing issue arising in general applications is that the length of each `corr.yx` vector may differ, making them more difficult to store in the same output matrix. The reason for this is that the length of each `corr.yx` vector depends on the number of CCM libraries computed in Step 1, which in turn depends on the embedding parameters, which may differ across the corresponding response variables. We circumvent this by fixing the row dimension of the storage matrix `corr.mat.x` for loop k to the library size (`lib.size`) computed for the corresponding response variable in loop j. There are NA elements added to the end of a `corr.yx` vector whose length is less than this library size. This occurs when `CCM_boot(multispatialCCM)` computes correlation coefficients for fewer libraries than (`lib.size`). In this case, we can remove the NA elements by adjusting the row length of the storage matrix `corr.mat.x` to the original length of the `corr.yx` vectors. We call this new matrix `corr.x`. Step 3a exports `corr.x` for each response series. For clarity, we rename, for example, the matrix for which v is the response variable: `v_as response variable`:

```
v   w
0   0.025397823
0   0.024285006
0   0.03187018
0   0.02467108
0   0.042724333
.   .

.   .
0   0.740892847
0   0.74243995
0   0.742783186
0   0.741667912
0   0.74375636
```

The first column contains zeros because we do not cross-map the response variable with itself.

Step 4 prepares the CCM results to be portrayed in a network diagram. First, we convert the `corr.mat.x` matrices into a *weighted adjacency matrix*. Causal interaction

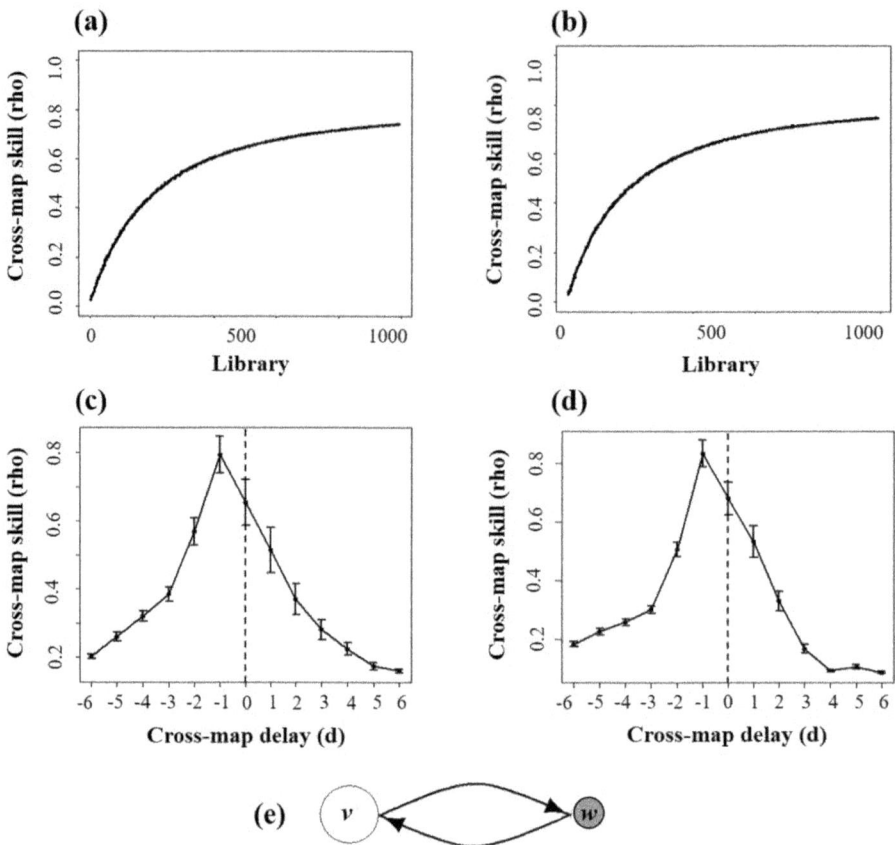

Fig. 8.2 Convergent cross mappings for model.1. (a) The cross mapping v xmap w correctly detects that w drives v in the first equation. (b) The cross mapping w xmap v correctly detects that v drives w in the second equation. (c) The delayed cross mapping v xmap w confirms the result in (a), since the maximum average correlation coefficient occurs at a negative delay $(= -1)$. (d) The delayed cross mapping w xmap v confirms the result in (b), since the maximum average correlation coefficient occurs at a negative delay $(= -1)$. (e) The network plot portrays the bi-causal interaction between v and w.

is indicated if a cross mapping generates correlation coefficients converging along a plateau to a high value. We approximate a convergence plateau by selecting the final 1-frac points along the curve, where frac is a parameter required by ccm_udf. We average the convergence points for each cross mapping and store them in the wam matrix exported in Step 4a. The time series serving as response variables are listed along the rows of wam and those taking their turn as forcing variables along the columns. For this illustration, the wam matrix is

```
     v     w
v   0.00  0.72
w   0.73  0.00
```

There are zeros along the main diagonal because we do not cross-map variables with themselves. The nonzero element in the first row (0.72) is the estimated convergent correlation coefficient for v xmap w (testing whether w forces v), and that in the second row (0.73) corresponds to w xmap v (testing whether v forces w).

Step 5 runs the user-defined function for CCM (`ccm_udf`).

Finally, Step 6 converts the generated `wam` matrix into an adjacency matrix amenable to network plotting. We assign a scale of integers to each element in the `wam` matrix to indicate its strength in indicating a causal interaction. In this illustration, we assign elements a '1' if they are greater than or equal to 0.65, and a '0' otherwise; consequently, each nonzero element of the `wam` in our illustration receives a '1' in the adjacency matrix:

```
  v  w
v 0  1
w 1  0
```

In network plots (to be discussed in Section 8.4), the variables appear as nodes in a diagram and causal interactions as arrows between interactive variables. In plotting the `adjacency` matrix in our illustration, for example, a bi-directional arrow will appear between v and w.

8.2.1 False Positives

Sugihara *et al.* (2012) noted that CCM could give false positives in the case of 'generalized synchrony'. False positives occur when CCM incorrectly detects a unidirectional causal relationship between two variables that does not exist. Generalized synchrony occurs between two variables when, for example, x strongly forces y, but y does not interact with x in the other direction. Consequently, system dynamics are synchronized to x, and CCM may misconstrue this as y forcing x (a false positive). Ye *et al.* (2015) illustrate generalized synchrony with `model.2`:

$$v(t+1) = v(t)[3.8 - 3.8v(t)]$$
$$w(t+1) = w(t)[3.1 - 3.1w(t) - 0.8v(t)]$$
(8.2)

There is strong forcing of $v(t)$ on $w(t+1)$, with a delay of one period, but w does not interact with v in the first equation of (8.2). We use Code 8.1 to run CCM on this system with the commands `model.1<-FALSE` and `model.2<-TRUE`. CCM generates the following adjacency matrix:

```
  v  w
v 0  1
w 1  0
```

CCM falsely detects that w drives v in the first equation of (8.2) (we know this from the 1 in the upper right corner of the matrix). The plot of the cross mapping v xmap w also displays this false positive, since prediction skill levels off at a high value (Figure 8.3a). Alternatively, CCM correctly detects that v drives w in the second equation (Figure 8.3b).

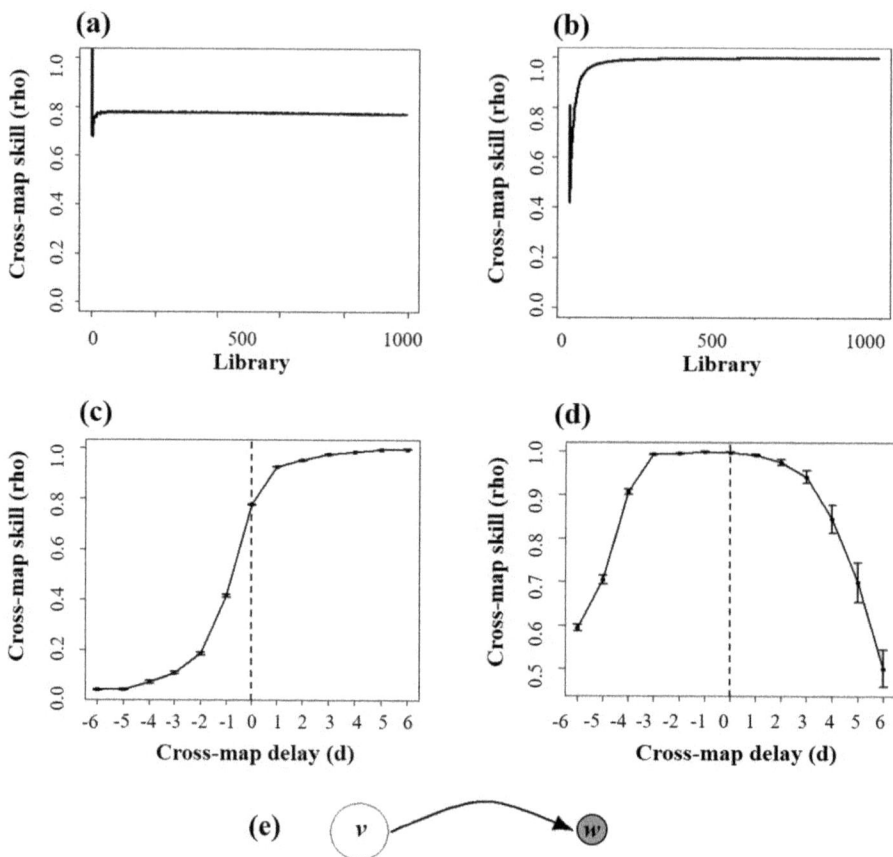

Fig. 8.3 Convergent cross mappings for model.2. (a) The cross mapping v xmap w falsely detects that w drives v in the first equation of (8.2). (b) The cross mapping w xmap v. correctly detects that v drives w in the second equation. (c) The delayed cross mapping v xmap w confirms that the result in (a) is a false positive, since the maximum average correlation coefficient occurs at a positive delay. (d) The delayed cross mapping w xmap v confirms the result in (b), since the maximum average correlation coefficient occurs at a negative delay. (e) The network plot shows that v forces w.

8.3   Extended (Delayed) Cross Convergent Mapping

Ye *et al.* (2015) formulated *extended (delayed) CCM* to provide a further causality test to screen for false positives in the face of generalized synchrony . Delayed CCM cross-maps over a spectrum of negative and positive delayed responses between driving and response variables to determine the delay for which CCM performs best. The method detects a negative delay in the true causal direction. For example, when applied to model.2, we will see that w xmap v performs best for the one-period negative delayed response formulated between the response variable w and the driving variable v in the second equation. The response variable best predicts past values of the driving variable. Viewing this interaction in the opposite direction, we will see that v xmap w performs best for a forward delayed response. The driving variable v best predicts future values of the response variable w. Consider a simple example: person A angers B, but B has a 'slow fuse' and does not respond for two days. To detect the true causal interaction in this relationship (i.e. how A's behaviour causes B's anger) requires that we go two days into the past. Since B does not cause A's behaviour, going into the past does not help us understand A. Detecting A's impact on the relationship requires that we go into the future.

Code 8.2 formulates a user-defined function `ccm_delay_udf` to run delayed CCM following Ye *et al.* (2015). Code 8.2 uses these results to revise the adjacency matrix computed for CCM to remove false positives. For data, we first select `model.1` and solve for time series v and w. We next set the parameters required by `ccm_delay_udf`. These include the embedding parameters required to reconstruct an attractor from a response variable, the desired number of random libraries to be used in CCM (*n.lib*) and the spectrum of delays over which to run CCM, in particular, the numbers of backward delays (*num.bd*) and forward delays (*num.fd*). The user-defined function `ccm_delay_udf` also requires input of a forcing variable A and a response variable B when we eventually run it. This performs the cross mapping: B xmap A.

```
#Code 8.2  User-defined function for delayed CCM (logistic model example)
rm(list=ls(all=TRUE))

#Define working directories
dir.udf<-as.character("C:/User Defined Functions")
dir.results<-as.character("C:/Results")

#Load user-defined functions
setwd(dir.udf)
dump("embed_udf", file="embed_udf.R")
source("embed_udf.R")

library(multispatialCCM)
library(tseriesChaos)
library(plotrix)

#Select model to run
model.1<-TRUE    #TRUE means run model.1
model.2<-FALSE   #TRUE means run model.2

#Solve ODE system to generate data
system<-function(v0,w0,nstep,a,b,c,d){
vt<-numeric();wt<-numeric()
v<-v0;w<-w0
vt[1]<-v;wt[1]<-w
for(i in 2:nstep){
 if(model.1){v1<-v*(a-(a*v)-(b*w)) #System (1)
   w1<-w*(c-(c*w)-(d*v))}
 if(model.2){v1<-v*(a-(a*v)) #System (2)
   w1<-w*(c-(c*w)-(d*v))}
 v<-v1;w<-w1
 vt[i]<-v;wt[i]<-w
 } #end loop i
solutions<-cbind(vt,wt)
return(solutions)
```

```
} #end user-defined function

if(model.1){solutions<-system(0.2,0.4,1000,3.78,0.07,3.77,0.08)} #run System (1)
if(model.2){solutions<-system(0.2,0.4,1000,3.8,0,3.1,0.8)} #run System (2)
v<-solutions[,1];w<-solutions[,2]
data<-cbind(v,w)
n.vars<-ncol(data)  #number of time series vectors
names<-colnames(data)

#Set parameters
E<-2  #embedding dimension
tau<-1  #embedding delay
n.lib<-20  #number of random libraries
num.bd<-6  #number of backward delays
num.fd<-6  #number of forward delays
delay.vec<-c(-(num.bd:1),0,(1:num.fd))  #vector of delays
delay_length<-length(delay.vec)
alpha<-0.05  #parameter used to update adjacency matrix

#Formulate user-defined function 'ccm_delay_udf' to run delayed ccm

ccm.delay<-function(A,B,E,tau,n.lib,num.bd,num.fd){

corr_0<-matrix(0,n.lib,1)  #store results from ccc without delay
corr_b_mat<-matrix(0,num.bd,n.lib) #store results with backward delays
corr_f_mat<-matrix(0,num.fd,n.lib) #store results with forward delays

for(i in 1:n.lib) {  #loop through randomly selected libraries

  #Step 1: Randomly select library length
  lib.min<-tau*(E-1)+(E+1) #minimum possible library size
  lib.max<-length(x)-E+2   #maximum possible library size
  lib.adj<-(lib.max-lib.min)/4 #avoid overly small libraries
  set.seed<-500
  lib<-as.integer(runif(1,(lib.min+lib.adj),lib.max))

  #Step 2: ccm without delay
  ccm.0<-CCM_boot(A,B,E,tau,DesiredL=lib,iterations=100)
  corr.0<-ccm.0$rho  #ccm correlation coefficient for each library
  #print("corr.0");print(corr.0)

  #Step 3: ccm with backward delays
  corr_b_vec<-matrix(0,num.bd,1)
  for(j in 1:num.bd){
    #print("j");print(j)
    embed<-embedd(B,2,j)  #put backward delay into response variable
    B.b<-embed[,2]  #response variable
    A.b<-A[1:length(B.b)]  #driving variable
    ccm.b<-CCM_boot(A.b,B.b,E,tau,DesiredL=lib,iterations=100)
    corr.b<-ccm.b$rho  #ccm correlation coefficients for each library
    #print("corr.b");print(corr.b)
    corr_b_vec[j,]<-corr.b
  } #end loop j

  #Step 4:  ccm with forward delays
  corr_f_vec<-matrix(0,num.fd,1)
  for(k in 1:num.fd){
    #print("k");print(k)
    embed<-embedd(A,2,k) #put forward delay into driving variable
    A.f<-embed[,2]
```

```
    B.f<-B[1:length(A.f)]
    ccm.f<-CCM_boot(A.f,B.f,E,tau,DesiredL=lib,iterations=100)
    corr.f<-ccm.f$rho  #ccm correlation coefficients for each library
    #print("corr.f");print(corr.f)
    corr_f_vec[k,]<-corr.f
    #print("corr_f_vec");print(corr_f_vec)
  }  #end loop k

  #Step 5: Store results over all libraries
    #Correlation coefficients without delay
  corr_0[i,]<-corr.0

    #Matrix of correlation coefficients across libraries (columns)
      #for backward delays (rows)
  corr_b_mat[,i]<-corr_b_vec  #corr coeff for backward delays

    #Matrix of correlation coefficients across libraries (columns)
      #for forward delays (rows)
  corr_f_mat[,i]<-corr_f_vec  #corr coeff for forward delays

} #end loop i

#Step 6: Average correlation coefficients across all libraries
  #No delay
corr.ave.0<-mean(corr_0)
corr.sd.0<-sd(corr_0)  #standard deviations
lb0<-corr.ave.0-corr.sd.0 #lower confidence bounds
ub0<-corr.ave.0+corr.sd.0 #upper confidence bounds
v0<-cbind(lb0,corr.ave.0,ub0)

  #Backward delays: v.b is a matrix whose 1st column is lower confidence
    #bounds, 2nd column is average correlations and 3rd column is upper
    #confidence bounds. The rows are backward time delays.
corr.ave.b<-apply(corr_b_mat,1,mean) #average across libraries (rows)
corr.sd.b<-apply(corr_b_mat,1,sd) #standard deviation across libraries
lb.b<-corr.ave.b-corr.sd.b #lower confidence bound for each delay
ub.b<-corr.ave.b+corr.sd.b #upper confidence bound for each delay
v.b<-cbind(lb.b,corr.ave.b,ub.b) #collect results in matrix
v.b<-v.b[nrow(v.b):1,]  #reverse time ordering (rows) for plot

  #Forward delays: v.f is a matrix whose 1st column is lower confidence
    #bounds, 2nd column is average correlations and 3rd column is upper
    #confidence bounds. The rows are forward time delays.
corr.ave.f<-apply(corr_f_mat,1,mean) #average across libraries (rows)
corr.sd.f<-apply(corr_f_mat,1,sd) #standard deviation across libraries
lb.f<-corr.ave.f-corr.sd.f #lower confidence bound for each delay
ub.f<-corr.ave.f+corr.sd.f #upper confidence bound for each delay
v.f<-cbind(lb.f,corr.ave.f,ub.f)

#'results' is a matrix whose 1st column is lower confidence bounds, 2nd
  #column is average correlations and 3rd column is upper confidence bounds.
  #The rows are delays starting with the most negative delay and ending with
  #the largest positive delay.
results.yx<-rbind(v.b,v0,v.f)
return(results.yx)

}  #end user-defined function ccm_delay_udf

#Step 7: Run user-defined function 'ccm_delay_udf'
```

```
#Loop j specifies each variable in turn as the response process (x),
#and loop k iterates over the remaining variables as forcing (driving)
#processes (y)

#Storage matrix for loop j
hold.j1<-matrix(0,n.vars,n.vars)
hold.j2<-matrix(0,n.vars,n.vars)

for(j in 1:n.vars){  #response (driven) process (x)
  print("j");print(j)
  x<-data[,j]
  #embed response process x
  results.embed<-embed_udf(x)
  E<-2  #E<-results.embed[[2]]
  tau<-1  #tau<-results.embed[[1]]

  #Storage matrices for loop k
  hold.k1<-matrix(0,delay_length,n.vars)
  hold.k2<-matrix(0,1,n.vars)
  hold.k3<-matrix(0,1,n.vars)

  for(k in 1:n.vars){  #forcing (driving) process (y)
    if(j==k) next  #don't cross map to same attractor
    print("k");print(k)
    y<-data[,k]

#Step 7a: Run user-defined function ccm_delay_udf
    results.yx<-ccm.delay(A=y,B=x,E,tau,n.lib,num.bd,num.fd)
    ave.yx<-results.yx[,2] #average correlation coefficients for each ccm delay
    print("ave.yx");print(ave.yx)
    hold.k1[,k]<-na.omit(ave.yx)  #matrix storing ave.yx vectors for jth response variable
    lb.yx<-results.yx[,1]  #lower confidence bound for each ccm delay
    ub.yx<-results.yx[,3]  #upper confidence bound for each ccm delay

#Step 7b: Plot delayed cross-mapping skill with confidence intervals
    setwd(dir.results)
    plotCI(ave.yx,ui=ub.yx,li=lb.yx,xaxt="n",xlab="cross map delay (d)",
    ylab="cross map skill (p)",lwd=2,font=2,font.lab=2,cex.axis=1.2,
    cex.lab=1.2)  #x xmap y
    abline(v=(num.bd+1),lty=2,lwd=2)
    lines(ave.yx,lwd=2)  #Put line through points
    #Custom x-axis showing backward and forward delays, side=1 is x-axis,
      #"at" defines 'tick-mark' numbers
    xaxis<-seq(1:(num.bd+num.fd+1))-(num.bd+1)
    axis(side=1,at=1:(num.bd+num.fd+1),labels=xaxis,font=2,cex.axis=1.2)

    #Export plots
    file<-as.character("")
    names.var<-as.matrix(names)
    file<- as.character(paste(substr(file,1,nchar(file)-4),
      names.var[j],"_delay_xmap_",names.var[k],".png", sep=""))
    dev.copy(png,filename=paste(file,".png",sep=""))
    dev.off()

  #Step 8: Compute parameters used below to cull false positives from ccm adjacency matrix

    #Step 8a: Criterion 1. Identify ccm delay in ave.yx with maximum averaged correlation
    #coefficient
    o<-order(ave.yx,decreasing=TRUE) #Order indices of ave.yx in decreasing order
    print("o");print(o)
```

```
delay.vec.o<-delay.vec[o]  #Applying ordering to delay.vec (vector of ccm delays)
print("delay.vec.o");print(delay.vec.o)
delay.o<-delay.vec.o[1]  #delay giving max correlation coefficient for cross mapping
print("delay.o");print(delay.o)
hold.k2[,k]<-delay.o

#Step 8b: Criterion 2. Compute critical value 'cv.yx' determining whether there is an
  #alpha % difference between the maximum and minimum values of ave.yx
cv.yx<-max(ave.yx)-(1+alpha)*min(ave.yx)
print("cv.yx");print(cv.yx)
hold.k3[,k]<-cv.yx
} #end loop k

#Step 9: Export 'hold.k1' matrix associated with each response variable j.
  #First, insert 'delay.vec' (the backward and forward delays for the run,
  #e.g., -2,-1,0,1,2) as the first column of hold.k1.
hold.k1a<-cbind(delay.vec,hold.k1)
print("hold.k1a");print(hold.k1a)
setwd(dir.results)
file<-as.character("")
names.var<-as.matrix(names)
names.var1<-c("delay",names.var)
file<- as.character(paste(substr(file,1,nchar(file)-4),
  "delayed_ccm_",names.var[j],"_as response variable",".csv", sep=""))
print(file)
write.table(hold.k1a,file,col.names=names.var1,
          row.names=FALSE,sep=",")

hold.j1[j,]<-hold.k2 #driving variable in column
hold.j2[j,]<-hold.k3
} #endloop j

#Step 10: Export hold.j1 matrix renamed 'delay_max'
row.names <- c(names)  #Insert row and column labels in hold.j1
col.names<-c("",names)
mat1<-cbind(row.names,hold.j1)
mat2<-rbind(col.names,mat1)
write.table(mat2,"delay_max.csv",col.names=FALSE,row.names=FALSE,sep=",")

#Step 11:  Cull false positives from adjacency matrix computed by ccm.
adjacency<-matrix(c(0,1,1,0),2,byrow=T) #adjacency matrix from ccm
adjacency_delay<-as.matrix(replace(adjacency,hold.j1>0&hold.j2>0,0))
print("adjacency_delay");print(adjacency_delay)

  #Step 11a:  Export adjacency_delay matrix
row.names <- c(names)
col.names<-c("",names)
mat1<-cbind(row.names,adjacency_delay)
mat2<-rbind(col.names,mat1)
write.table(mat2,"adjacency_delay.csv",col.names=FALSE,row.names=FALSE,sep=",")
```

In Code 8.2, we formulate `ccm_delay_udf` with a loop i that iterates over $n.lib = 20$ libraries. In Step 1, the length of each library is randomly selected as a draw from a uniform distribution over an interval of lengths $[lib.min + lib.adj, lib.max]$, depending on the embedding parameters and the length of the time series (Clark, 2015). We use the command `set.seed<-500` to generate a reproducible random series with the command `runif`. We add an adjustment factor $lib.adj$ to the lower bound $lib.min$ so that random library lengths are sufficiently long for CCM convergence. In each iteration of loop i, the code is broken into running CCM without delay (Step 2), CCM with backward delays (Step 3) and CCM with forward delays (Step 4).

We run CCM without delay with the response series $B(t) = (B_1, B_2, \ldots, B_N)$ cross mapping the forcing series $A(t) = (A_1, A_2, \ldots, A_N)$, where N is the length of each series. We run CCM for a single period backward delay with the delayed response series $B(t) = (B_2, \ldots, B_N)$ cross mapping the forcing series $A(t) = (A_1, A_2, \ldots, A_{N-1})$, where the observations B_1 and A_N are lost to the delay. More generally, Step 3 formulates $num.bd$ backward-delayed cross mappings with loop j, where the jth iteration cross maps $B(t) = (B_j, \ldots, B_N)$ against $A(t) = (A_1, A_2, \ldots, A_{N-j})$. We obtain the response series for the jth backward delay by embedding B with embedding dimension 2 and embedding delay j. In other words, we forward delay B to cross map against unadjusted A. Loop j runs each backward-delayed cross mapping for given random library i.

In Step 4, we run CCM for $num.fd$ forward-delayed cross mappings with loop k. We follow the same logic as for backward-delayed CCM, except that we now embed the forcing series A with embedding dimension 2 and embedding delay k. This results in a kth forward-delayed cross mapping of $B(t) = (B_1, \ldots, B_{N-k})$ against $A(t) = (A_k, A_2, \ldots, A_N)$.

Step 5 stores the correlation coefficients associated with these delayed cross mappings across the $n.lib = 20$ libraries. For illustration, we display output for `model.1` and cross mapping v xmap w. The vector `corr_0` stores the correlation coefficients for CCM without delay:

```
[,1]        [,2]        [,3]        . . . [,18]     [,19]     [,20]
0.7318048   0.6842870   0.6408820   . . . 0.6285427 0.5306342 0.6568170
```

The matrix `corr_b_mat` stores the correlation coefficients for the $num.bd = 6$ backward-delayed CCM (rows) across $n.lib = 20$ libraries (columns):

```
          [,1]        [,2]        [,3] . . . [,20]
[1,]  0.8504884   0.8189843   0.7880906 . . . 0.7994096
[2,]  0.6140139   0.5910237   0.5657833 . . . 0.5765907
[3,]  0.4051244   0.3945313   0.3826577 . . . 0.3876936
[4,]  0.3392846   0.3312874   0.3219471 . . . 0.3217417
[5,]  0.2745237   0.2698294   0.2610379 . . . 0.2638132
[6,]  0.2118138   0.2054358   0.1990440 . . . 0.2054633
```

The matrix `corr_f_mat` does the same for $num.fd = 6$ forward delays:

```
          [,1]        [,2]        [,3]        . . . [,20]
[1,]  0.5945762   0.5458374   0.4983071   . . . 0.5182710
[2,]  0.4243567   0.3884807   0.3616811   . . . 0.3724237
[3,]  0.3135171   0.2994315   0.2780292   . . . 0.2777889
[4,]  0.2456939   0.2344776   0.2227594   . . . 0.2279646
[5,]  0.1837078   0.1816485   0.1805506   . . . 0.1754439
[6,]  0.1662510   0.1622380   0.1615639   . . . 0.1597572
```

In Step 6, we average correlation coefficients across libraries for the various cross mappings and compute lower and upper confidence bounds. For CCM without delay, we average the `corr_0` vector of correlation coefficients, and compute the standard deviation to compute a confidence interval: *average ± standard deviation*. We do this for backward- and forward-delayed CCM by using the **apply** command to compute these statistics for each row of the corresponding storage matrices. We collect these results in a matrix **results** whose rows run from the most distant backward delay to the most distant forward delay: $num.bd, \ldots, -1, 0, 1, \ldots, num.fd$, and whose columns contain the lower confidence bound, the average correlation coefficient **corr.ave** and the upper confidence bound. The **results** matrix for **model.1** and cross mapping v xmap w is

```
       lower bound  corr.ave   upper bound
 [1,]  0.1963255    0.2028461  0.2093667
 [2,]  0.2491501    0.2611578  0.2731655
 [3,]  0.3056558    0.3211606  0.3366654
 [4,]  0.3649121    0.3849238  0.4049355
 [5,]  0.5303948    0.5696387  0.6088826
 [6,]  0.7398272    0.7934090  0.8469908
 [7,]  0.5873050    0.6539842  0.7206634
 [8,]  0.4491039    0.5156663  0.5822286
 [9,]  0.3266287    0.3707956  0.4149624
[10,]  0.2530907    0.2819927  0.3108947
[11,]  0.2069644    0.2254551  0.2439457
[12,]  0.1641586    0.1745370  0.1849155
[13,]  0.1554694    0.1614644  0.1674594
```

In Step 7, we run the user-defined function **ccm.delay_udf** with outer loop j iterating over the response series and inner loop k iterating over each series as the forcing series. This runs **ccm_delay_udf** for **model.1** four times. The first iteration of loop j fixes $B = v$ as the response variable and $A = v$ and $A = w$ in turn as the forcing variables to perform cross mappings v xmap v and then v xmap w. The second iteration of loop j fixes $B = w$ as the response variable and again iterates over $A = v$ and $A = w$ as the forcing variables to perform cross mappings w xmap v and then w xmap w. The average correlation coefficients for the delayed cross mappings associated with a given response variable are stored in the matrix **hold.k1** and exported to an external *.csv* file (Step 9). The exported file name when, for example, v is fixed as the response variable is **delayed_ccm_v_as response variable**:

```
delay v w
-6     0   0.202846085
-5     0 0.261157803
-4     0 0.321160592
-3     0 0.384923817
-2     0 0.569638704
-1     0 0.793408998
0      0 0.65398417
1      0 0.515666256
2      0 0.370795572
3      0 0.281992704
4      0 0.225455063
5      0 0.174537014
6      0 0.161464402
```

The second column contains zeros since **ccm.delay_udf** does not cross-map a variable against itself. In Step 7b, cross-mapping skills (i.e. average correlation coefficients with confidence bounds) for computed cross mappings are plotted against the delay (d) and exported. The plots for `model.1` accurately detect the one-period delayed interaction running between v and w in both equations (Figure 8.2c,d).

In Step 8, we turn to the job of culling false positives from the adjacency matrix computed in `ccm_udf`. We formulate two rules for culling a detected interaction as a false positive. First, the maximum average correlation coefficient for the associated delayed cross mapping occurs at a positive delay. Second, the maximum average correlation for a positive delay is significantly larger than the values for the other delays. We want to avoid declaring a false positive when, for example, the maximum value differs from the minimum value by a minuscule order of magnitude, or, in other words, when the delayed CCM plot for a given cross mapping is relatively flat. There are several ways of deciding what constitutes 'significantly larger' – we propose a possible rule in Step 8b.

Step 8a identifies the delay yielding the maximum average correlation coefficient for a given cross mapping. For example, in the `delayed_ccm_v_as response variable` matrix already mentioned, the maximum value (0.793408998) occurs at a delay equal to -1, consistent with `model.1`. The maximum delay values for the cross mappings associated with a given response variable are stored in the storage vector `hold.k2`. For this illustration, the vector of maximum delay values associated with v serving as the response variable is 0, -1, where the first element is zero because v is not cross-mapped with itself, and the second element is the value for v xmap w. The `hold.j1` matrix (renamed `delay_max` when exported in Step 9) stores (as rows) the vector of maximum delay values associated with each response variable. For this illustration, `delay_max` is

```
       v    w
v   0  -1
w  -1   0
```

Step 8b proposes a rule for determining whether there is a significant difference between the maximum and minimum average correlation coefficients for a given cross mapping: `cv.yx<-max(ave.yx)-(1+alpha)*min(ave.yx)`. When the parameter `cv.yx` is positive, there is an *alpha* percentage difference between the maximum and minimum values, which is deemed significant. We selected *alpha* = 0.05 when setting parameters at the beginning of Code 8.2. The `hold.j2` matrix stores the critical values for each cross mapping. For this illustration, we have

```
       v       w
v   0.000   0.644
w   0.750   0.000
```

We see that each cross mapping exceeds the 5% difference tolerance level, since each element of the matrix is positive. For example, the critical value for the cross mapping v xmap w is $0.644 > 0$. The `delay_max` and `hold.j2` matrices are exported in Steps 9 and 10.

If the criteria in Steps 8a and 8b are both met, Step 11 removes a '1' from the CCM `adjacency` matrix and replaces it with a '0' in the revised `adjacency_delay` matrix with the R command `replace()`. Since delayed CCM does not detect a false positive when applied to `model.1`, the `adjacency` and `adjacency_delay` matrices remain the same.

Finally, we use Code 8.2 to solve `model.2` to illustrate how the code works to detect a false positive. The plot for v xmap w (Figure 8.3c) accurately detects a false positive (i.e. that w does not force v in the first equation), since the peak correlation coefficient occurs at a positive delay. Since this peak value exceeds the next largest value by at least 5%, the `adjacency_delay` matrix zeros out this interaction:

```
  v  w
v 0  0
w 1  0
```

Alternatively, the delayed CCM plot for w xmap v accurately detects that v forces w in the second equation, since the maximum average correlation coefficient occurs at a negative delay (Figure 8.3d).

8.4 Network Plots

We can use network plots to graphically summarize causal interactions identified in adjacency matrices with the R package `igraph` (Csardi, 2015). Time series variables appear as **nodes** whose size and colour can represent binary attributes (such as gender) or continuous attributes (such as age or size). When the adjacency matrix has an element equal to 1, the network plot generates an *edge* (line) from the forcing to the response variable *node*. *Directed edges* have arrows indicating the direction of interaction. *Weighted edges* indicate the strength of interaction by edge width or colour.

Code 8.3 draws a network plot from imported adjacency and attribute matrices. An *attribute* matrix contains data on binary and continuous attributes of node variables. For this illustration, we generate these matrices inside the code. We first generate the adjacency matrix computed in Code 8.1 with the `model.1` equations:

```
  v w
v 0 1
w 1 0
```

The nodes in the network plot are labelled 'v' and 'w'. We next generate a sample **attribute** matrix as a **dataframe** object:

```
variable binary continuous
    v       M        10
    w       F         5
```

The binary attribute assigns gender with 'M' for male and 'F' for female (second column), and the continuous attribute assigns age (third column).

```
#Code 8.3  Network plots from adjacency matrices
rm(list=ls(all=TRUE))

#Working Directories
dir.data<-"C:/Users/Ray/Documents/R_files/Data"
dir.results<-"C:/Users/Ray/Documents/R_files/Results"

#Libraries
library(igraph)

#Import adjacency matrix: The columns are drivers of a given variable (row)
#setwd(dir.data)
```

```
#adjacency<-read.csv("adjacency.csv",header=TRUE,row.names=1,check.names=FALSE)
#data<-as.matrix(adjacency)
#names<-colnames(data)

#Make adjacency matrix
adjacency<-matrix(c(0,1,1,0),2,byrow=T)
rownames(adjacency,do.NULL=FALSE);rownames(adjacency)<-c("v","w")
colnames(adjacency,do.NULL=FALSE);colnames(adjacency)<-c("v","w")
data<-adjacency
names<-colnames(data)

#Import attribute matrix 'a': The first column contains variable names, the
 #second column contains some binary attribute (e.g. gender) and the third
 #column measures on some continuous attribute (e.g. size).
#a<-read.csv("attributes.csv")

#Make attribute data frame
variable<-c('v','w')
binary<-c('M','F')
continuous<-c(10,5)
a<-data.frame(variable,binary,continuous)

#Step 1: Compute edge density--Ratio of edges to total possible edges in adjacency matrix
#edge_density(net.wgt)  #compute with igraph command
edge.density<-(sum(rowSums(data)))/((ncol(data)*nrow(data))-length(diag(data)))

#Step 2:  Select plot options
 #Type of attribute matrix
no_attribute<-FALSE  #TRUE: attribute matrix not imported
binary<-FALSE        #TRUE: attribute matrix has only binary column
continuous<-FALSE    #TRUE: attribute matrix has only continuous column
both<-TRUE           #TRUE: attribute matrix has both binary and continuous columns

 #Edge width and color
edge_width<-FALSE #TRUE: edge width denotes strength of relationship (else: width fixed)
edge_color<-FALSE  #TRUE: edge color denotes strength of relationship (else: monochrome)

 #Node (vertex) size
#node.size=TRUE  #TRUE: node size adapts to continuous attribute (else: fixed)

#Step 3: Describe and set network plot configuration parameters
#vertex.label.font=1 [1 (regular), 2 (bold), 3 (italic), 4 (bold italic)]

#vertex.size" = size of circle around node in plot

#vertex.color=""  node color (some possibilities--red,pink,blue,light blue,
                 #green,light green,yellow,black,grey,light grey)

#attribute_binary_symbol<-c("","") #letters used for attribute_binary attribute

#attribute_binary_color<-c("","") node color reflecting attribute_binary attribute
                    #some possibilities--red,pink,blue,light blue,green,
                    #light green,yellow,black,grey,light grey

#frac.weight=1  default fixed edge width (set larger to increase width)

#edge.color=""  #edge color (lines between nodes)

#weighted=TRUE #edge thickness reflects strength of relationship
        #=NULL #uniform edge thickness
```

```
#edge.curved=TRUE    curves edges between nodes
           #=FALSE

#mode="directed"  #arrows on edges
                   #2 arrowed edges if edge.curved=TRUE

#layout=layout.kamada.kawai  (type of network plot)
  #Other possibilities:layout.auto,layout.random,layout.circle,layout.sphere,
  #layout.fruchterman.reingold,layout.spring,layout.reingold.tilford,
  #layout.fruchterman.reingold.grid,lyaout.lgl,layout.star

#Set parameters
vertex.label.font=2
vertex.size=30 #(fixed node size, increase with 30*2)
label.font.size=3
vertex.color="white"
attribute_binary_symbol<-c("M","F")
attribute_binary_color<-c("white","grey")
frac.weight=3
edge.arrow.size=2  #set arrow size to edge size
edge.color="black"
weighted=TRUE
width.adj=1
edge.curved=TRUE
mode="directed"
layout=layout.star

#Step 4:  Run network plot
data.t<-t(data) #take transpose of data so arrows come out right
net.wgt=graph.adjacency(data.t,mode,weighted,diag=FALSE)
V(net.wgt)$label.cex<-label.font.size  #font size for node label

if(no_attribute){ #commands if attribute matrix not imported

  #Plot commands when edge width denotes strenth of relationship
  if(edge_width){
    plot.igraph(net.wgt,vertex.label.color="black",vertex.label.font=vertex.label.font,
      layout=layout,vertex.color=vertex.color,vertex.size=vertex.size,
      edge.color=edge.color,edge.width=E(net.wgt)$weight/width.adj,
      edge.curved=edge.curved,edge.arrow.size=edge.arrow.size)} else{ #fixed edge
    plot.igraph(net.wgt,vertex.label.color="black",vertex.label.font=vertex.label.font,
      layout=layout,vertex.color=vertex.color,vertex.size=vertex.size,
      edge.color=edge.color,edge.width=frac.weight,edge.curved=edge.curved,
      edge.arrow.size=edge.arrow.size)
  } #end if(edge_width)

  #Plot commands when edge color denotes strength of relationship (3 levels)
  if(edge_color){
    E(net.wgt)$color <- ifelse(E(net.wgt)$weight ==1, "light grey",
      ifelse(E(net.wgt)$weight ==2, "dark grey", "black"))
    plot.igraph(net.wgt,vertex.label.color="black",vertex.label.font=vertex.label.font,
      layout=layout,vertex.color=vertex.color,vertex.size=vertex.size,
      edge.color=E(net.wgt)$color,edge.width=E(net.wgt)$weight/3,edge.curved=edge.curved,
      edge.arrow.size=edge.arrow.size)
    #Include plot legend defining edge colors
    colors<- cbind("light grey", "dark grey", "black")
    labels1<-cbind("0.60-0.79", "0.89-0.89","0.90-1.00")
    legend(-2.5,-1,legend=labels1, col=colors, pch=19,title="")
  } #end if(edge_color)
} #end if(no_attribute)
```

```
if(binary){
  attribute_binary<-a[,2]

  #Reflecting binary attribute by node color
  nodecolor=as.character(attribute_binary[match(V(net.wgt)$name,names)])
  nodecolor=gsub(attribute_binary_symbol[1],attribute_binary_color[1],nodecolor)
  nodecolor=gsub(attribute_binary_symbol[2],attribute_binary_color[2],nodecolor)

  if(edge_width){
    plot.igraph(net.wgt,vertex.label.color="black",vertex.label.font=vertex.label.font,
      layout=layout,vertex.color=nodecolor,vertex.size=vertex.size,
      edge.color=edge.color,edge.width=E(net.wgt)$weight/width.adj,
      edge.curved=edge.curved,edge.arrow.size=edge.arrow.size)} else{ #fixed edge
    tkplot(net.wgt,vertex.label.color="black",vertex.label.font=vertex.label.font,
      layout=layout,vertex.color=nodecolor,vertex.size=vertex.size,
      edge.color=edge.color,edge.width=frac.weight,edge.curved=edge.curved,
      edge.arrow.size=edge.arrow.size)
  } #end if(edge_width)

  #Plot commands when edge color denotes strength of relationship (3 levels)
  if(edge_color){
    E(net.wgt)$color <- ifelse(E(net.wgt)$weight ==1, "light grey",
      ifelse(E(net.wgt)$weight ==2, "dark grey", "black"))
    plot.igraph(net.wgt,vertex.label.color="black",vertex.label.font=vertex.label.font,
      layout=layout,vertex.color=nodecolor,vertex.size=vertex.size,
      edge.color=E(net.wgt)$color,edge.width=E(net.wgt)$weight/3,edge.curved=edge.curved,
      edge.arrow.size=edge.arrow.size)
    #Include plot legend defining edge colors
    colors<- cbind("light grey", "dark grey", "black")
    labels1<-cbind("0.60-0.79", "0.89-0.89","0.90-1.00")
    legend(-2.5,-1,legend=labels1, col=colors, pch=19,title="")
  } #end if(edge_color)
} #end if(binary)

if(continuous){
  attribute_continuous<-a[,3]

  #Reflecting continuous attribute by node size
  node.size<-setNames(attribute_continuous,V(net.wgt)$name)*5

  if(edge_width){
    plot.igraph(net.wgt,vertex.label.color="black",vertex.label.font=vertex.label.font,
      layout=layout,vertex.color=vertex.color,vertex.size=as.matrix(node.size),
      edge.color=edge.color,edge.width=E(net.wgt)$weight/width.adj,
      edge.curved=edge.curved,edge.arrow.size=edge.arrow.size)} else{ #fixed edge
    plot.igraph(net.wgt,vertex.label.color="black",vertex.label.font=vertex.label.font,
      layout=layout,vertex.color=vertex.color,vertex.size=as.matrix(node.size),
      edge.color=edge.color,edge.width=frac.weight,edge.curved=edge.curved,
      edge.arrow.size=edge.arrow.size)
  } #end if(edge_width)

  #Plot commands when edge color denotes strength of relationship (3 levels)
  if(edge_color){
    E(net.wgt)$color <- ifelse(E(net.wgt)$weight ==1, "light grey",
      ifelse(E(net.wgt)$weight ==2, "dark grey", "black"))
    plot.igraph(net.wgt,vertex.label.color="black",vertex.label.font=vertex.label.font,
      layout=layout,vertex.color=vertex.color,vertex.size=as.matrix(node.size),
      edge.color=E(net.wgt)$color,edge.width=E(net.wgt)$weight/3,edge.curved=edge.curved,
      edge.arrow.size=edge.arrow.size)
    #Include plot legend defining edge colors
```

```
      colors<- cbind("light grey", "dark grey", "black")
      labels1<-cbind("0.60-0.79", "0.89-0.89","0.90-1.00")
      legend(-2.5,-1,legend=labels1, col=colors, pch=19,title="")
   } #end if(edge_color)
} #end if(continuous)

if(both){
  attribute_binary<-a[,2];attribute_continuous<-a[,3]

  #Reflecting binary attribute by node color
  nodecolor=as.character(attribute_binary[match(V(net.wgt)$name,names)])
  nodecolor=gsub(attribute_binary_symbol[1],attribute_binary_color[1],nodecolor)
  nodecolor=gsub(attribute_binary_symbol[2],attribute_binary_color[2],nodecolor)

 #Reflecting continuous attribute by node size
  node.size<-setNames(attribute_continuous,V(net.wgt)$name)*5

  if(edge_width){
    plot.igraph(net.wgt,vertex.label.color="black",vertex.label.font=vertex.label.font,
      layout=layout,vertex.color=nodecolor,vertex.size=as.matrix(node.size),
      edge.color=edge.color,edge.width=E(net.wgt)$weight/width.adj,
      edge.curved=edge.curved,edge.arrow.size=edge.arrow.size)} else{ #fixed edge
    plot.igraph(net.wgt,vertex.label.color="black",vertex.label.font=vertex.label.font,
      layout=layout,vertex.color=nodecolor,vertex.size=as.matrix(node.size),
      edge.color=edge.color,edge.width=frac.weight,edge.curved=edge.curved,
      edge.arrow.size=edge.arrow.size)
  } #end if(edge_width)

  #Plot commands when edge color denotes strength of relationship (3 levels)
  if(edge_color){
    E(net.wgt)$color <- ifelse(E(net.wgt)$weight ==1, "light grey",
      ifelse(E(net.wgt)$weight ==2, "dark grey", "black"))
    plot.igraph(net.wgt,vertex.label.color="black",vertex.label.font=vertex.label.font,
      layout=layout,vertex.color=nodecolor,vertex.size=as.matrix(node.size),
      edge.color=E(net.wgt)$color,edge.width=E(net.wgt)$weight/3,edge.curved=edge.curved,
      edge.arrow.size=edge.arrow.size)
    #Include plot legend defining edge colors
    colors<- cbind("light grey", "dark grey", "black")
    labels1<-cbind("0.60-0.79", "0.89-0.89","0.90-1.00")
    legend(-2.5,-1,legend=labels1, col=colors, pch=19,title="")
  } #end if(edge_color)
} #end if(both)

dev.copy(png,"adjacency.png")
dev.off()
```

Step 1 of Code 8.3 computes a statistic commonly used to characterize the amount of information in a network plot: *edge density*. The density of interactions is calculated as the ratio of edges to total edges possible. We compute this from the `adjacency` matrix by dividing the elements equal to 1 by the total number of elements in the matrix less the number of diagonal elements. For this illustration, the edge density is computed as: $2/(4-2) = 1$. There is bi-causal interaction between the only two nodes in the diagram.

Step 2 formats the network plot to information in the attribute matrix. If we input `no_attribute<-TRUE`, then the code is directed to commands creating the plot without an attribute matrix. The code performs the same function for an attribute matrix containing either binary or continuous attributes or both. If we input `edge_width<-TRUE` or `edge_color<-TRUE`, edges display the strength of an interaction by their width or colour, respectively. Finally, if we input `node.size<-TRUE`, node size adapts to the continuous attribute. For example, variables of greater age can be represented with larger nodes.

Step 3 describes and then sets configuration parameters for the network plots. For example, the `layout` parameter allows us to select from a number of possible network plot layouts. Finally, Step 4 creates the network plot with commands described in the documentation to the R package `igraph` (Csardi, 2015). Replacing the `plot.igraph` command with `tkplot` generates an interactive layout.

The network plot derived from the solutions of `model.1` portrays the bi-causal interaction between v and w (Figure 8.2e). Consistent with the sample attribute matrix and the specified configuration parameters, the binary variable is indicated with node size, and one of the nodes is shaded to demonstrate how another characteristic of interest might be portrayed. The network plot computed from the solutions of `model.2` accurately portrays only a single edge running from the forcing variable v to the response variable w (Figure 8.3e).

8.5 Real-World Application

A key question in the epidemiological literature is whether outbreaks of infectious diseases interact. Using linear correlation and regression techniques, Rohani *et al.* (2003) found evidence of interactive disease dynamics between measles and whooping cough in five European cities. Since correlation and regression results do not imply causation, and linear techniques are not reliable when applied to data governed by nonlinear dynamics, we apply CCM to test for multistrain dynamics in disease data first diagnosed with deterministic nonlinear dynamic structure. In particular, we test for interactive dynamics between between weekly cases of scarlet fever and measles in New York City in the pre-vaccine period 1915–1947 (1722 observations) (Source: Project Tycho, http://www.tycho.pitt.edu/index.php). These data are found in the `disease_1915.csv` file in the `C:/Data` directory of the book's web materials.

We initially applied the Toeplitz method of singular spectrum analysis (window = 520) to each mean-adjusted time series to test for structured variation (i.e. a *signal*). We isolated a strong signal in the scarlet fever series accounting for 72% of variation in the series from the mean. The signal is dominated by a strong typical annual cycle (62%) in conjunction with fainter five-year (6%) and semiannual (4%) cycles, where the percentages in parentheses measure the relative strengths of component cycles in accounting for overall variation. We isolated a signal of moderate strength in the measles series accounting for about 60% of variation in the series (window = 520). The signal is dominated by a relatively strong annual cycle (30%), a 2.5-year cycle (13%), a 2-year cycle (9%) and a 37-month cycle (7%). Pre-vaccine annual cycling in measles was largely driven by school terms, since school-age children were the core group for transmission (Bjørnstad *et al.*, 2002). Plots of the two disease series (grey curves) and their respective isolated signals (black curves) give strong visual evidence that the signals successfully capture much of the observed variation in the data (Fig-

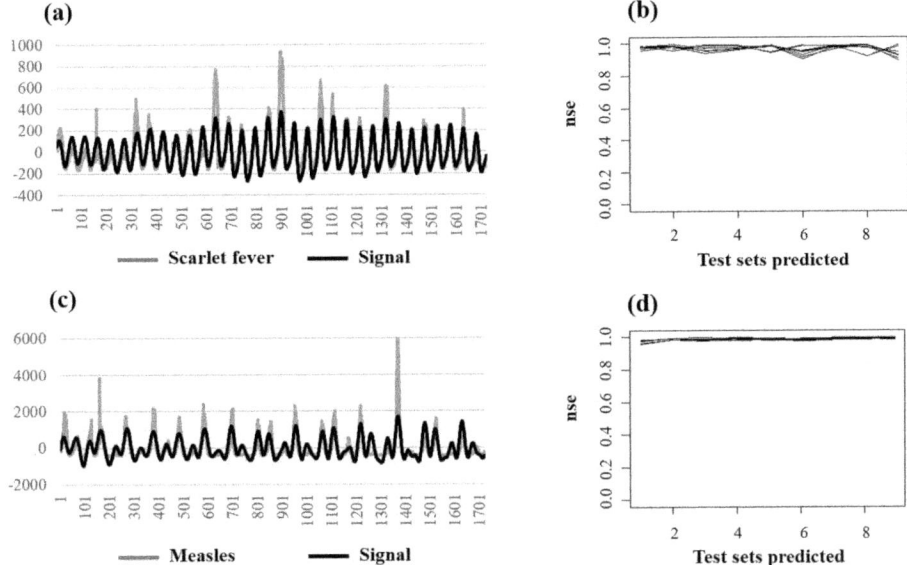

Fig. 8.4 Diagnostics of scarlet fever and measles time series data. Plots of disease data (grey curves) and isolated signals (black curves) for scarlet fever (a) and measles (c) show that signals capture much of the observed variability in the data. Nonlinear cross prediction plots for scarlet fever (b) and measles (d) provide evidence that signals are stationary for purposes of nonlinear analysis.

ure 8.4a, c). Probing for nonstationarity with nonlinear cross prediction (Code 6.15) provides strong evidence that each disease signal is stationary for purposes of nonlinear time series analysis, since each signal segment is highly skilled at cross-predicting the others (Figure 8.4b, d).

We reconstructed low-dimensional attractors from each disease signal exhibiting nonrepeating (aperiodic) oscillations (Figure 8.5). We tested the null hypothesis that an attractor is reconstructed from a signal most likely generated by linear stochastic dynamics with *AAFT* and *PPS* surrogate time series ensembles. We selected nonlinear prediction skill as the discriminating statistic, and conducted an upper-tailed test to reject the null hypothesis only when the attractor reconstructed from the disease signal predicted with more skill than surrogate attractors. We set the probability of false rejection at $\alpha = 0.05$ and the rank order test parameter at $k = 5$. This generated $(k/\alpha) - 1 = 99$ surrogate data vectors. We rejected the null hypothesis if the Nash–Sutcliffe coefficient of efficiency (nse) measured from a signal attractor was among the k largest nse values measured from the surrogate attractors.

We weakly rejected the null hypothesis for the scarlet fever signal since the predictive skill for the signal attractor ($nse = 0.99$) barely exceeded the upper limit computed for the attractors reconstructed from both AAFT ($nse = 0.98$) and PPS ($nse = 0.97$) surrogates. Similarly, we weakly rejected the null hypothesis for the measles signal

(a) **(b)**

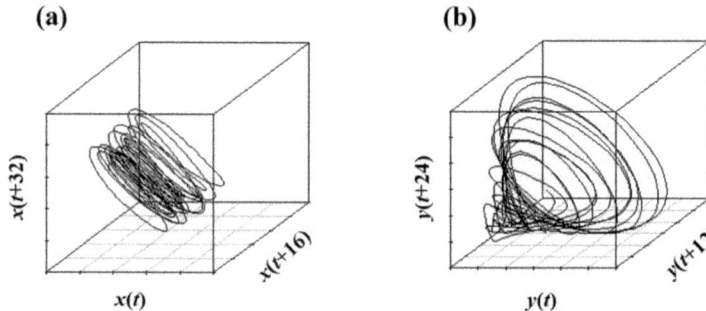

Fig. 8.5 Low-dimensional attractors reconstructed from scarlet fever (a) and measles (b) signals.

since the predictive skill for the signal attractor ($nse = 0.99$) barely exceeded the upper limits for both AAFT ($nse = 0.97$) and PPS ($nse = 0.96$) surrogates.

   In summary, these diagnostics provide empirical evidence of low-dimensional attractors reconstructed from relatively strong stationary disease signals potentially generated by nonlinear deterministic dynamics, and consequently justify the application of CCM to detect causal interactions between the scarlet fever and measles time series. Code 8.4 combines Codes 8.1 and 8.2 to run CCM and delayed CCM in sequence. The code initially imports the time series data, and sets parameters required to run the user-defined functions for CCM (`ccm_udf`) and delayed CCM (`ccm_delay_udf`). The code runs `ccm_udf` to generate the **adjacency** matrix (as in Code 8.1), and then runs `ccm_delay_udf` to generate the **adjacency_delay** matrix culling false-positive interactions (as in Code 8.2).

```
#Code 8.4  R code to run CCM followed by delayed CCM
rm(list=ls(all=TRUE))

#Define working directories
dir.udf<-as.character("C:/Users/Ray/Documents/R_files/Code/User Defined Functions")
dir.data<-as.character("C:/Users/Ray/Documents/R_files/Data")
dir.results<-as.character("C:/Users/Ray/Documents/R_files/Results")

#Load user-defined functions
setwd(dir.udf)
dump("embed_udf", file="embed_udf.R")
source("embed_udf.R")
dump("ccm_udf", file="ccm_udf.R")
source("ccm_udf.R")
dump("ccm_delay_udf", file="ccm_delay_udf.R")
source("ccm_delay_udf.R")

#Load libraries
library(tseriesChaos)
library(multispatialCCM)
library(plotrix)

#Load data from an external .csv file
setwd(dir.data)
```

```
ts<-read.csv("disease_1915.csv")
data<-ts[,c(4,6)]
#data<-data[1:300,]
data<-na.omit(data)
data<-as.matrix(data)
names<-colnames(data)
n.vars<-ncol(data)

#Parameter required by ccm_udf
#'frac' is a parameter used in ccm_udf to compute the 'weighted
#adjacency matrix'. (1-frac) sets the fraction of points used to calculate
#the asymptote of the ccm curve (plotting the correlation coefficient
#against the library of points on the attractor used to cross predict
#the other variables.
frac<-0.66

#Parameters required by ccm_delay_udf
n.lib<-20  #number of random libraries
num.bd<-6  #number of backward delays
num.fd<-6  #number of forward delays
delay.vec<-c(-(num.bd:1),0,(1:num.fd))  #vector of delays
delay_length<-length(delay.vec)
alpha<-0.05  #parameter used to update adjacency matrix

#Step 1: Run ccm_udf
results_ccm<-ccm_udf(data,frac)
wam<-results_ccm[[1]] #weighted adjacency matrix

#Step 2:  Convert the 'wam' matrix into a scaled adjacency matrix 'adjacency'
  #by assigning a scale of integers to the approximated asymptotes in 'wam'. For
  #example, asymptotes greater than 0.65 are assigned 1 with 0 otherwise.
mat.1<-wam
for(row in 1:nrow(wam)){#row
   for(col in 1:nrow(wam)){#column
       if(mat.1[row,col]>=0.65){mat.1[row,col]=1}
       if(mat.1[row,col]<0.65){mat.1[row,col]=0}
   } #end col
} #end row
#Reading file: Driving variable listed in column.
adjacency<-mat.1

#Export 'adjacency' matrix after inserting row and column labels
row.names <- c(names)
col.names<-c("",names)
mat1<-cbind(row.names,adjacency)
mat2<-rbind(col.names,mat1)
write.table(mat2,"adjacency.csv",col.names=FALSE,row.names=FALSE,sep=",")

#Step 3: Run user-defined function 'ccm_delay_udf'

  #Loop j specifies each variable in turn as the response process (x),
  #and loop k iterates over the remaining variables as forcing (driving)
  #processes (y)

#Storage matrix for loop j
hold.j1<-matrix(0,n.vars,n.vars)
hold.j2<-matrix(0,n.vars,n.vars)

for(j in 1:n.vars){  #response (driven) process (x)
  print("j");print(j)
```

```
x<-data[,j]
#embed response process x
results.embed<-embed_udf(x)
E<-results.embed[[2]]
tau<-results.embed[[1]]

#Storage matrices for loop k
hold.k1<-matrix(0,delay_length,n.vars)
hold.k2<-matrix(0,1,n.vars)
hold.k3<-matrix(0,1,n.vars)

for(k in 1:n.vars){  #forcing (driving) process (y)
  if(j==k) next  #don't cross-map to same attractor
  print("k");print(k)
  y<-data[,k]

#Step 3a: Run user-defined function ccm_delay_udf
  results.yx<-ccm.delay(A=y,B=x,E,tau,n.lib,num.bd,num.fd)
  ave.yx<-results.yx[,2] #average correlation coefficients for each ccm delay
  print("ave.yx");print(ave.yx)
  hold.k1[,k]<-na.omit(ave.yx)  #matrix storing ave.yx vectors for jth response variable
  lb.yx<-results.yx[,1]  #lower confidence bound for each ccm delay
  ub.yx<-results.yx[,3]  #upper confidence bound for each ccm delay

#Step 3b: Plot delayed cross-mapping skill with confidence intervals
  setwd(dir.results)
  plotCI(ave.yx,ui=ub.yx,li=lb.yx,xaxt="n",xlab="cross map delay (d)",
  ylab="cross map skill (p)",lwd=2,font=2,font.lab=2,cex.axis=1.2,
  cex.lab=1.2)  #x xmap y
  abline(v=(num.bd+1),lty=2,lwd=2)
  lines(ave.yx,lwd=2)  #Put line through points
  #Custom x-axis showing backward and forward delays, side=1 is x-axis,
    #"at" defines 'tick-mark' numbers
  xaxis<-seq(1:(num.bd+num.fd+1))-(num.bd+1)
  axis(side=1,at=1:(num.bd+num.fd+1),labels=xaxis,font=2,cex.axis=1.2)

  #Export plots
  file<-as.character("")
  names.var<-as.matrix(names)
  file<- as.character(paste(substr(file,1,nchar(file)-4),
    names.var[j],"_delay_xmap_",names.var[k],".png", sep=""))
  dev.copy(png,filename=paste(file,".png",sep=""))
  dev.off()

#Step 4: Compute parameters used below to cull false positives from ccm adjacency matrix
  #Step 4a: Criterion 1. Identify ccm delay in ave.yx with maximum averaged correlation
  #coefficient
  o<-order(ave.yx,decreasing=TRUE) #Order indices of ave.yx in decreasing order
  print("o");print(o)
  delay.vec.o<-delay.vec[o]  #Applying ordering to delay.vec (vector of ccm delays)
  print("delay.vec.o");print(delay.vec.o)
  delay.o<-delay.vec.o[1]  #delay giving max correlation coefficient for cross mapping
  print("delay.o");print(delay.o)
  hold.k2[,k]<-delay.o

  #Step 4b: Criterion 2. Compute critical value 'cv.yx' determining whether there is an
    #alpha % difference between the maximum and minimum values of ave.yx
  cv.yx<-max(ave.yx)-(1+alpha)*min(ave.yx)
  print("cv.yx");print(cv.yx)
  hold.k3[,k]<-cv.yx
```

```
} #end loop k

#Step 5: Export 'hold.k1' matrix associated with each response variable j.
  #First, insert 'delay.vec' (the backward and forward delays for the run,
  #e.g., -2,-1,0,1,2) as the first column of hold.k1.
hold.k1a<-cbind(delay.vec,hold.k1)
print("hold.k1a");print(hold.k1a)
setwd(dir.results)
file<-as.character("")
names.var<-as.matrix(names)
names.var1<-c("delay",names.var)
file<- as.character(paste(substr(file,1,nchar(file)-4),
  "delayed_ccm_",names.var[j],"_as response variable",".csv", sep=""))
print(file)
write.table(hold.k1a,file,col.names=names.var1,
          row.names=FALSE,sep=",")

hold.j1[j,]<-hold.k2  #driving variable in column
hold.j2[j,]<-hold.k3
} #endloop j

#Step 6: Export hold.j1 matrix renamed 'delay_max'
row.names <- c(names)  #Insert row and column labels in hold.j1
col.names<-c("",names)
mat1<-cbind(row.names,hold.j1)
mat2<-rbind(col.names,mat1)
write.table(mat2,"delay_max.csv",col.names=FALSE,row.names=FALSE,sep=",")

#Step 7:  Cull false positives from adjacency matrix computed by ccm.
adjacency_delay<-as.matrix(replace(adjacency,hold.j1>0&hold.j2>0,0))
print("adjacency_delay");print(adjacency_delay)

  #Step 7a:  Export adjacency_delay matrix
row.names <- c(names)
col.names<-c("",names)
mat1<-cbind(row.names,adjacency_delay)
mat2<-rbind(col.names,mat1)
write.table(mat2,"adjacency_delay.csv",col.names=FALSE,row.names=FALSE,sep=",")
```

CCM results provide evidence that measles moderately forces scarlet fever in the cross mapping *scarlet fever* xmap *measles*, since the correlation coefficients converge to 0.7 (Figure 8.6a). Scarlet fever strongly forces measles in the cross mapping *measles* xmap *scarlet fever*, since correlation coefficients converge to almost 1 (Figure 8.6b). Delayed CCM does not detect a false positive in either direction, since the maximum average correlation coefficients occur at negative delays on relative flat graphs (Figure 8.6c, d). The network plot portrays the bi-directional forcing (Figure 8.6e).

8.6 Detecting Change Points

CCM requires variables with continuous time records so that we can empirically reconstruct the required shadow phase spaces. However, real-world systems are often heterogeneous in that some essential variables may have irregular records reflecting only occasional activity. How can we empirically test, for example, whether a continuous record of environmental degradation is connected to a sporadically discharged

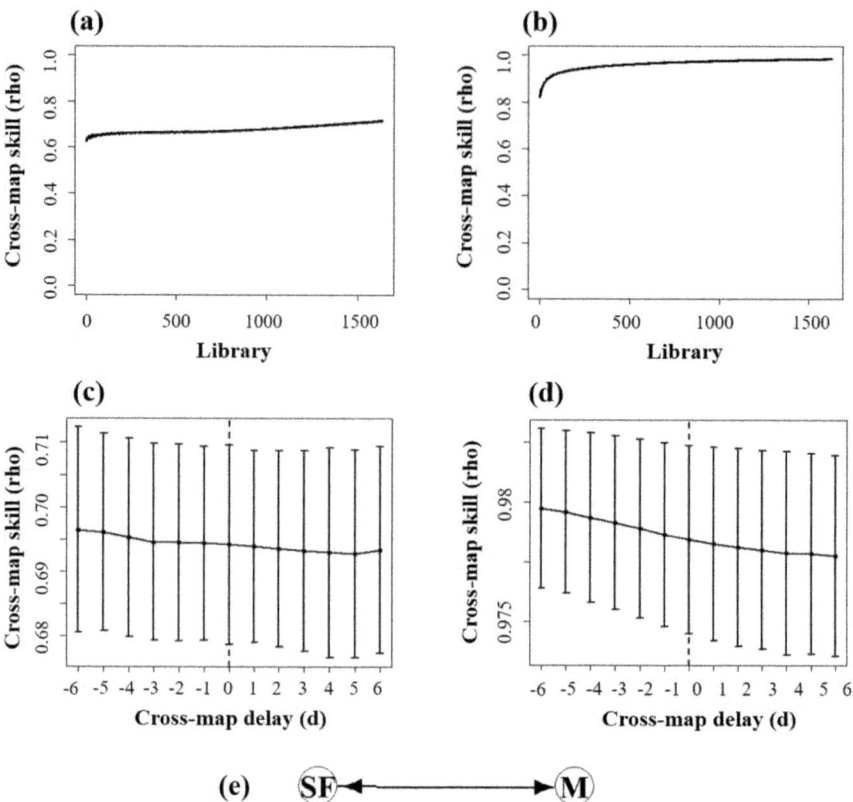

Fig. 8.6 Detecting causality in disease data. (a) CCM detects that measles moderately drives scarlet fever in the cross mapping *scarlet fever* xmap *measles*. (b) CCM detects that scarlet fever strongly drives measles in the cross mapping *measles* xmap *scarlet fever*. (c, d) Delayed CCM does not detect a false positive in either direction, since the maximum average correlation coefficients occur at negative delays on relatively flat graphs. (e) The network plot shows bi-directional forcing.

effluent? In this case, we search for evidence of *change points* indicating abrupt structural changes in environmental dynamics corresponding to the timing of discharge spikes. In this chapter, we focus on the *singular spectrum transformation (SST)* approach to detecting change points, which is based on singular spectrum analysis (SSA) (Chapter 6) (Itoh and Marwan, 2013; Moskvina and Zhigljavsky, 2003).

We illustrate the mechanics of SST change-point detection with a synthetic time series generated by modifying eqns (15) and (16) of Itoh and Marwan (2013):

$$Y(t) = \alpha t + A\sin\left(\frac{2\pi}{T}t\right) + \varepsilon_t \tag{8.3a}$$

with

$$\alpha = \begin{cases} 0.25 & (1 \leq t \leq 200) \\ -0.15 & (200 < t \leq 550) \\ 0.05 & (550 < t \leq 1000) \end{cases}$$

$$(8.3b)$$

$$T = \begin{cases} 24 & (1 \leq t \leq 700) \\ 48 & (700 < t \leq 550) \end{cases}$$

This is a sine function with linear trend and Gaussian noise $\varepsilon \sim N(0,1)$. The trend component shifts twice in time periods $t = 200$ and $t = 550$, and the frequency of the oscillatory component decreases at $t = 700$ (Figure 8.7a). Code 8.5 runs eqns (8.3) for $N = 1000$ periods and exports the resulting time series in the file **ts.csv**.

```
#Code 8.5 Generate sine test function for SST change-point detection
#Adapted from equations 15-16 in Itoh and Marwin (2013)
rm(list=ls(all=TRUE))

dir.results<-as.character("C:/Results")

N<-1000 #time periods
#First segment (t1=1:200)
t1<-1:200  #time steps
alpha<-0.25  #linear trend
T<-24 #period
phi<-0  #phase
A<-100 #amplitude
a<-1 #number of cycles completed over T
frequency<-a/T;frequency
#set.seed(3)
#Add Gaussian noise
ts.1<-(alpha*t1)+A*sin(((a*2*pi/T)*t1))

#Second segment (t2=201:550) Increase amplitude
t2<-201:550  #time steps
alpha<--0.15
ts.2<-(alpha*t2)+A*sin(((a*2*pi/T)*t2))

#Third segment (t3=551:700) Increase frequency
t3<-551:700  #time steps
alpha<-0.05
ts.3<-(alpha*t3)+A*sin(((a*2*pi/T)*t3))

#Fourth segment (t3=701:1000) Increase frequency
t4<-701:1000  #time steps
T<-48
ts.4<-(alpha*t4)+A*sin(((a*2*pi/T)*t4))

#Entire test time series
ts<-c(ts.1,ts.2,ts.3,ts.4)
plot(ts,type='l')

#Export test series
setwd(dir.results)
write.table(ts,sep=",","ts.csv")
```

We first apply SSA to decompose the time series into constituent trend and oscillatory components with Codes 6.9 and 6.11 (Chapter 6). As a quick refresher, the

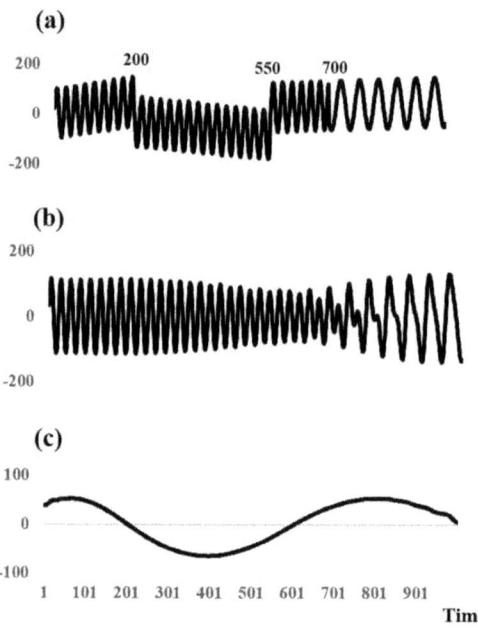

Fig. 8.7 SSA decomposition of eqns (8.3). (a) The test series has two change points in a linear trend component (at periods 200 and 550) and one change point in an oscillatory (sine) component (at period 700). (b) Reconstructed oscillatory component composed of 24-period and 48-period oscillations (eigentriplets 1, 2, 5 and 6). (c) Reconstructed trend component (eigentriplets 3 and 4).

time series in converted into a trajectory matrix whose first column is the time series and whose remaining columns are its delayed copies. SSA runs an eigensystem decomposition of the trajectory matrix that expresses it as a sum of new matrices. Each new matrix is computed as the product of eigensystem components (an eigentriplet), including a left eigenvector that characterizes dynamic patterns in the data (Ide and Inoue, 2005). Change-point detection operates off of these left eigenvectors (as we shall explain). SSA proceeds by dividing the new matrices into groups corresponding to trend, oscillations and noise (unstructured variation) components in the data, after which the matrix groups are reconverted into time series vectors of the respective components.

For the time series generated by eqns (8.3), we run the R package Rssa (Korobeynikov *et al.*, 2015) to isolate the expected 24-period (eigentriplets 1 and 2) and 48-period (eigentriplets 5 and 6) oscillations, accounting respectively for 50% and 10% of the total variation in the series, and the trend component (eigentriplets 3 and 4), accounting for 25% of the total variation (Figure 8.7b, c). We use an SSA window length of 500 periods (half of the length of the time series). We store the reconstructed time series components in the file sin_cp_ssa.csv (which can be found in the web materials for this book).

Code 8.6 formulates a user-defined function `sst_udf` to perform SST change-point detection following Itoh and Marwan (2013). The function requires input of a time series x and three parameters b, g and k, which we shall define shortly. We apply `sst_udf` to detect the change point built into the combined oscillatory component (eigentriplets 1, 2, 5 and 6), and the two change points in the trend component (eigentriplets 3 and 4). We import the file `sin_cp_ssa.csv` and work first with the oscillatory component (`x<-ts$oscillations`) by commenting out the trend data.

```
#Code 8.6  User-defined function for SST change-point detection
rm(list=ls(all=TRUE))

#Define working directories
dir.udf<-as.character("C:/User Defined Functions")
dir.data<-as.character("C:/Data")

#Load R packages
library(Rssa)   #singular spectrum analysis
library(ppls)   #normalize to unit vector
library(psych) #trace of matrix

#Load data
setwd(dir.data)
ts<-read.csv("sin_cp_ssa.csv")
x<-ts$oscillations
#x<-ts$trend

#Detection parameters
b<-90  #change-point window (oscillations)
#b<-50  #change-point window (trend)
g<-0     #forward delay before future part begins
k<-2   #number of left eigenvectors to use in projection

#User-defined function for SST change-point detection (sst_udf)
sst<-function(x,b,g,k){
 #Step 1: Partition oscillatory component into 'past' and 'future' time series with CP window
  windows<-length(x)-(2*b)-g+1  #number of windows
  hold.i<-matrix(0,windows,1)
  for (i in 1:windows) {
  xp<-x[i:(i+b-1)] #past time series
  #print("xp");print(xp)
  xf<-x[(i+b+g):(i+(2*b)+g-1)]  #future series
  #print("xf");print(xf)
 #Step 2:  Run Rssa on past and future time series
  L_sst<-(length(xp)+1)%/%2 #default takes integer half of time series length
  sp<-ssa(xp,L=L_sst,kind="toeplitz")
  sf<-ssa(xf,L=L_sst,kind="toeplitz")
  vp<-calc.v(sp,idx=1:L_sst) #left eigenvectors (columns) for past series
  vf<-calc.v(sf,idx=1:L_sst) #left eigenvectors (columns) for future series
  vp_norm<-apply(vp,2,normalize.vector)  #normalize vp to unit vector
  vf_norm<-apply(vf,2,normalize.vector)  #normalize vf to unit vector
 #Step 3:  Compute CP score (z) (Okayasu et al. 2012)
  vp_k<-vp_norm[,1:k]  #hyperplane with first k left eigenvectors for past series
  vf_1<-vf_norm[,1]  #left eigenvector associated with largest singular value for future series
  ip_vec<-t(vf_1)%*%vp_k #project vf_1 onto vp_k
  z<-1-sum(ip_vec^2)  #Itoh and Marwan 2013
  hold.i[i,]<-z
  } #end loop i
z<-list(hold.i)
```

```
  return(z)
} #end user-defined function sst

z<-sst(x,b,g,k)
z<-unlist(z)
plot(z,type='l',ylim=c(0,1),ylab="CP score",cex.lab=1.5,cex.axis=1.5,font=2,lwd=2)

#Identify windows giving peak z scores
max.1<-which(z==max(z[1:300]));max.1
max.2<-which(z==max(z[300:800]));max.2
```

In Step 1, we partition the oscillatory component into past and future series whose lengths are set by the change-point window b. For example, given a general reference time t_r, the past series begins at $t = t_r - b$ and ends at the period before the reference time, $t = t_r - 1$. The future series begins with a possible forward delay of g periods at time $t = t_r + g$ and ends at $t = t_r + g + b - 1$ (Figure 8.8a). The earliest possible reference time is $t_r = b + 1$, since the past series begins at the first observation. The associated future series begins g periods later and ends at $t = 2b + g$ (Figure 8.8b). Finally, the latest possible reference point is $t_r = N - b - g + 1$, since the future series ends at the final observation N (Figure 8.8c). There are $N - 2b - g + 1$ feasible reference times (windows), which we calculate by subtracting the earliest from the latest possible reference time plus 1. Loop i iterates over the feasible reference times to generate the corresponding past and future series for each.

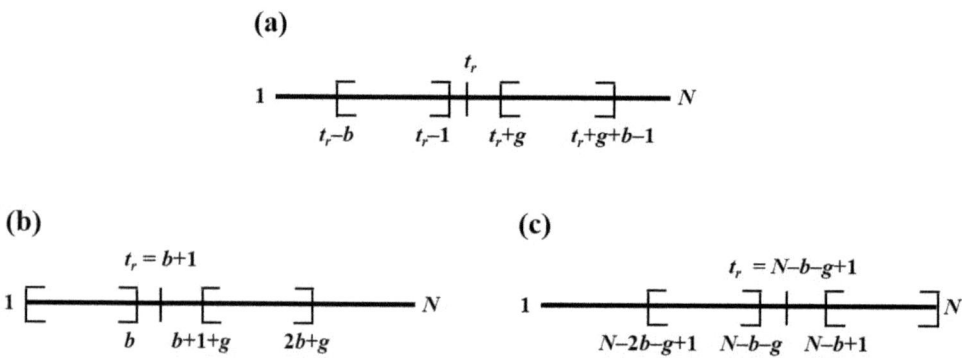

Fig. 8.8 Partitioning a time series into past and future series with the SST window parameter b. (a) Time boundaries on the windows defining past and future series for a general reference point t_r. (b) Time boundaries associated with the earliest possible reference point $t_r = b + 1$. (c) Time boundaries associated with the latest possible reference point $t_r = N - b - g + 1$.

In Step 2 of Code 8.6, we determine whether the dynamics characterizing the past series have abruptly changed in the future series. In particular, we run SSA on each series with window length L_sst set to half of the length of the partitioned series. We next retrieve the left eigenvectors from decomposition of the past series (characterizing past dynamics) and the future series (characterizing future dynamics) with the command calc.v(). From decomposition of the past series, we select the left eigenvectors from the eigentriplets associated with the k largest singular values: $(vp_1, \ldots, vp_k)$. These left eigenvectors span a hyperplane vp_k given by

$$vp\_k = \text{span}\{vp_1, \ldots, vp_k\} \tag{8.4}$$

From decomposition of the future series, we select the left eigenvector from the eigen-triplet associated with the largest singular value: vf_1. Then, with one last bit of linear algebra, we can determine whether the dynamics have shifted from the past series to the future series. Specifically, we project vf_1 (representing future behaviour) onto the hyperplane vp_k (representing past behaviour) by taking the inner product:

$$ip\_vec = (vf\_1)^{\mathrm{T}} vp\_k \tag{8.5}$$

where superscript T denotes the transpose. This projection allows us, in Step 3, to estimate the distance between the vector vf_1 and the hyperplane vp_k using a metric such as that proposed by Itoh and Marwan (2013):

$$z = 1 - \sum_{i=1}^{k} (ip\_vec)^2 \tag{8.6}$$

The metric z, referred to as the *change-point (CP) score*, falls in the interval $0 \leq z \leq 1$. The CP score spikes at times (*change points*) when eqn (8.6) measures a relatively large distance, thereby indicating a break between past and future dynamic behaviour.

Loop i repeats Steps 2 and 3 for each reference time in the feasible interval:

$$b + 1 \leq t_r \leq N - b - g + 1 \tag{8.7}$$

This results in a time series of CP scores over the $N - 2b - g + 1$ windows. To identify the reference time t_r to which a particular CP score corresponds, we add the SST window parameter b to the window index in loop i. For example, we note in Figure 8.8b that the first (earliest) window corresponds to $t_r = b + 1$.

The computed CP scores for the oscillatory component of eqns (8.3) are shown in (Figure 8.9a, black curve). We set the SST window at $b = 90$ and the forward delay parameter at $g = 0$. We use the left eigenvectors associated with the two largest singular values $k = 2$ in the SSA decomposition of the past series to span the hyperplane vp_k (eqn (8.4)). We detect a change point in frequency at $t = 720$ (since the CP score is 1), which is in the neighbourhood of the change point built into the equations ($t = 700$). Turning to the trend component of eqns (8.3), we set the SST window at $b = 50$, the forward delay parameter at $g = 0$ and the number of spanning left eigenvectors at $k = 2$. CP scores reach 1 at two change points occurring at $t = 179$ and $t = 581$, which are in the neighbourhoods of those built into the equation: $t = 200$ and $t = 550$ (Figure 8.9b: black curve).

There are no precise rules for setting the change-point parameters L_sst, b, g and k (Itoh and Marwan, 2013). One option is to run the change-point algorithm for wide intervals of parameter values (Itoh and Marwan, 2013). For example, one study found that the analysis remains stable over a wide range of the SST window parameter b

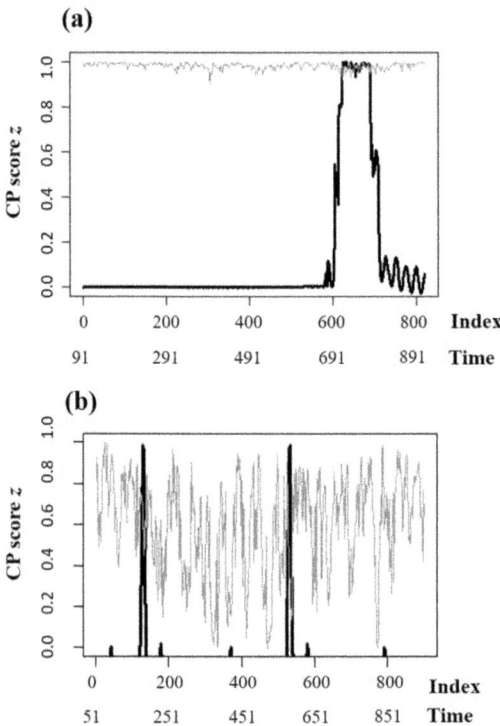

Fig. 8.9 CP scores for oscillatory and trend components (black curves) with point-by-point 95% confidence limits (grey curves). (a) The change point in the oscillatory component is detected in period 720. The associated CP score exceeds the upper 95% confidence level $(1 > 0.95)$. (b) The change points in the trend component are detected in periods 179 and 581. The associated CP scores exceed the upper 95% confidence limits: period 179 $(0.98 > 0.83)$, period 581 $(0.98 > 0.89)$.

(Ide and Inoue, 2005). Another option is to set parameter values that bring plotted CP scores into sharper focus (Itoh and Kurths, 2011).

Code 8.7 tests the statistical significance of the time vector of CP scores with surrogate data (Chapter 7). In particular, we use AAFT surrogates to test the null hypothesis that CP scores are generated by a linear stochastic process, and consequently that high scores occur randomly rather than indicating structural change points (Itoh and Marwan, 2013). Our object is to provide point-by-point 95% upper confidence levels for CP scores. If the CP score rests above the upper confidence level, then we reject the null hypothesis, and we deem the associated time period to be a statistically significant structural change point.

We proceed by loading the oscillatory and trend components from the SSA decomposition of eqns (8.3) and their respective change-point parameters b, g and k. In Step 1, we run the user-defined function for change-point detection sst_udf on the oscillatory component.

```
#Code 8.7  Change-point detection with point-by-point upper confidence level
rm(list=ls(all=TRUE))

#Define working directories
dir.udf<-as.character("C:/Users/Ray/Documents/R_files/Code/User Defined Functions")
dir.data<-as.character("C:/Users/Ray/Documents/R_files/Data")

#Load user-defined function
setwd(dir.udf)
dump("sst_udf", file="sst_udf.R")
source("sst_udf.R")

library(fractal)  #generate linear stochastic surrogates

#Load data
setwd(dir.data)
ts<-read.csv("sin_cp_ssa.csv")
x<-ts$oscillations  #ssa reconstruction of oscillatory component
#x<-ts$trend  #ssa reconstruction of trend

#Detection parameters
b<-90   #change-point window (oscillations)
#b<-50  #change-point window (trend)
g<-0    #forward delay before future part begins
windows<-length(x)-(2*b)-g+1  #change-point windows
k<-2 #number of left eigenvectors to use in projection

#Step 1:  Run change-point detection on ssa component
z.x<-sst(x,b,g,k)  #CP score (z)
z.x<-unlist(z.x)

#Step 2:  Surrogate data testing for point-by-point 95% confidence limit
 #Step 2a: Number of surrogates and significance level
k=2  #used to set number of surrogates for one- and two-sided tests
alpha=0.05  #significance level
nsurr=(k/alpha)-1   #surrogates for one-sided hypothesis test

 #Step 2b: CP scores for surrogate data vectors
#Each column of matrix 'z.surr' contains CP scores for a surrogate data vector
z.surr<-matrix(0,windows,nsurr)
for (i in 1:nsurr){
surr_out<-surrogate(x,method="aaft")
surr<-surr_out[1:length(surr_out)]
z<-sst(surr,b,g,k)
z<-unlist(z)
print('z');print(z)
z.surr[,i]<-z
}
print('z.surr');print(z.surr)

 #Step 2c: User-defined function for rank order test on a vector of CP scores
rank_udf<-function(x,k){
z.high<-sort(x)[length(x)-(k-1)]  #lower limit on k largest values
return<-z.high
}

z.high<-apply(z.surr,1,rank_udf,k=k) #apply rank_udf to each column of z.surr

#Step 3: Plot of CP scores (z) for each window with 95% confidence upper limit
plot(z.x,type="l",xlab= "window",ylab="CP score (z)",ylim=c(0,1),
```

```
cex.lab=1.5,las=1,cex.axis=1.5,font=2,lwd=3)
lines(z.high,col="dark gray") #plot regression line

#Step 4: Identify window giving peak CP score (z) and 95% confidence limit
max<-which(z.x==max(z.x[300:800]));max  #window giving max CP score
z.x.max<-z.x[max]  #max CP score
z.high.max<-z.high[max] #corresponding 95% upper limit
max;z.x.max;z.high.max
```

In Step 2, we generate surrogate data vectors to run an upper-tailed hypothesis test using nonparametric rank order statistics . In Step 2a, we compute the required number of surrogate vectors to run the test at the $alpha = 0.05$ significance level. Setting $k = 2$ generates $(k/alpha) - 1 = 39$ surrogate data vectors. In each time period, we rank the CP scores for the surrogate vectors in descending order, and reject the null hypothesis if the CP score computed from the trend or oscillation component of the observed data is among the $k = 2$ largest of the descending surrogate scores.

In Step 2b, we partition each surrogate vector into past and future series, run the change-point user-defined function `sst_udf` on each partitioned series, and compute CP scores through time. This results in an output matrix `z.surr` whose columns contain the CP scores for each surrogate vector and whose rows contain the scores in each period across surrogate vectors.

In Step 2c, we write a user-defined function `rank_udf` to sort a vector in descending order, and pick out the value that must be exceeded for a selected significance level. We then apply `rank_udf` to each row of `z.surr` with the R command `apply`. This results in a time series of point-by-point 95% upper confidence limits `z.high` that a CP score must exceed to be statistically significant.

The 95% confidence limits for the oscillatory and trend components associated with eqns (8.3) are plotted in Figure 8.9 (grey curves). The CP score for the change point detected in the oscillatory component (period 720) exceeds the upper 95% confidence level ($1 > 0.95$). Similarly, the CP scores for the change points detected in the trend component (periods 179 and 581) exceed the upper 95% confidence limits: $0.98 > 0.83$ and $0.98 > 0.89$, respectively.

8.7 Detecting Tipping Points

We now consider a probabilistic approach to detecting causality motivated by a continuous record having a weak signal due to a high degree of noise. Again, we search for evidence of abrupt structural changes in the continuous record corresponding to spikes in sporadic records, but now rely on *histograms* instead of SSA to identify *tipping points* (Livina *et al.*, 2011; Lenton and Livina, 2016). A histogram is a bar chart measuring frequencies with which intervals of observations (*bins*) occur in the time series. Histograms destroy temporal patterns in the time series, but this becomes less costly as patterns account for relatively smaller fractions of total variability. The logic behind relying on histograms to detect system dynamics is that bins containing stable equilibria occur relatively more frequently in the time series, as displayed by relatively higher peaks. For example, the histogram of a time series converging to a stable equilibrium has a single spiked peak over the corresponding bin. A tipping point occurs, for example, when a histogram transitions from a single peak to multiple peaks reflecting cyclical dynamics. At its root, the strategy for detecting tipping points is to compute histograms for a sliding window through the time series, count histogram peaks in each window, and identify times when peaks abruptly change in number.

We illustrate the mechanics of tipping-point detection with a synthetic time series (*ts*) generated by incorporating an abrupt shift in dynamics into the Lorenz equations at time $t = 100$. For $t \leq 1000$, the Lorenz equations are calibrated for a stable focus equilibrium (*sigma* $= 10$, *beta* $= \frac{8}{3}$, *rho* $= 5$), while for $1000 < t \leq 3000$), they are calibrated for the chaotic butterfly attractor (*sigma* $= 10$, *beta* $= \frac{8}{3}$, *rho* $= 28$) (Figure 8.10a).

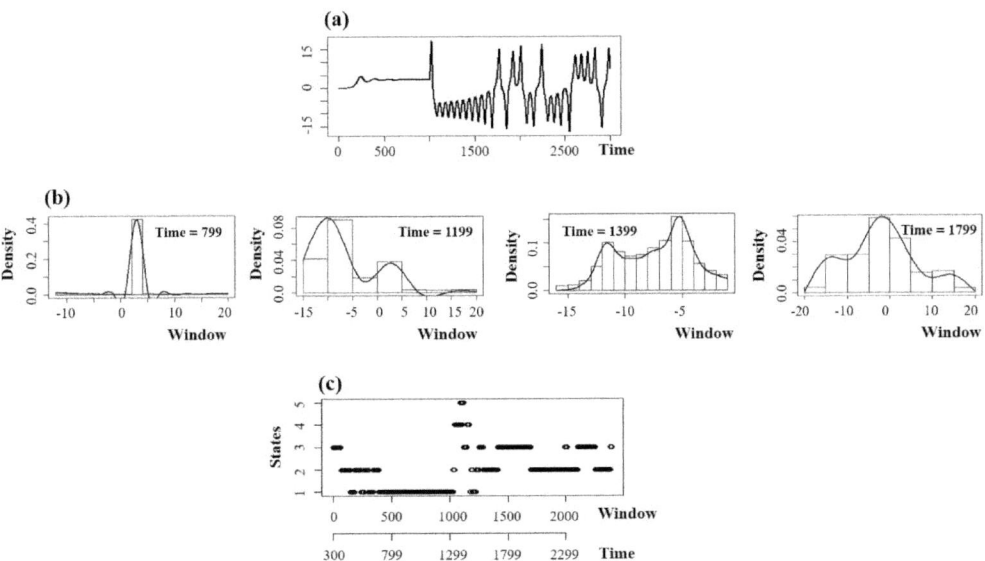

Fig. 8.10 Detecting tipping points. (a) Lorenz series constructed with a tipping point at $t = 1000$ between stable-focus and chaotic regimes. (b) Histograms with fitted spline-regression curves were computed for sliding windows of 600 points to gauge how histogram peaks (measuring state numbers) varied across windows. The plots show histograms for four windows in the sequence. (c) Number of states computed for each window with corresponding real time on the second horizontal axis. Wide-ranging multiple states in initial times ($0 < t \leq 500$) and later times ($1000 < t \leq 1400$) are due to transitional dynamics as the system converges to a stable focus and a chaotic attractor, respectively. Once the system has settled into the chaotic attractor, histograms assume a broad single peak with flat shoulders (right-most plot in (b)). The number of states alternates between three and two (when one of the shoulders disappears for a time).

We use Code 8.8 to detect the tipping point ($t = 1000$). Roughly following Livina *et al.* (2011) and Lenton and Livina (2016), the code computes histograms for a sliding window through the time series and makes a movie of how the histograms change over the series. It also fits a spline regression to the histogram bars to use for counting histogram peaks (and therefore the most frequently visited states). Finally, the code plots the number of states for each window across time.

```
#Code 8.8 Tipping-point detection
rm(list=ls(all=TRUE))
options(scipen=999) #disable scientific notation

#Directories
dir.udf<-as.character("C:/User Defined Functions")
dir.results<-as.character("C:/Results")

library(animation)
library(graphics) #histograms
ff.loc<-"C:/User Defined Functions/ffmpeg"

#Load user-defined functions
setwd(dir.udf)
dump("lorenz_udf", file="lorenz_udf.R");source("lorenz_udf.R")

#Generate Lorenz data with user-defined function 'lorenz_udf'
 #First segment, x1:  Stable focus
lorenz_results<-lorenz_udf(x0=0.01,y0=0.01,z0=0.01,sigma=10,beta=8/3,rho=5,
 t.end=10,delta=0.01)
x1<-lorenz_results[[1]];y1<-lorenz_results[[2]];z1<-lorenz_results[[3]]
x1<-x1[1:length(x1)]
 #Second segment, x2:  Chaos
lorenz_results<-lorenz_udf(x0=x1[length(x1)],y0=y1[length(y1)],z0=z1[length(z1)],
 sigma=10,beta=8/3,rho=28,t.end=20,delta=0.01)
x2<-lorenz_results[[1]];y2<-lorenz_results[[2]];z2<-lorenz_results[[3]]
 #Time series, ts
ts<-c(x1,x2)
plot(ts,type='l',lwd=2,cex.axis=1.3,cex.lab=1.3,font=2,xlab="time")

#Window parameters
b<-300  #length of each side of reference midpoint
win.length<-2*b #window length
#Histogram parameters
breaks<-11  #number of breaks in histogram

#Animation parameter: interval (seconds) between frames
interval<-0.005

#Step 1: Make movie of histograms through windowed segments of data
setwd(dir.results)
saveVideo(
 expr={
 ani.record(reset=TRUE) #clear history before recording
 ani.options(interval=interval) #set interval between frames (seconds)
 nwin<-length(ts)-(2*b)+1 #number of windows
 hold.i<-matrix(0,nwin,1)  #storage matrix for number of states in windowed histograms
 for (i in 1:nwin) {
  #Step 1a: Partition time series into windows
  xp<-ts[i:(i+b-1)] #past time series
  xf<-ts[(i+b):(i+(2*b)-1)]  #future series
  x.win<-c(xp,xf)

  #Step 1b: Run histogram and compute spline-regression line for each window
  histogram<-hist(x.win,breaks=breaks,freq=FALSE,plot=TRUE,lwd=1,cex.axis=1.3,cex.lab=1.3,font=2)
  par(new=T)
  density<-unlist(histogram$density)
```

```
y<-density
x<-1:length(y)
f <- splinefun(x, y, method="natural") #fit spline regression to histogram densities
splinecoef <- get("z", envir = environment(f)) #regression coefficients
#spline-regression curve
curve(f(x),1,length(x),lwd = 2.5,xaxt='n',yaxt='n',ann=FALSE,ylim=c(min(y),max(y)))
par(new=F)
ani.record()  #record the current frame

#Step 1c: Preprocess spline-regression values
y.spl<-f(x, deriv = 0) #spline-regression values (one for each histogram bar)
y.spl[y.spl<0]<-0  #make negative values zero
y.spl[(y.spl/max(y.spl))<0.05]<-0 #remove 'noise' floor
y.feasible<-which(y.spl>0)  #pre-processed values

#Step 1d: Compute number of states for each window
 #Only one feasible spline-regression value--front bar
if((length(y.feasible)==1)&(y.feasible==1)) {states=1}
 #Only one feasible spline-regression value--single interior bar
if((length(y.feasible)==1)&(f(y.feasible, deriv = 2)<0)) {states=1}
 #Multiple interior bars
if(length(y.feasible)>1) {
  derivative.2<-f(x, deriv = 2) #2nd derivative of spline-regression line
  #remove 2nd derivatives for endpoint bars
  derivative.2a<-derivative.2[y.feasible][2:(length(y.feasible)-1)]
  signs<-sign(derivative.2a) #count number of sign changes in second derivative
   #over histogram bars
  ip<-sum(signs[-1] != signs[-length(derivative.2a)]) #number of inflection points
  states<-1+(ip%/%2)  #number of states for ith histogram
} #end if
hold.i[i,]<-states  #number of states in ith windowed histogram
} #end loop i
states<-hold.i  #states in windowed histograms (vector)
},
video.name="histogram.mp4",
ffmpeg=ff.loc
)  #end saveVideo


window<-1:nwin
time<-b+(window-1)  #window midpoints

#Plot number of states with two x axes. The upper axis is the window number and the
 #lower axis is the corresponding time (= b + (window-1)).
plot(states,xlab="",cex.axis=1.3,cex.lab=1.3,font=2)
axis(1, at = c(0, 500, 1000, 1500, 2000), labels = c("300", "799", "1299", "1799","2299"),
 cex.axis=1.3,cex.lab=1.3,font=2,line = 3)  #line=3 is lines between two x axes
```

After generating the Lorenz time series, Code 8.8 requires input of parameters governing construction of the sliding windows, the histograms and the movie. Following the logic for constructing a sliding window in Code 8.7, the parameter b fixes the window length on either side of a reference time, so that the entire length is *2b*. The **breaks** parameter determines the number of bars in the histogram. A histogram constructed with a relatively large number of breaks gives a more detailed display of peaks, but the trade-off is that relatively minor peaks may reflect noise more than systematic dynamics. We can experiment to find break numbers that bring major peaks into focus. Finally, the **interval** parameter fixes the elapsed time between histogram frames in the movie.

In Step 1, we use the `saveVideo` command in the `animation` library (Xie, 2017) to film the movie of sequenced histograms. Loop i iterates through the number of sliding windows calculated with the command `nwin<-length(ts)-2b+1`. We partition the time series into windows (Step 1a) and then compute a histogram for the ith window with the `hist()` command (Step 1b). Setting `freq=FALSE` results in histogram bars measuring probability densities instead of frequencies. We further use the `splinefun()` command to fit a spline regression to histogram peaks. We plot the spline regression against the histogram with the command `curve()` and record the iterated sequence of histograms and regression lines with the command `ani.record()`. Figure 8.10b shows a sequence of histograms at times before and after the built-in tipping point at $t = 1000$. Before the tipping point, histograms display a single peak corresponding to stable-focus dynamics (left-most plot). In the neighbourhood of the tipping point, histograms transition towards the shape characterizing the classic chaotic butterfly attractor, namely a single broad peak with relatively flat shoulders (right-most plot).

In Step 1c, we begin the process of using the spline regression to count histogram peaks. We first retrieve the spline regression values with the command `y.spl<-f(x, deriv = 0)` and set negative values to zero. We next remove the (*noise*) floor that we observe, for example, on either side of the single peak in Figure 8.10b (left-most plot). The noise floor contains minor peaks generated by numerical error, that if counted, would greatly exaggerate the true number of states. Removing this noise floor can be done in several ways; we do it by setting regression points less than 5% of the maximum value to zero.

In Step 1d we count the number of histogram peaks for the cases of a corner peak (i.e. when the first bin is the highest), and interior peaks. We identify interior peaks by counting inflection points ip as the sign changes in the second derivatives of the spline-regression function retrieved with the command `derivative.2<-f(x,deriv=2)`. We then calculate the number of states reflected in histogram i as $states = 1 + ip/2$. Finally, we plot the number of states in each windowed histogram with horizontal axes for the window number and the corresponding real time (calculated as $time = b + window - 1$).

Figure 8.10c shows this plot generated for the Lorenz series in this illustration. Time starts at 300 as the midpoint of the first window. For initial times $300 < t \le 500$, the plot shows multiple transitional states as dynamics converge towards the stable-focus equilibrium. The plot then displays a single state corresponding to the stable focus. As windows begin to cover the tipping point ($t = 1000$), the plot reflects multiple states as the dynamics transition to the chaotic butterfly attractor. Once windows settle into the attractor, histograms assume a single broad peak with rounded shoulders. The plot reflects this by alternating between three and two states (when one of the shoulders disappears for a time). The plot succeeds in identifying the tipping point built into the time series.

8.8   Summary

In this chapter, we have presented empirical causal-detection methods for the case in which the tested variables have continuous records and another case in which some have irregular-sporadic records.

Convergent cross mapping (CCM) requires that variables have the continuous time records needed to reconstruct the shadow phase spaces used to measure causal interaction. Some practitioners rightly question how we can count on phase space reconstruction when 'it remains challenging to measure (and even define) noise and non-equilibrium behavior in natural systems' and given uncertainty over whether 'systems [are] approximately deterministic' (Colby and Baskerville, 2016). These concerns underlie the importance of applying CCM to signal-processed time series data diagnosed with deterministic, low-dimensional and nonlinear dynamics. Fortunately, we can make this diagnosis in a statistically rigorous manner with nonlinear time series methods. If we fail to detect evidence of deterministic dynamics in data, we should turn to more appropriate empirical causality tests. When it comes to CCM (and any other method): *If the shoe doesn't fit, don't wear it*!

In the final two sections, we have investigated methods designed to test whether variables with continuous records exhibit abrupt structural changes in dynamic behaviour corresponding to the irregular timing of variables exhibiting only occasional activity. If we uncover a strong signal in the continuous record, we can use a structural change-point detection approach based on singular spectrum analysis. Alternatively, if we detect a weak noisy signal, we can turn to a probabilistic tipping-point detection approach based on histograms constructed from the observed data. Armed with these methods, we can detect causal interactions when confronted with a broad array of heterogeneous historic records.

9
Phenomenological Modelling

`In solving a problem of this sort, the grand thing is
to be able to reason backward.'
                    Arthur Conan Doyle, A Study in Scarlet

9.1   Introduction

Phenomenological models mathematically describe relationships among empirically observed phenomena without attempting to explain the underlying mechanisms (Hilborn
and Mangel, 1997). Within the context of NLTS , phenomenological modelling goes
beyond phase space reconstruction to extract equations governing real-world system
dynamics from a single or multiple observed time series (Baker *et al.*, 1996; Dong *et al.*,
2015; Gouesbet and Maquet, 1992; Breeden and Hubler, 1990; Brunton *et al.*, 2016).
Brunton *et al.* (2016) explain that phenomenological modelling allows us to '[distill]
physical models of dynamical processes from big data . . . to extrapolate the dynamics
beyond the attractor where they were sampled and constructed' (p. 1). Data-driven
phenomenological modelling is a valuable intermediary between empirical observation and theory, because reliable explanation depends on accurate descriptions of how
things actually work .

   In this chapter, we illustrate phenomenological modelling by showing how it works
'in reverse' to extract the Lorenz equations from chaotic time series solutions: first from
all three solutions and then from just one. Long-run dynamics of the Lorenz model
are captured by the famous butterfly attractor, which successful phenomenological
representations replicate.

9.2   Components of a Phenomenological Model

Phenomenological models in NLTS are systems of first-order ordinary differential equations (ODEs), often expressed as polynomial expansions of interacting observed variables:

$$\frac{dx}{dt} = \dot{x} = \alpha_1 + \alpha_2 z + \alpha_3 y + \alpha_4 yz + \alpha_5 x + \alpha_6 xz + \alpha_7 xy + \alpha_8 xyz$$

$$\frac{dy}{dt} = \dot{y} = \beta_1 + \beta_2 z + \beta_3 y + \beta_4 yz + \beta_5 x + \beta_6 xz + \beta_7 xy + \beta_8 xyz \qquad (9.1)$$

$$\frac{dz}{dt} = \dot{z} = \gamma_1 + \gamma_2 z + \gamma_3 y + \gamma_4 yz + \gamma_5 x + \gamma_6 xz + \gamma_7 xy + \gamma_8 xyz$$

Nonlinear Time Series Analysis with R. Ray Huffaker, Marco Bittelli and Rodolfo Rosa, Oxford University Press (2017).
© Ray Huffaker, Marco Bittelli and Rodolfo Rosa. DOI: 10.1093/oso/9780198782933.001.0001

The system (9.1), for example, is composed of three ODEs expressed as third-degree polynomials in three observed time series variables (x_t, y_t and z_t), where the time series are the isolated signals and α_i, β_i and γ_i ($i = 1, \ldots, 8$) are unknown fixed coefficients. System dimensionality is guided by the embedding dimension estimated in phase space reconstruction, and system variables are selected with convergent cross mapping . The order of each polynomial is selected so that the system faithfully reproduces empirically reconstructed dynamics.

In this chapter, we construct R codes that formulate a phenomenological model from data on multiple observed variables (Code 9.7) and a single observed variable (Code 9.8). Each of these codes is composed of a sequence of user-defined functions that (1) approximate the time derivatives $\dot{x}$, $\dot{y}$ and $\dot{z}$ with finite-differencing techniques, (2) compute polynomial expansions of selected order, (3) estimate unknown coefficients with linear regression techniques (since polynomial ODEs are linear in their coefficients), (4) compute the goodness of fit of the estimated ODEs and (5) numerically solve the estimated ODE system for simulated values of the observed variables. We construct the code in stages by investigating each user-defined function in its entirety and then collapsing it to a *run* statement when investigating the next. The time series used to illustrate each user-defined functions are the solutions to the Lorenz equations generated by the user-defined function `lorenz_udf` (Code 6.16). We can then present the final codes (9.7) and (9.8) as a parsimonious sequence of run statements for the component user-defined functions.

9.3   Approximation of Derivatives with Finite Differences

Code 9.1 formulates a user-defined function `diff` that requires inputs of a time series x_t and an integration step δ to generate finite-difference approximations of time derivatives associated with the time series . We apply the following numerical recipe for computing *fourth-order centred finite differences* (Gouesbet and Maquet, 1992):

$$\dot{x}_t = \frac{8(x_{t+1} - x_{t-1}) - (x_{t+2} - x_{t-2})}{12\delta} + \vartheta(\delta^4) \tag{9.2}$$

where $\vartheta(\delta^4 t)$ is the truncation error. The first centred difference is computed in the third period ($t = 3$) since going two periods back in the equation takes us to the first observation, and we cannot go further back than that. Similarly, the last finite difference is computed two periods before the final period so that going two periods forward in the equation takes us to the final period. Consequently, the first two and last two observations are lost to differencing. The code uses this information to equate the length of the time series and differences.

```
#Code 9.1 Specify user-defined function to compute fourth-order centred finite differences
#(finite_differences_udf.R)

rm(list=ls(all=TRUE))   #remove objects between runs

#Load user-defined functions
setwd("C:/User Defined Functions")
dump("lorenz_udf", file="lorenz_udf.R")
```

```
source("lorenz_udf.R")

#Generate Lorenz data with user-defined function 'lorenz_udf'
lorenz_results<-lorenz_udf(x0=1,y0=1,z0=1,sigma=10,beta=8/3,rho=28,
 t.end=50,delta=0.001)
x<-lorenz_results[[1]];y<-lorenz_results[[2]];z<-lorenz_results[[3]]

#Set parameters for finite differences
delta<-0.001   #integration step

diff<-
  function(x,delta){  #formula for equally spaced grid
  #first centered observation is x.ts(3)
  x1<-x[3:(length(x)-2)] #first two and last two observations lost
  x_plus1<-x[4:length(x)]   #compute finite-differencing terms
  x_plus2<-x[5:length(x)]
  x_minus1<-x[2:length(x)]
  x_minus2<-x[1:length(x)]
  g<-length(x1);length(x_plus2)<-g;length(x_plus1)<-g  #make terms f equal length
  length(x_minus1)<-g;length(x_minus2)<-g
  #Compute 4th-order centred difference for equally spaced data points
  x.diff<-(-x_plus2+(8*x_plus1)-(8*x_minus1)+x_minus2)/(12*delta)

  #Make time series compatible length to finite difference
  g<-length(x.diff)
  #Time series loses first and last two observations (4 total).
  #To make same length as *.diff, remove two observations from
  #each end.
  x<-x[3:length(x)]   #remove 1st two observations
  length(x)<-g   #remove last two observations
  return<-list(x,x.diff)
} #end user-defined function

diff_x<-diff(x,delta)  #run user-defined function
x<-diff_x[[1]];x.diff<-diff_x[[2]]
```

Table 9.1 shows the output from Code 9.1 for the first six time periods (in the first column), the time series (in the second column) and the computed finite difference (in the third column).

9.4   Multivariate Polynomial Expansions

The *standard multivariate polynomial function* is

$$F(x_1, x_2, \ldots, x_m) = \sum_{n=1}^{N_P} A_n \, x_1^{\alpha_1} x_2^{\alpha_2} \cdots x_m^{\alpha_m} \tag{9.3}$$

where m is the number of variables in the expansion (three in the Lorenz system) and A_n are fixed unknown coefficients (Dong *et al.*, 2015). The number of monomials in the expansion is calculated as $N_p = (M_p)^m$, where M_p sets the range of monomial (individual) terms. For example, the number of monomials in each polynomial expansion in the system (9.1) is $N_p = (M_p = 2)^{(m=3)} = 8$, and monomial terms contain x_n^0 and x_n^1 ($n = 1, 2, 3$).

Table 9.1 Code 9.1 output: finite differences.

Time	x	$x.diff$
1	1.000516	0.5144436
2	1.001158	0.7676426
3	1.002051	1.0182859
4	1.003193	1.2664660
5	1.004583	1.5122737
6	1.006217	1.7557983

The powers taken by the monomial terms are calculated as follows:

$$\alpha_1 = \text{Int}\big\{(n-1)/M_p^{m-1}\big\}$$

$$\alpha_2 = \text{Int}\big\{[(n-1) - \alpha_1 M_p^{m-1}]/M_p^{m-2}\big\}$$

$$\alpha_3 = \text{Int}\big\{[(n-1) - \alpha_1 M_p^{m-1} - \alpha_2 M_p^{m-2}]/M_p^{m-3}\big\}$$

$$\alpha_4 = \text{Int}\big\{[(n-1) - \alpha_1 M_p^{m-1} - \alpha_2 M_p^{m-2} - \alpha_3 M_p^{m-3}]/M_p^{m-4}\big\} \tag{9.4}$$

$$\vdots$$

$$\alpha_m = \text{Int}\big\{(n-1) - \alpha_1 M_p^{m-1} - \alpha_2 M_p^{m-2} \cdots \alpha_{m-1} M_p\big\}$$

where Int{ } rounds down to the nearest integer (Dong *et al.*, 2015). The powers taken by the eight monomial terms in the system (9.1) are shown in Table 9.2.

Table 9.2 Powers taken by monomial terms.

n	α_1	α_2	α_3
1	0	0	0
2	$\text{Int}\{1/4\} = 0$	$\text{Int}\{[1 - (0)(4)]/2\} = \text{Int}\{1/2\} = 0$	$\text{Int}\{[1 - (0)(4)] - (0)(2)/1\} = \text{Int}\{1\} = 1$
3	$\text{Int}\{2/4\} = 0$	$\text{Int}\{[2 - (0)(4)]/2\} = \text{Int}\{1\} = 1$	$\text{Int}\{[2 - (0)(4)] - (1)(2)/1\} = \text{Int}\{0\} = 0$
4	$\text{Int}\{3/4\} = 0$	$\text{Int}\{[3 - (0)(4)]/2\} = \text{Int}\{3/2\} = 1$	$\text{Int}\{[3 - (0)(4)] - (1)(2)/1\} = \text{Int}\{1\} = 1$
5	$\text{Int}\{4/4\} = 1$	$\text{Int}\{[4 - (1)(4)]/2\} = \text{Int}\{0\} = 0$	$\text{Int}\{[4 - (1)(4)] - (0)(2)/1\} = \text{Int}\{0\} = 0$
6	$\text{Int}\{5/4\} = 1$	$\text{Int}\{[5 - (1)(4)]/2\} = \text{Int}\{1/2\} = 0$	$\text{Int}\{[5 - (1)(4)] - (0)(2)/1\} = \text{Int}\{1\} = 1$
7	$\text{Int}\{6/4\} = 1$	$\text{Int}\{[6 - (1)(4)]/2\} = \text{Int}\{1\} = 1$	$\text{Int}\{[6 - (1)(4)] - (1)(2)/1\} = \text{Int}\{0\} = 0$
8	$\text{Int}\{7/4\} = 1$	$\text{Int}\{[7 - (1)(4)]/2\} = \text{Int}\{3/2\} = 1$	$\text{Int}\{[7 - (1)(4)] - (1)(2)/1\} = \text{Int}\{1\} = 1$

The resulting standard polynomial expansion is

$$F(x_1, x_2, \ldots, x_m) = A_1 x_1^0 x_2^0 x_3^0 + A_2 x_1^0 x_2^0 x_3^1 + A_3 x_1^0 x_2^1 x_3^0 + A_4 x_1^0 x_2^1 x_3^1$$

$$+ A_5 x_1^1 x_2^0 x_3^0 + A_6 x_1^1 x_2^0 x_3^1 + A_7 x_1^1 x_2^1 x_3^0 + A_8 x_1^1 x_2^1 x_3^1$$

$$= A_1 + A_2 x_3 + A_3 x_2 + A_4 x_2 x_3$$

$$+ A_5 x_1 + A_6 x_1 x_3 + A_7 x_1 x_2 + A_8 x_1 x_2 x_3 \qquad (9.5)$$

Code 9.2 formulates a user-defined function `polynomial_udf` that computes polynomial expansions. It requires input of the embedding dimension m (i.e. the number of variables in the expansion) and the polynomial range parameter M_p. The function includes several conditional statements to accommodate the polynomial ranges $M_p = 2, 3, 4$ ($m = 2, 3$), $M_p = 2, 3$ ($m = 4$) and $M_p = 2$ ($m = 5$). It is straightforward (although tedious) to augment the code to accommodate other combinations. Since we are working with data generated by the Lorenz equations, we set $m = 3$ and $M_p = 2$. The user-defined function first specifies the monomial terms generated by selected combinations of m and M_p to use as labels in an output matrix. The function is then composed of two loops. Loop n computes the powers in each monomial term using eqns (9.4). The output of this loop is a matrix containing powers like those computed in Table 9.2. Loop i then populates the monomial terms with observed time series data. Finally, the function relies on the R package `mpoly` (Kahle, 2015) to express the monomial terms symbolically.

```
#Code 9.2 Specify user-defined function to create multivariate polynomial expansion and
#populate monomial terms with observed time series data for three variables
#(polynomial_udf.R)

rm(list=ls(all=TRUE))   #remove objects between runs

#Directories
dir.udf<-as.character("C:/User Defined Functions")

#Load user-defined functions
setwd(dir.udf)
dump("lorenz_udf", file="lorenz_udf.R")
source("lorenz_udf.R")
dump("finite_differences_udf", file="finite_differences_udf.R")
source("finite_differences_udf.R")

#Generate Lorenz data with user-defined function 'lorenz_udf'
lorenz_results<-lorenz_udf(x0=1,y0=1,z0=1,sigma=10,beta=8/3,rho=28,
 t.end=50,delta=0.001)
x<-lorenz_results[[1]];y<-lorenz_results[[2]];z<-lorenz_results[[3]]

#Set parameters for finite differences and polynomial expansion
delta<-0.001  #integration step
m<-3  #number of variables
Mp<-2  #polynomial range#Set parameters

#Finite differences
#Response variables: approximation of derivatives with finite differences
#Response variable:  user-defined function to compute finite differences
```

```
diff_x<-diff(x,delta)  #run user-defined function for x
diff_y<-diff(y,delta)
diff_z<-diff(z,delta)
x<-diff_x[[1]];y<-diff_y[[1]];z<-diff_z[[1]]  #times series same length as differences
x.diff<-diff_x[[2]];y.diff<-diff_y[[2]];z.diff<-diff_z[[2]] #finite differences

#User-defined function to calculate polynomial
#User-defined function to create multivariate polynomial expansion and populate
#monomial terms with observed time series data

polynomial_udf<-function(m,Mp){
  library(mpoly) #symbolic polynomial package

  #Names of monomial terms in columns of data matrix
  #Two Variables
  if(m==2&&Mp==2) {col.names<-c("constant","y","x","xy")}
  if(m==2&&Mp==3) {col.names<-c("constant","y","y2","x","xy","xy2","x2","x2y","x2y2")}
  if(m==2&&Mp==4) {col.names<-c("constant","y","y2","y3","x","xy","xy2","xy3","x2","x2y",
    "x2y2","x2y3","x3","x3y","x3y2","x3y3")}

  #Three variables
  if(m==3&&Mp==2) {col.names=c("constant","z","y","yz","x","xz","xy","xyz")}
  if(m==3&&Mp==3) {col.names=c("constant","z","z2","y","yz","yz2","y2","y2z","y2z2","x",
    "xz","xz2","xy","xyz","xyz2","xy2","xy2z","xy2z2","x2","x2z","x2z2","x2y","x2yz",
    "x2y2z2","x2y2","x2y2z2")}
  if(m==3&&Mp==4) {col.names=c("constant","z","z2","z3","y","yz","yz2","yz3","y2","y2z",
    "y2z2","y2z3","y3","y3z","y3z2","y3z3","x","xz","xz2","xz3","xy","xyz","xyz2","xyz3",
    "xy2","xy2z","xy2z2","xy2z3","xy3","xy3z","xy3z2","xy3z3","x2","x2z","x2z2","x2z3",
    "x2y","x2yz","x2yz2","x2yz3","x2y2","x2y2z","x2y2z2","x2y2z3","x2y3","x2y3z","x2y3z2",
    "x2y3z3","x3","x3z","x3z2","x3z3","x3y","x3yz","x3yz2","x3yz3","x3y2","x3y2z","x3y2z2",
    "x3y2z3","x3y3","x3y3z","x3y3z2","x3y3z3")}

  #Four variables
  if(m==4&&Mp==2) {col.names=c("constant","w","z","zw","y","yw","yz","yzw","x","xw","xz",
    "xzw","xy","xyw","xyz","xyzw")}
  if(m==4&&Mp==3) {col.names=c("constant","w","w2","z","zw","zw2","z2","z2w","z2w2","y",
    "yw","yw2","yz","yzw","yzw2","yz2","yz2w","yz2w2","y2","y2w","y2w2","y2z","y2zw",
    "y2zw2","y2z2","y2z2w","y2z2w2","x","xw","xw2","xz","xzw","xzw2","xz2","xz2w","xz2w2",
    "xy","xyw","xyw2","xyz","xyzw","xyzw2","xyz2","xyz2w","xyz2w2","xy2","xy2w","xy2w2",
    "xy2z","xy2zw","xy2zw2","xy2z2","xy2z2w","xy2z2w2","x2","x2w","x2w2","x2z","x2zw",
    "x2zw2","x2z2","x2z2w","x2z2w2","x2y","x2yw","x2yw2","x2yz","x2yzw","x2yzw2","x2yz2",
    "x2yz2w","x2yz2w2","x2y2","x2y2w","x2y2w2","x2y2z","x2y2zw","x2y2zw2","x2y2z2",
    "x2y2z2w","x2y2z2w2")}

  #Five variables
  if(m==5&&Mp==2) {col.names<-c("constant","v","w","wv","z","zv","zw","zwv","y","yv","yw",
    "ywv","yz","yzv","yzw","yzwv","x","xv","xw","xwv","xz","xzv","xzw","xzwv","xy","xyv",
    "xyw","xywv","xyz","xyzv","xyzw","xyzwv")}

  Np<-Mp^m  #number of monomials in polynomial expansion

  #Compute alphas: powers of x, y, z, w, v in monomials
  alpha1<-matrix(0,Np);alpha2<-matrix(0,Np);alpha3<-matrix(0,Np)  #storage matrices for loop
  alpha4<-matrix(0,Np);alpha5<-matrix(0,Np)
    for(n in 1:Np){  #loop through monomial terms
      alpha1a<-floor((n-1)/(Mp^(m-1)))
      alpha2a<-floor(((n-1)-(alpha1a*(Mp^(m-1))))/(Mp^(m-2)))
      alpha3a<-floor(((n-1)-(alpha1a*(Mp^(m-1)))-(alpha2a*(Mp^(m-2))))/(Mp^(m-3)))
      alpha4a<-floor(((n-1)-(alpha1a*(Mp^(m-1)))-(alpha2a*(Mp^(m-2)))-
        (alpha3a*(Mp^(m-3))))/(Mp^(m-4)))
```

```
    alpha5a<-floor((((n-1)-(alpha1a*(Mp^(m-1)))-(alpha2a*(Mp^(m-2)))-
      (alpha3a*(Mp^(m-3)))-(alpha4a*(Mp^(m-4))))/(Mp^(m-5)))
    alpha1[n]<-alpha1a  #load storage matrices
    alpha2[n]<-alpha2a
    alpha3[n]<-alpha3a
    alpha4[n]<-alpha4a
    alpha5[n]<-alpha5a
  }  #end loop

#'powers matrix': rows are powers of x (1st column), y (2nd column), z (3rd column),
#w (4th column) and v (5th column). Columns are monomial terms (mono).
powers<-cbind(alpha1,alpha2,alpha3,alpha4,alpha5)

#Populate monomial terms with time series data
termList2<-matrix(0,1) #store monomial terms in 'mpoly' language
data<-matrix(0,length(x),Np) #store monomials with observed time series
  for(i in 1:Np){ #loop through monomial terms
    if(m==2){ #2 variables
      termList<-list(c(x=powers[i,1],y=powers[i,2],coef=1))
      mono<-mpoly(termList)
      mono.data<-x^powers[i,1]*y^powers[i,2]}
    if(m==3){ #3 variables
      termList<-list(c(x=powers[i,1],y=powers[i,2],z=powers[i,3],coef=1))
      mono<-mpoly(termList)
      mono.data<-x^powers[i,1]*y^powers[i,2]*z^powers[i,3]}
    if(m==4){ #4 variables
      termList<-list(c(x=powers[i,1],y=powers[i,2],z=powers[i,3],w=powers[i,4],coef=1))
      mono<-mpoly(termList)
      mono.data<-x^powers[i,1]*y^powers[i,2]*z^powers[i,3]*w^powers[i,4]}
    if(m==5){ #5 variables
      termList<-list(c(x=powers[i,1],y=powers[i,2],z=powers[i,3],w=powers[i,4],v=powers[i,5],
        coef=1))
      mono<-mpoly(termList)
      mono.data<-x^powers[i,1]*y^powers[i,2]*z^powers[i,3]*w^powers[i,4]*v^powers[i,5]}
    data[,i]<-mono.data
    termList2[i]<-mono  #store mono in mpoly language
  } #end loop
 colnames(data)<-c(col.names)  #apply monomial names to data matrix
 return<-list(termList2,col.names,data)  #output
} #end user-defined function

poly.exp<-polynomial_udf(m,Mp)  #run user-defined function
col.names<-poly.exp[[2]]  #monomial terms
data<-poly.exp[[3]]        #populated data matrix
```

The output for Code 9.2 reports the monomial terms populated with observed time series data, as shown in Table 9.3 for the first six time periods.

9.5 Estimating System Coefficients: Ordinary Least Squares

The system (9.1) is linear in the coefficients, so a single-equation *ordinary least squares (OLS)* regression is feasible. Code 9.3 specifies a user-defined function `ols_udf` that runs OLS on a single ODE with inputs of the dependent (or response) variable `dep`, the matrix containing data on regressors `exog`, the embedding dimension m (i.e. the number of independent variables in the regression) and the polynomial range parameter M_p. The response variable is computed finite differences (Code 9.1) and the regressors

Table 9.3 Code 9.2 output: monomial terms populated with observed time series data.

Time	Constant	z	y	yz	x	xz	xy	xyz
1	1	0.997	1.05	1.05	1.00	0.997	1.05	1.05
2	1	0.995	1.08	1.07	1.00	0.996	1.08	1.07
3	1	0.994	1.03	1.10	1.00	0.996	1.11	1.10
4	1	0.992	1.13	1.12	1.00	0.995	1.13	1.12
5	1	0.991	1.16	1.44	1.00	0.995	1.16	1.15
6	1	0.989	1.18	1.17	1.01	0.995	1.19	1.18

are the populated monomial terms (Code 9.2). We illustrate the use of Code 9.3 to fit the first ODE in the Lorenz system in the chaotic range:

$$\dot{x} = 10(y - x) \tag{9.6}$$

Accordingly, we set $m = M_p = 2$ in Code 9.3.

```
#Code 9.3 Ordinary least squares (OLS) regression

rm(list=ls(all=TRUE))   #remove objects between runs

#Directories
dir.udf<-as.character("C:/User Defined Functions")

#Load user-defined functions
setwd(dir.udf)
dump("lorenz_udf", file="lorenz_udf.R")
source("lorenz_udf.R")
dump("finite_differences_udf", file="finite_differences_udf.R")
source("finite_differences_udf.R")
dump("polynomial_udf", file="polynomial_udf.R")
source("polynomial_udf.R")

#Generate Lorenz data with user-defined function 'lorenz_udf'
lorenz_results<-lorenz_udf(x0=1,y0=1,z0=1,sigma=10,beta=8/3,rho=28,
 t.end=50,delta=0.001)
x<-lorenz_results[[1]];y<-lorenz_results[[2]];z<-lorenz_results[[3]]

#Set parameters for finite differences and polynomial expansion
delta<-0.001   #integration step
m<-2   #number of variables
Mp<-2   #polynomial range

#Response variable:  user-defined function to compute finite differences
diff_x<-diff(x,delta)   #run user-defined function for x
diff_y<-diff(y,delta)
diff_z<-diff(z,delta)
x<-diff_x[[1]];y<-diff_y[[1]];z<-diff_z[[1]]   #times series same length differences
x.diff<-diff_x[[2]] #finite differences

#Regressors: user-defined function to compute multivariate polynomial expansion
poly.exp<-poly(m,Mp)
#Data check: compare order of terms in 'mpoly' with column names
```

```
termList2<-poly.exp[[1]]   #monomial terms
poly1<-mpoly(termList2);term<-terms(poly1);term
col.names<-poly.exp[[2]];col.names
#Monomials populated with data
data1<-poly.exp[[3]]   #populated data matrix
data2<-data1[complete.cases(data1),]   #remove last row with na's
data<-data2[,-1] #remove 1st column of 1's (regression packages don't require it)

#User-defined function to run OLS
ols_udf<-function(dep,exog,m,Mp){
data.reg<-cbind(dep,exog);data.reg<-as.data.frame(data.reg) #create data frame
if(m==2) {
  if(Mp==2){
    eqn <- dep~x+y+xy
  }  #end Mp=2

  if(Mp==3){
    eqn <- dep~x+y+xy+x2+y2+x2y+xy2+x2y2
  }   #end Mp=3

  if(Mp==4) {
    eqn <- dep~x+y+xy+x2+y2+x2y+xy2+x2y2+x3+y3+x3y+xy3+x2y3+x3y2+x3y3
  }  #end Mp=4
} #end if m=2

if(m==3) {
  if(Mp==2){
    eqn <- dep~z+y+yz+x+xz+xy+xyz
  }   #end Mp=2

  if(Mp==3){
    eqn <- dep~z+z2+y+yz+yz2+y2+y2z+y2z2+x+xz+xz2+xy+xyz+xyz2+xy2+xy2z+xy2z2+x2+x2z+x2z2+
      x2y+x2yz+x2yz2+x2y2+x2y2z+x2y2z2
  }   #end Mp=3

  if(Mp==4) {
    eqn <- dep~z+z2+z3+y+yz+yz2+yz3+y2+y2z+y2z2+y2z3+y3+y3z+y3z2+y3z3+x+xz+xz2+xz3+
      xy+xyz+xyz2+xyz3+xy2+xy2z+xy2z2+xy2z3+xy3+xy3z+xy3z2+xy3z3+x2+x2z+x2z2+x2z3+x2y+
      x2yz+x2yz2+x2yz3+x2y2+x2y2z+x2y2z2+x2y2z3+x2y3+x2y3z+x2y3z2+x2y3z3+x3+x3z+x3z2+
      x3z3+x3y+x3yz+x3yz2+x3yz3+x3y2+x3y2z+x3y2z2+x3y2z3+x3y3+x3y3z+x3y3z2+x3y3z3
  }   #end Mp=4
} #end if m=3

if(m==4) {
  if(Mp==2){
    eqn <- dep~w+z+zw+y+yw+yz+yzw+x+xw+xz+xzw+xy+xyw+xyz+xyzw
  }  #end if Mp=2

  if(Mp==3){
    eqn <- dep~w+w2+z+zw+zw2+z2+z2w+z2w2+y+yw+yw2+yz+yzw+yzw2+yz2+yz2w+yz2w2+y2+y2w+
      y2w2+y2z+y2zw+y2zw2+y2z2+y2z2w+y2z2w2+x+xw+xw2+xz+xzw+xzw2+xz2+xz2w+xz2w2+xy+xyw+
      xyw2+xyz+xyzw+xyzw2+xyz2+xyz2w+xyz2w2+xy2+xy2w+xy2w2+xy2z+xy2zw+xy2zw2+xy2z2+xy2z2w+
      xy2z2w2+x2+x2w+x2w2+x2z+x2zw+x2zw2+x2z2+x2z2w+x2z2w2+x2y+x2yw+x2yw2+x2yz+x2yzw+x2yzw2+
      x2yz2+x2yz2w+x2yz2w2+x2y2+x2y2w+x2y2w2+x2y2z+x2y2zw+x2y2zw2+x2y2z2+x2y2z2w+x2y2z2w2
  }   #end if Mp = 3
}   #end if m=4

if(m==5) {
  eqn <- dep~v+w+wv+z+zv+zw+zwv+y+yv+yw+ywv+yz+yzv+yzw+yzwv+x+xv+xw+xwv+xz+xzv+xzw+xzwv+
    xy+xyv+xyw+xywv+xyz+xyzv+xyzw+xyzwv
```

```
}  #end if m=5

fit<-lm(eqn,data.reg)
coeffs<-as.numeric(coef(fit))
table<-summary(fit)

results<-list(coeffs,table)
return(results)
} #end user-defined function

results<-ols_udf(x.diff,data,m,Mp)
coeffs_ols<-results[[1]]
table<-results[[2]]
```

The output for Code 9.3 includes the summary table with regression statistics. OLS regression perfectly fits the Lorenz equation (9.6) with coefficient of determination $R^2 = 1$.

9.6   Estimating System Coefficients: Regularized Regression Methods

Estimating phenomenological models with real-world data may introduce statistical problems of multicollinearity and overfitting. *Multicollinearity* arises when independent variables (*regressors*) are highly linearly correlated. Estimated coefficients become unreliable, because they can shift erratically from small to large values with small changes in the data. Identifying redundant regressors is difficult, resulting in overly complex phenomenological models with an excessive number of coefficients. These models may fit the data perfectly, since even absurd and incorrect models can do so by including enough coefficients relative to sampled data. However, they are *overfitted* in the sense that they account for random instead of structured variation in sampled data and thus cannot be reliably used to extrapolate dynamic structure outside of the sample. Polynomial models are particularly susceptible to multicollinearity and overfitting.

Regularized (or *penalized*) regression techniques offer potential remedies. These techniques purposefully introduce bias into estimation (meaning that the estimator's expected value differs from the estimated coefficient's true value) to reduce the variability of estimated coefficients and to select essential regressors. In particular, *ridge regression* reduces variability by shrinking the size of the coefficients of highly correlated regressors towards each other. It does so by imposing an ℓ_2 constraint that restricts the sum of squares of estimated coefficients in OLS estimation. Alternatively, the *least absolute shrinkage and selection operator (LASSO)* reduces overfitting by simultaneously estimating coefficients and selecting essential regressors. Regressors whose estimated coefficients are below given thresholds are driven from the model, which tends to select one of the highly correlated regressors while discarding the others. LASSO does this by imposing an ℓ_1 constraint restricting the sum of the absolute value of regression coefficients in OLS estimation (Tibshirani, 1996). Brunton *et al.* (2016) have emphasized the value of regularized regression techniques , such as LASSO, in phenomenological modelling of real-world nonlinear dynamical systems that may be governed by only a few relevant terms. These techniques are geared to generate parsimonious models that balance accuracy with model complexity to avoid overfitting.

The R package `glmnet` (Friedman *et al.*, 2015) runs Ridge regression, LASSO or a combination of the two as controlled by an *elastic-net mixing parameter* α. Ridge regression is run for $\alpha = 0$, LASSO for $\alpha = 1$ (the default) or a mixed approach bridging the gap between the two approaches for intermediate α values. Running a mixed model can increase numerical stability (by reducing variability due to extreme correlations) while retaining the model selection benefits of LASSO.

Adaptive LASSO can increase the consistency of LASSO (meaning that the probability that the estimated coefficient approaches its true value converges to 1 as the sample size increases indefinitely) by computing adaptive weights differentially penalizing coefficients in the ℓ_1 constraint. Adaptive LASSO has *oracle* properties; in other words, it performs as well as if the true underlying model had been given in advance (Zou, 2006). The `adaptive.weights` routine in the R package `MESS` (Ekstrom, 2015) computes weights as

$$1/|\widehat{\beta}|^{nu}$$

that is, as the inverse of the absolute value of the estimated coefficient raised to the *nu*th power, where *nu* is a tuning parameter.

Code 9.4 formulates a user-defined function `lasso_udf` to estimate the coefficients for a single ODE with LASSO using the R package `glmnet` or Adaptive LASSO by computing adaptive weights with the R routine `adaptive.weights`. The user-defined function `lasso_udf` requires input of the matrix `exog` of time series observations on exogenous variables, the vector `dep` of observations on the response variable, the elastic-net mixing parameter α, the parameter *weight* that runs Adaptive LASSO regression when set to TRUE and the Adaptive LASSO non-negative tuning parameter *nu*.

For this illustration, we again fit the first ODE in the Lorenz system (eqn 9.6) in the chaotic range. We set the elastic-net mixing parameter to $\alpha = 1$ to run unmixed LASSO and set *weight* = TRUE to run Adaptive LASSO with tuning parameter *nu* = 1.

```
#Code 9.4 LASSO regression (lasso_udf.R)

rm(list=ls(all=TRUE))   #remove objects between runs

#Directories
dir.udf<-as.character("C:/User Defined Functions")

#Load user-defined functions
setwd(dir.udf)
dump("lorenz_udf", file="lorenz_udf.R")
source("lorenz_udf.R")
dump("finite_differences_udf", file="finite_differences_udf.R")
source("finite_differences_udf.R")
dump("polynomial_udf", file="polynomial_udf.R")
source("polynomial_udf.R")
dump("ols_udf", file="ols_udf.R")
source("ols_udf.R")

#Generate Lorenz data with user-defined function 'lorenz_udf'
lorenz_results<-lorenz_udf(x0=1,y0=1,z0=1,sigma=10,beta=8/3,rho=28,
 t.end=50,delta=0.001)
x<-lorenz_results[[1]];y<-lorenz_results[[2]];z<-lorenz_results[[3]]
```

```
#Set parameters
delta<-0.001  #integration step
m<-2 #number of equations
Mp<-2 #polynomial range
alpha<-1 #glmnet elastic-net parameter
weight<-TRUE #Run Adaptive LASSO by computing adaptive weights
nu<-1 #Adaptive LASSO non-negative tuning parameter

#Response variable: user-defined function to compute finite differences
diff_x<-diff(x,delta)  #run user-defined function for x
diff_y<-diff(y,delta)
diff_z<-diff(z,delta)
x<-diff_x[[1]];y<-diff_y[[1]];z<-diff_z[[1]]  #times series same length differences
x.diff<-diff_x[[2]] #finite differences

#Regressors: user-defined function to compute multivariate polynomial expansion
poly.exp<-poly(m,Mp)
#Data check: compare order of terms in 'mpoly' with column names
termList2<-poly.exp[[1]]  #monomial terms
poly1<-mpoly(termList2);term<-terms(poly1);term
col.names<-poly.exp[[2]];col.names
#Monomials populated with data
data1<-poly.exp[[3]]  #populated data matrix
data2<-data1[complete.cases(data1),]  #remove last row with na's
data<-data2[,-1] #remove 1st column of 1's (regression packages don't require it)

#User-defined function to run OLS
results_ols<-ols_udf(x.diff,data,m,Mp)
coeffs_ols<-results_ols[[1]]
table_ols<-results_ols[[2]]

#User-defined function to run LASSO
lasso_udf<-function(exog,dep,alpha,weight,nu){
library(glmnet) #for lasso
library(MESS) #for adaptive weights

if(weight){
weights<-adaptive.weights(exog, dep, nu, weight.method = c("multivariate"))
fit<-glmnet(exog,dep,alpha=alpha,penalty.factor=weights$weights)} else {
fit<-glmnet(exog,dep,alpha=alpha)}

plot(fit) #coefficient path against constraint
plot(fit,xvar="dev",label=TRUE) #coefficient path against fraction deviance
table<-print(fit) #number of variables, %deviation, lambda
coef.all<-coef(fit)  #coefs for all lambda

#Model selection: cross validation
#Choose regression for given value of lambda):
#lambda.min is value at minimal mean squared error (min MSE)
#lambda.1se is value whose mean squared error is within one standard error of min MSE
#Make sure to give same run specifications in cv.glmnet() as in glmet()

if(weight){
cv<-cv.glmnet(exog,dep,alpha=alpha,penalty.factor=weights$weights)} else {
cv<-cv.glmnet(exog,dep,alpha=alpha)}

plot (cv) #MSE against log(lambda)
lambda.min<-cv$lambda.min #retrieve deviance explained for coef.min
lambda.min<-as.numeric(format(lambda.min, scientific=TRUE,digits=4))
lambda.1se<-cv$lambda.1se #retrieve deviance explained for coef.1se
```

```
lambda.1se<-as.numeric(format(lambda.1se, scientific=TRUE,digits=4))

#Coefficients
coef.min<-coef(fit, s=lambda.min)
coef.min<-round(as.vector(coef.min),digits=4)
coef.1se<-coef(fit, s=lambda.1se)
coef.1se<-round(as.vector(coef.1se),digits=4)
coeffs<-cbind(coef.min,coef.1se)
rownames(coeffs)<-col.names

#Retrieve fraction of explained deviance for different coefficient groups
index.min<-which(table[,3]==lambda.min)  #coef.min
dev.min<-table[,2][index.min];dev.min<-round(as.vector(dev.min),digits=2)
index.1se<-which(table[,3]==lambda.1se)  #coef.1se
dev.1se<-table[,2][index.1se];dev.1se<-round(as.vector(dev.1se),digits=2)
dev<-rbind(dev.min,dev.1se)

results<-list(coeffs,dev)
return(results)
} #end user defined function

results_lasso<-lasso_udf(data,x.diff,alpha,weight,nu)
coeffs_lasso<-results_lasso[[1]]
dev_lasso<-results_lasso[[2]]
```

The R package `glmnet` estimates several coefficient vectors associated with different values of a parameter λ measuring the extent to which the ℓ_1 constraint is binding on OLS estimation. For example, $\lambda = 0$ returns unconstrained OLS estimated coefficients. The R routine `cv.glmnet` uses *cross validation* to select the value of λ associated with the LASSO model providing the best fit to the data on a *mean squared error (MSE)* basis. MSE is the average of squared deviations between finite differences computed from the data and those predicted by the estimated regression equation, where n is the sample size. The routine reports cross-validated models associated with the value of λ giving minimum MSE, `lambda.min`, and the λ whose MSE is within one standard error of the minimum MSE, namely `lambda.1se`. The user-defined function `lasso_udf` retrieves the estimated coefficients for the two cross-validated models and reports the percentage of variation in the response variable that each explains (*explained deviation*).

The output of Code 9.4 reports the estimated coefficients for the cross-validated models, as shown in Table 9.4. We observe the bias that LASSO deliberately enters into estimation since the estimated coefficients for both models deviate from the true coefficients in the Lorenz equation (9.6). Since the Lorenz data do not exhibit multicollinearity, OLS performs well (returning unbiased estimates), and LASSO is not needed in this particular case to correct for associated problems. Later in this chapter, we will see a case where LASSO is needed.

9.7 Goodness of Fit

We generalize the cross-validation technique used by the R package `glmnet` to assess the *goodness of fit* of estimated phenomenological models. How well do estimated models perform in predicting finite differences computed by Code 9.1? The problem with assessing performance from a single calculated value of the mean squared error,

Table 9.4 Code 9.4 output: LASSO coefficient estimates.

Monomial term	*coef.min*	*coef.1se*
Constant	0.053	0.053
z	0.00	0.00
y	9.68	9.68
yz	0.00	0.00
x	−9.64	−9.64
xz	0.00	0.00
xy	0.00	0.00
xyz	0.00	0.00

for example, is subjectively determining which values indicate an acceptable fit when values can range from zero to infinity. We attack this problem by assessing the statistical significance of the calculated measure. We follow Ritter and Munoz-Carpena (2013) by calculating (1) bootstrapped confidence intervals from a distribution of the performance measure obtained by resampling the computed finite differences and (2) probability values that the performance measure exceeds given thresholds representing an acceptable fit. We again adopt the dimensionless *Nash–Sutcliffe coefficient of efficiency (nse)* .

Code 9.5 formulates a user-defined function `goodfit_3` to statistically assess goodness of fit in ODE systems with three equations. It requires inputs of the following: the calculated finite differences ($x.dep$, $y.dep$ and $z.dep$); the information required to simulate the differences with the phenomenological model, including the observed time series (x, y and z), the fitted parameters (xp, yp and zp) and the polynomial range parameter (M_p); and *ci*. If we set `ci<-TRUE`, a 95% bootstrapped confidence interval is calculated for the computed *nse* in a subsequent conditional statement. The confidence interval tells us that 95% of the *nse* values computed in repeated sampling would rest in the computed interval. Computed *nse* values are increasingly statistically significant as they rest in tighter confidence intervals. We next select the regression method with the conditional statement `ols<-TRUE` to run OLS, or `ols<-FALSE` to run LASSO. For this illustration, we select the Adaptive LASSO estimated coefficients associated with `lambda.min`.

```
#Code 9.5 User-defined function to calculate goodness of fit with three variables
rm(list=ls(all=TRUE))   #remove objects between runs

#Directories
dir.udf<-as.character("C:/User Defined Functions")

#Load user-defined functions
setwd(dir.udf)
dump("lorenz_udf", file="lorenz_udf.R")
source("lorenz_udf.R")
dump("finite_differences_udf", file="finite_differences_udf.R")
source("finite_differences_udf.R")
```

```
dump("polynomial_udf", file="polynomial_udf.R")
source("polynomial_udf.R")
dump("ols_udf", file="ols_udf.R")
source("ols_udf.R")
dump("lasso_udf", file="lasso_udf.R")
source("lasso_udf.R")

#Generate Lorenz data with user-defined function 'lorenz_udf'
lorenz_results<-lorenz_udf(x0=1,y0=1,z0=1,sigma=10,beta=8/3,rho=28,
 t.end=50,delta=0.001)
x<-lorenz_results[[1]];y<-lorenz_results[[2]];z<-lorenz_results[[3]]

#Set parameters
delta<-0.001  #integration step
m<-3 #number of equations
Mp<-2 #polynomial range
alpha<-1 #glmnet mixing parameter
weight<-TRUE #If 'TRUE' run Adaptive LASSO
nu<-1 #Adaptive LASSO non-negative tuning parameter
ci<-FALSE #95% confidence intervals on calculated Nash-Sutcliffe measure

#Select regression method
ols<-FALSE  #TRUE=run ols, FALSE=run lasso

#Response variable:  User-defined function to compute finite differences
diff_x<-diff(x,delta)  #run user-defined function for x
diff_y<-diff(y,delta)
diff_z<-diff(z,delta)
x<-diff_x[[1]];y<-diff_y[[1]];z<-diff_z[[1]]  #times series same length differences
x.diff<-diff_x[[2]];y.diff<-diff_y[[2]];z.diff<-diff_z[[2]] #finite differences

#Regressors: User-defined function to compute multivariate polynomial expansion
poly.exp<-poly(m,Mp)
#Data Check: compare order of terms in 'mpoly' with column names
termList2<-poly.exp[[1]]  #monomial terms
poly1<-mpoly(termList2);term<-terms(poly1);term
col.names<-poly.exp[[2]];col.names
#Monomials populated with data
data1<-poly.exp[[3]]  #populated data matrix
data2<-data1[complete.cases(data1),]  #remove last row with na's
data<-data2[,-1] #remove 1st column of 1's (regression packages don't require it)

#Regression
if(ols){#User-defined function to run OLS
results_ols_x<-ols_udf(x.diff,data,m,Mp)  #x.diff
xp.ols<-results_ols_x[[1]]
table_ols_x<-results_ols_x[[2]]

results_ols_y<-ols_udf(y.diff,data,m,Mp)  #y.diff
yp.ols<-results_ols_y[[1]]
table_ols_y<-results_ols_y[[2]]

results_ols_z<-ols_udf(z.diff,data,m,Mp)  #z.diff
zp.ols<-results_ols_z[[1]]
table_ols_z<-results_ols_z[[2]]

xp<-xp.ols;yp<-yp.ols;zp<-zp.ols  #OLS parameters

} else
```

```
{#User-defined function to run LASSO
results_lasso_x<-lasso_udf(data,x.diff,alpha,weight,nu)
xp.min<-as.vector(results_lasso_x[[1]][,1])
xp.1se<-as.vector(results_lasso_x[[1]][,2])
dev_x<-results_lasso_x[[2]]

results_lasso_y<-lasso_udf(data,y.diff,alpha,weight,nu)
yp.min<-as.vector(results_lasso_y[[1]][,1])
yp.1se<-as.vector(results_lasso_y[[1]][,2])
dev_y<-results_lasso_y[[2]]

results_lasso_z<-lasso_udf(data,z.diff,alpha,weight,nu)
zp.min<-as.vector(results_lasso_z[[1]][,1])
zp.1se<-as.vector(results_lasso_z[[1]][,2])
dev_z<-results_lasso_z[[2]]

xp<-xp.min;yp<-yp.min;zp<-zp.min #LASSO min coefficients
#xp<-xp.1se;yp<-yp.1se;zp<-zp.1se #LASSO 1se coefficients
}


#User-defined function to assess goodness of fit
goodfit_3<-function(x.dep,y.dep,z.dep,x,y,z,xp,yp,zp,Mp,ci) {
if(Mp==2){
#Regression parameters
x1=xp[1];x2=xp[2];x3=xp[3];x4=xp[4];x5=xp[5];x6=xp[6];x7=xp[7];x8=xp[8]
y1=yp[1];y2=yp[2];y3=yp[3];y4=yp[4];y5=yp[5];y6=yp[6];y7=yp[7];y8=yp[8]
z1=zp[1];z2=zp[2];z3=zp[3];z4=zp[4];z5=zp[5];z6=zp[6];z7=zp[7];z8=zp[8]
#Regression equations
dx <- x1+(x2*z)+(x3*y)+(x4*y*z)+(x5*x)+(x6*x*z)+(x7*x*y)+(x8*x*y*z)
dy <- y1+(y2*z)+(y3*y)+(y4*y*z)+(y5*x)+(y6*x*z)+(y7*x*y)+(y8*x*y*z)
dz <- z1+(z2*z)+(z3*y)+(z4*y*z)+(z5*x)+(z6*x*z)+(z7*x*y)+(z8*x*y*z)
} #end Mp=2

if(Mp==3){
#Regression parameters
x1=xp[1];x2=xp[2];x3=xp[3];x4=xp[4];x5=xp[5];x6=xp[6];x7=xp[7];x8=xp[8];x9=xp[9];
x10=xp[10];x11=xp[11];x12=xp[12];x13=xp[13];x14=xp[14];x15=xp[15];x16=xp[16];
x17=xp[17];x18=xp[18];x19=xp[19];x20=xp[20];x21=xp[21];x22=xp[22];x23=xp[23];
x24=xp[24];x25=xp[25];x26=xp[26];x27=xp[27];y1=yp[1];y2=yp[2];y3=yp[3];y4=yp[4];
y5=yp[5];y6=yp[6];y7=yp[7];y8=yp[8];y9=yp[9];y10=yp[10];y11=yp[11];y12=yp[12];
y13=yp[13];y14=yp[14];y15=yp[15];y16=yp[16];y17=yp[17];y18=yp[18];y19=yp[19];
y20=yp[20];y21=yp[21];y22=yp[22];y23=yp[23];y24=yp[24];y25=yp[25];y26=yp[26];
y27=yp[27];z1=zp[1];z2=zp[2];z3=zp[3];z4=zp[4];z5=zp[5];z6=zp[6];z7=zp[7];
z8=zp[8];z9=zp[9];z10=zp[10];z11=zp[11];z12=zp[12];z13=zp[13];z14=zp[14];
z15=zp[15];z16=zp[16];z17=zp[17];z18=zp[18];z19=zp[19];z20=zp[20];
z21=zp[21];z22=zp[22];z23=zp[23];z24=zp[24];z25=zp[25];z26=zp[26];z27=zp[27]
#Regression equations
dx <- x1+(x2*z)+(x3*z^2)+(x4*y)+(x5*y*z)+(x6*y*z^2)+(x7*y^2)+(x8*y^2*z)+(x9*y^2*z^2)+
    (x10*x)+(x11*x*z)+(x12*x*z^2)+(x13*x*y)+(x14*x*y*z)+(x15*x*y*z^2)+(x16*x*y^2)+
    (x17*x*y^2*z)+(x18*x*y^2*z^2)+(x19*x^2)+(x20*x^2*z)+(x21*x^2*z^2)+(x22*x^2*y)+
    (x23*x^2*y*z)+(x24*x^2*y*z^2)+(x25*x^2*y^2)+(x26*x^2*y^2*z)+(x27*x^2*y^2*z^2)
dy <- y1+(x2*z)+(y3*z^2)+(y4*y)+(y5*y*z)+(y6*y*z^2)+(y7*y^2)+(y8*y^2*z)+(y9*y^2*z^2)+
    (y10*x)+(y11*x*z)+(y12*x*z^2)+(y13*x*y)+(y14*x*y*z)+(y15*x*y*z^2)+(y16*x*y^2)+
    (y17*x*y^2*z)+(y18*x*y^2*z^2)+(y19*x^2)+(y20*x^2*z)+(y21*x^2*z^2)+(y22*x^2*y)+
    (y23*x^2*y*z)+(y24*x^2*y*z^2)+(y25*x^2*y^2)+(y26*x^2*y^2*z)+(y27*x^2*y^2*z^2)
dz <- z1+(z2*z)+(z3*z^2)+(z4*y)+(z5*y*z)+(z6*y*z^2)+(z7*y^2)+(z8*y^2*z)+(z9*y^2*z^2)+
    (z10*x)+(z11*x*z)+(z12*x*z^2)+(z13*x*y)+(z14*x*y*z)+(z15*x*y*z^2)+(z16*x*y^2)+
    (z17*x*y^2*z)+(z18*x*y^2*z^2)+(z19*x^2)+(z20*x^2*z)+(z21*x^2*z^2)+(z22*x^2*y)+
    (z23*x^2*y*z)+(z24*x^2*y*z^2)+(z25*x^2*y^2)+(z26*x^2*y^2*z)+(z27*x^2*y^2*z^2)
} #end Mp=3
```

```
if(Mp==4){
#Parameters
x1=xp[1];x2=xp[2];x3=xp[3];x4=xp[4];x5=xp[5];x6=xp[6];x7=xp[7];
x8=xp[8];x9=xp[9];x10=xp[10];x11=xp[11];x12=xp[12];x13=xp[13];x14=xp[14];
x15=xp[15];x16=xp[16];x17=xp[17];x18=xp[18];x19=xp[19];x20=xp[20];x21=xp[21];
x22=xp[22];x23=xp[23];x24=xp[24];x25=xp[25];x26=xp[26];x27=xp[27];x28=xp[28];
x29=xp[29];x30=xp[30];x31=xp[31];x32=xp[32];x33=xp[33];x34=xp[34];x35=xp[35];
x36=xp[36];x37=xp[37];x38=xp[38];x39=xp[39];x40=xp[40];x41=xp[41];x42=xp[42];
x43=xp[43];x44=xp[44];x45=xp[45];x46=xp[46];x47=xp[47];x48=xp[48];x49=xp[49];
x50=xp[50];x51=xp[51];x52=xp[52];x53=xp[53];x54=xp[54];x55=xp[55];x56=xp[56];
x57=xp[57];x58=xp[58];x59=xp[59];x60=xp[60];x61=xp[61];x62=xp[62];x63=xp[63];
x64=xp[64];y1=yp[1];y2=yp[2];y3=yp[3];y4=yp[4];y5=yp[5];y6=yp[6];y7=yp[7];
y8=yp[8];y9=yp[9];y10=yp[10];y11=yp[11];y12=yp[12];y13=yp[13];y14=yp[14];
y15=yp[15];y16=yp[16];y17=yp[17];y18=yp[18];y19=yp[19];y20=yp[20];y21=yp[21];
y22=yp[22];y23=yp[23];y24=yp[24];y25=yp[25];y26=yp[26];y27=yp[27];y28=yp[28];
y29=yp[29];y30=yp[30];y31=yp[31];y32=yp[32];y33=yp[33];y34=yp[34];y35=yp[35];
y36=yp[36];y37=yp[37];y38=yp[38];y39=yp[39];y40=yp[40];y41=yp[41];y42=yp[42];
y43=yp[43];y44=yp[44];y45=yp[45];y46=yp[46];y47=yp[47];y48=yp[48];y49=yp[49];
y50=yp[50];y51=yp[51];y52=yp[52];y53=yp[53];y54=yp[54];y55=yp[55];y56=yp[56];
y57=yp[57];y58=yp[58];y59=yp[59];y60=yp[60];y61=yp[61];y62=yp[62];y63=yp[63];
y64=yp[64];z1=zp[1];z2=zp[2];z3=zp[3];z4=zp[4];z5=zp[5];z6=zp[6];z7=zp[7];
z8=zp[8];z9=zp[9];z10=zp[10];z11=zp[11];z12=zp[12];z13=zp[13];z14=zp[14];
z15=zp[15];z16=zp[16];z17=zp[17];z18=zp[18];z19=zp[19];z20=zp[20];z21=zp[21];
z22=zp[22];z23=zp[23];z24=zp[24];z25=zp[25];z26=zp[26];z27=zp[27];z28=zp[28];
z29=zp[29];z30=zp[30];z31=zp[31];z32=zp[32];z33=zp[33];z34=zp[34];z35=zp[35];
z36=zp[36];z37=zp[37];z38=zp[38];z39=zp[39];z40=zp[40];z41=zp[41];z42=zp[42];
z43=zp[43];z44=zp[44];z45=zp[45];z46=zp[46];z47=zp[47];z48=zp[48];z49=zp[49];
z50=zp[50];z51=zp[51];z52=zp[52];z53=zp[53];z54=zp[54];z55=zp[55];z56=zp[56];
z57=zp[57];z58=zp[58];z59=zp[59];z60=zp[60];z61=zp[61];z62=zp[62];z63=zp[63];
z64=zp[64]
#Regression equations
dx <- x1+(x2*z)+(x3*z^2)+(x4*z^3)+(x5*y)+(x6*y*z)+(x7*y*z^2)+(x8*y*z^3)+(x9*y^2)+
    (x10*y^2*z)+(x11*y^2*z^2)+(x12*y^2*z^3)+(x13*y^3)+(x14*y^3*z)+(x15*y^3*z^2)+
    (x16*y^3*z^3)+(x17*x)+(x18*x*z)+(x19*x*z^2)+(x20*x*z^3)+(x21*x*y)+(x22*x*y*z)+
    (x23*x*y*z^2)+(x24*x*y*z^3)+(x25*x*y^2)+(x26*x*y^2*z)+(x27*x*y^2*z^2)+(x28*x*y^2*z^3)+
    (x29*x*y^3)+(x30*x*y^3*z)+(x31*x*y^3*z^2)+(x32*x*y^3*z^3)+(x33*x^2)+(x34*x^2*z)+
    (x35*x^2*z^2)+(x36*x^2*z^3)+(x37*x^2*y)+(x38*x^2*y*z)+(x39*x^2*y*z^2)+(x40*x^2*y*z^3)+
    (x41*x^2*y^2)+(x42*x^2*y^2*z)+(x43*x^2*y^2*z^2)+(x44*x^2*y^2*z^3)+(x45*x^2*y^3)+
    (x46*x^2*y^3*z)+(x47*x^2*y^3*z^2)+(x48*x^2*y^3*z^3)+(x49*x^3)+(x50*x^3*z)+
    (x51*x^3*z^2)+(x52*x^3*z^3)+(x53*x^3*y)+(x54*x^3*y*z)+(x55*x^3*y*z^2)+(x56*x^3*y*z^3)+
    (x57*x^3*y^2)+(x58*x^3*y^2*z)+(x59*x^3*y^2*z^2)+(x60*x^3*y^2*z^3)+(x61*x^3*y^3)+
    (x62*x^3*y^3*z)+(x63*x^3*y^3*z^2)+(x64*x^3*y^3*z^3)
dy <- y1+(y2*z)+(y3*z^2)+(y4*z^3)+(y5*y)+(y6*y*z)+(y7*y*z^2)+(y8*y*z^3)+(y9*y^2)+
    (y10*y^2*z)+(y11*y^2*z^2)+(y12*y^2*z^3)+(y13*y^3)+(y14*y^3*z)+(y15*y^3*z^2)+
    (y16*y^3*z^3)+(y17*x)+(y18*x*z)+(y19*x*z^2)+(y20*x*z^3)+(y21*x*y)+(y22*x*y*z)+
    (y23*x*y*z^2)+(y24*x*y*z^3)+(y25*x*y^2)+(y26*x*y^2*z)+(y27*x*y^2*z^2)+(y28*x*y^2*z^3)+
    (y29*x*y^3)+(y30*x*y^3*z)+(y31*x*y^3*z^2)+(y32*x*y^3*z^3)+(y33*x^2)+(y34*x^2*z)+
    (y35*x^2*z^2)+(y36*x^2*z^3)+(y37*x^2*y)+(y38*x^2*y*z)+(y39*x^2*y*z^2)+(y40*x^2*y*z^3)+
    (y41*x^2*y^2)+(y42*x^2*y^2*z)+(y43*x^2*y^2*z^2)+(y44*x^2*y^2*z^3)+(y45*x^2*y^3)+
    (y46*x^2*y^3*z)+(y47*x^2*y^3*z^2)+(y48*x^2*y^3*z^3)+(y49*x^3)+(y50*x^3*z)+
    (y51*x^3*z^2)+(y52*x^3*z^3)+(y53*x^3*y)+(y54*x^3*y*z)+(y55*x^3*y*z^2)+(y56*x^3*y*z^3)+
    (y57*x^3*y^2)+(y58*x^3*y^2*z)+(y59*x^3*y^2*z^2)+(y60*x^3*y^2*z^3)+(y61*x^3*y^3)+
    (y62*x^3*y^3*z)+(y63*x^3*y^3*z^2)+(y64*x^3*y^3*z^3)
dz <- z1+(z2*z)+(z3*z^2)+(z4*z^3)+(z5*y)+(z6*y*z)+(z7*y*z^2)+(z8*y*z^3)+(z9*y^2)+
    (z10*y^2*z)+(z11*y^2*z^2)+(z12*y^2*z^3)+(z13*y^3)+(z14*y^3*z)+(z15*y^3*z^2)+
    (z16*y^3*z^3)+(z17*x)+(z18*x*z)+(z19*x*z^2)+(z20*x*z^3)+(z21*x*y)+(z22*x*y*z)+
    (z23*x*y*z^2)+(z24*x*y*z^3)+(z25*x*y^2)+(z26*x*y^2*z)+(z27*x*y^2*z^2)+(z28*x*y^2*z^3)+
    (z29*x*y^3)+(z30*x*y^3*z)+(z31*x*y^3*z^2)+(z32*x*y^3*z^3)+(z33*x^2)+(z34*x^2*z)+
    (z35*x^2*z^2)+(z36*x^2*z^3)+(z37*x^2*y)+(z38*x^2*y*z)+(z39*x^2*y*z^2)+(z40*x^2*y*z^3)+
    (z41*x^2*y^2)+(z42*x^2*y^2*z)+(z43*x^2*y^2*z^2)+(z44*x^2*y^2*z^3)+(z45*x^2*y^3)+
```

```
(z46*x^2*y^3*z)+(z47*x^2*y^3*z^2)+(z48*x^2*y^3*z^3)+(z49*x^3)+(z50*x^3*z)+
(z51*x^3*z^2)+(z52*x^3*z^3)+(z53*x^3*y)+(z54*x^3*y*z)+(z55*x^3*y*z^2)+(z56*x^3*y*z^3)+
(z57*x^3*y^2)+(z58*x^3*y^2*z)+(z59*x^3*y^2*z^2)+(z60*x^3*y^2*z^3)+(z61*x^3*y^3)+
(z62*x^3*y^3*z)+(z63*x^3*y^3*z^2)+(z64*x^3*y^3*z^3)
} #end Mp=4

obs<-cbind(x.dep,y.dep,z.dep)  #observed
reg<-cbind(dx,dy,dz)  #regression
names<-colnames(reg)<-c("x.ode","y.ode","z.ode")

#Define bootstrapping statistic: The R bootstrapping routine works off a
#user-defined function that calculates a statistic to be measured in each sample.
library(boot)
nse_udf_boot<-function(data,indices){
    d <- data[indices,]
    z1<-d[,1];z2<-d[,2]
    num<-sum((z1-z2)^2)
    den<-sum((z1-mean(z1))^2)
    nse<-1-(num/den)
    return(nse)
} #end boot function

model=nse_udf_boot #see "b=boot()" below

#Loop i: iterates through each time series and corresponding ssa
#reconstruction.

#Output
#hold.i: Nash-Sutcliffe efficiency for each time series
#hold.ii[,i]: 95% bootstrapped confidence interval on mnse
#hold.iii[,i]: P-value that Nash-Sutcliffe is < 0.65
#hold.iv: 11 rows equal to fitness probability categories calculated

hold.i<-matrix(0,1,length(names))
hold.ii<-matrix(0,2,length(names))
hold.iii<-matrix(0,1,length(names))
hold.iv<-matrix(0,11,length(names))
    for(i in 1:length(names)){
        w1.obs<-obs[,i]
        w1.reg<-reg[,i]
        data1<-cbind(w1.obs,w1.reg)

        #Calculate Nash-Sutcliffe efficiency for each time series
        num<-sum((w1.reg-w1.obs)^2)
        den<-sum((w1.obs-mean(w1.obs))^2)
        nse<-1-(num/den)

        #Bootstrap Nash-Sutcliffe statistic
        b = boot(data=data1,statistic=model,R=5000)
        #b$t gives nse statistic for bootstraps (Rx1)

        #Compute confidence interval around nse.ssa if ci=TRUE in
        #function arguments. 'bca' adjusts for bias and skewness in
        #bootstrap distribution (Ritter and Munoz-Carpena, 2013, p.37)
        if(ci) {
        ci.out<-boot.ci(b, conf=0.95,type="bca")
        ci<-ci.out$bca[4:5]}

        #cumulative distribution function
        cdf<-ecdf(b$t)
```

```
    plot_cdf<-plot(cdf,main=names[i],ylab="cdf",xlab="nse")

    #compute cdf values
    tt <- seq(0,1, by = 0.001)
    cdf.val<-cdf(tt);#cdf.val<-unlist(cdf.val)

    #Probabilities of nse fit intervals
    Prob_less_0<-cdf(0)                        #Prob(nse<0)
    Prob_0_0.9<-cdf(0.099)-cdf(0)              #Prob(0.0<nse<0.099)
    Prob_0.1_0.19<-cdf(0.199)-cdf(0.1)         #Prob(0.1<nse<0.199)
    Prob_0.2_0.29<-cdf(0.299)-cdf(0.2)         #Prob(0.2<nse<0.29)
    Prob_0.3_0.39<-cdf(0.399)-cdf(0.3)         #Prob(0.3<nse<0.399)
    Prob_0.4_0.49<-cdf(0.499)-cdf(0.4)         #Prob(0.4<nse<0.49)
    Prob_0.5_0.59<-cdf(0.599)-cdf(0.5)         #Prob(0.5<nse<0.599)
    Prob_0.6_0.69<-cdf(0.699)-cdf(0.6)         #Prob(0.6<nse<0.69)
    Prob_0.7_0.79<-cdf(0.799)-cdf(0.7)         #Prob(0.7<nse<0.799)
    Prob_0.8_0.89<-cdf(0.899)-cdf(0.8)         #Prob(0.8<nse<0.899)
    Prob_0.9_1<-cdf(1)-cdf(0.9)                #Prob(0.9<nse<1)

    fit.int1<-cbind(Prob_less_0,Prob_0_0.9,Prob_0.1_0.19,Prob_0.2_0.29,Prob_0.3_0.39,
        Prob_0.4_0.49,Prob_0.5_0.59,Prob_0.6_0.69,Prob_0.7_0.79,Prob_0.8_0.89,Prob_0.9_1)

    #Hypothesis testing  H0: nse < nse0 = 0.65; HA: nse >= nse0 = 0.65

    #Compute probability in favor of H0: Prob(nse < 0.65) (Ritter and
    #Munoz-Carpena, 2013, pp. 38-39)
    pval<-cdf(0.65)

  hold.i[,i]<-nse
  if(ci){hold.ii[,i]<-ci}
  hold.iii[,i]<-pval
  hold.iv[,i]<-fit.int1
} #end loop i

#1st row (Nash-Sutcliffe for each time series)
#2nd and 3rd rows are lower and upper 95% confidence bound if ci=TRUE
if(ci) {
results1<-rbind(hold.i,hold.ii,hold.iii,hold.iv)
results2<- matrix(results1,nrow=15,ncol=length(names),byrow=FALSE,dimnames=
  list(c("nse","95% low","95% high","P-value","Prob(nse<0)","Prob(0<nse<0.099)",
  "Prob(0.1<nse<0.199)","Prob(0.2<nse<0.299)","Prob(0.3<nse<0.399)",
  "Prob(0.4<nse<0.499)","Prob(0.5<nse<0.599)","Prob(0.6<nse<0.699)",
  "Prob(0.7<nse<0.799)","Prob(0.8<nse<0.899)","Prob(0.9<nse<1)"),
  c(names[1:length(names)]))))
} else {
results1<-rbind(hold.i,hold.iii,hold.iv)
results2<- matrix(results1,nrow=13,ncol=length(names),byrow=FALSE,dimnames=
  list(c("nse","P-value","Prob(nse<0)","Prob(0<nse<0.099)",
  "Prob(0.1<nse<0.199)","Prob(0.2<nse<0.299)","Prob(0.3<nse<0.399)",
  "Prob(0.4<nse<0.499)","Prob(0.5<nse<0.599)","Prob(0.6<nse<0.699)",
  "Prob(0.7<nse<0.799)","Prob(0.8<nse<0.899)","Prob(0.9<nse<1)"),
  c(names[1:length(names)]))))}

results<-list(dx,dy,dz,results2)
return(results)
} #end user-defined function

results_gf<-goodfit_3(x.diff,y.diff,z.diff,x,y,z,xp,yp,zp,Mp,ci)
fit_table<-results_gf[[4]]
```

The R package `boot` is used for *bootstrapped* resampling and the routine `boot.ci` is used to compute confidence intervals. To run `boot`, we formulate a user-defined function to compute *nse* for each resampling. We next set up a loop to run `boot` and `boot.ci` for each ODE in the phenomenological model separately, and use the R routine `ecdfstats` to compute an *empirical cumulative distribution function (CDF)* for the bootstrapped performance measure. We use the empirical CDF to compute the probabilities that the calculated *nse* is within various probability intervals, and the probability (*p*-value) that it exceeds the threshold value $nse \geq 0.65$. Small *p*-values support the decision to accept the alternative hypothesis that the computed *nse* indicates an acceptable fit. On a last note, if the routine `boot.ci` fails to produce a confidence interval, a remedy is to increase the number of bootstrap replications (i.e. *R* in the `boot()` command).

The output of Code 9.5 returns the results shown in Table 9.5, whose columns are the fitted ODEs and whose rows are the computed *nse*, the *p*-value that $nse \geq 0.65$ and the probabilities that *nse* rests in various intervals. The code also returns plotted CDFs for each fitted ODE (Figure 9.1). We observe that the ODEs estimated with Adaptive LASSO provide excellent fits to the computed finite differences (Code 9.1) from the Lorenz data, since the computed *nse* measures are almost 1 and the computed *p*-values are zero.

Table 9.5 Code 9.5 output: goodness-of-fit table for estimated ODEs.

	x.ode	*y.ode*	*z.ode*
nse	0.999	0.999	0.999
p-value ($nse \leq 0.65$)	0.000	0.000	0.000
Prob($nse < 0$)	0.000	0.000	0.000
Prob($0 < nse < 0.099$)	0.000	0.000	0.000
Prob($0.1 < nse < 0.199$)	0.000	0.000	0.000
Prob($0.2 < nse < 0.299$)	0.000	0.000	0.000
Prob($0.3 < nse < 0.399$)	0.000	0.000	0.000
Prob($0.4 < nse < 0.499$)	0.000	0.000	0.000
Prob($0.5 < nse < 0.599$)	0.000	0.000	0.000
Prob($0.6 < nse < 0.699$)	0.000	0.000	0.000
Prob($0.7 < nse < 0.799$)	0.000	0.000	0.000
Prob($0.8 < nse < 0.899$)	0.000	0.000	0.000
Prob($0.9 < nse < 1$)	1.000	1.000	1.000

9.8 Solution of Phenomenological Model

Code 9.6 uses the R package `deSolve` (Soetaert *et al.*, 2016) to solve the estimated phenomenological model for simulated solutions to the Lorenz equations. The code formulates a user-defined function `solve_3` that takes as inputs the information needed

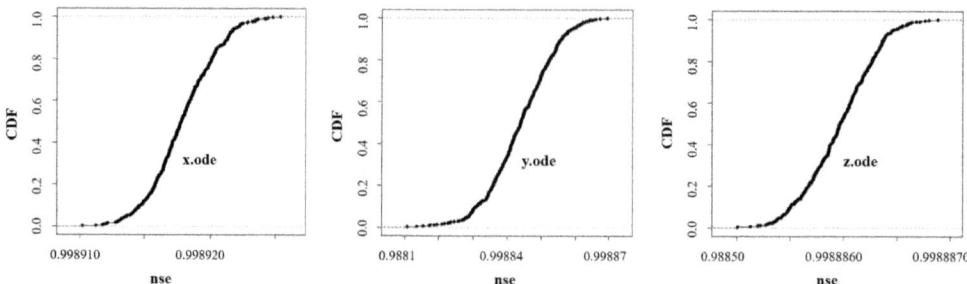

Fig. 9.1 Empirical cumulative distribution functions fit to bootlegged NSE performance measure.

to specify model ODEs, including the observed time series (x, y and z), the estimated model coefficients (xp, yp and zp) and the polynomial range parameter (M_p), together with the information needed to solve the model, including the time step required for integration `delta`, and the integration method (`method`). We select the `lsoda` integration method, one of several options available in `deSolve`. This method switches automatically from *nonstiff* to *stiff* integration methods when required for numerical stability. Finally, we solve the ODEs from initial conditions set equal to the values of the observed time series in the initial time period.

```
#Code 9.6 User-defined function to solve phenomenological Lorenz model
rm(list=ls(all=TRUE))  #remove objects between runs

#Directories
dir.udf<-as.character("C:/User Defined Functions")

#Load user-defined functions
setwd(dir.udf)
dump("lorenz_udf", file="lorenz_udf.R")
source("lorenz_udf.R")
dump("finite_differences_udf", file="finite_differences_udf.R")
source("finite_differences_udf.R")
dump("polynomial_udf", file="polynomial_udf.R")
source("polynomial_udf.R")
dump("ols_udf", file="ols_udf.R")
source("ols_udf.R")
dump("lasso_udf", file="lasso_udf.R")
source("lasso_udf.R")
dump("goodfit_3_udf", file="goodfit_3_udf.R")
source("goodfit_3_udf.R")

#Generate Lorenz data with user-defined function 'lorenz_udf'
lorenz_results<-lorenz_udf(x0=1,y0=1,z0=1,sigma=10,beta=8/3,rho=28,
  t.end=50,delta=0.01)
x<-lorenz_results[[1]];y<-lorenz_results[[2]];z<-lorenz_results[[3]]

#Set parameters
delta<-0.01  #integration step
m<-3  #number of equations
Mp<-2  #polynomial range
alpha<-1  #glmnet elastic-net mixing parameter
```

```
weight<-TRUE  #run adaptive LASSO
nu<-1 #Adaptive LASSO non-negative tuning parameter
ci<-FALSE #95% confidence intervals for NSE
method<-"lsoda"

#Select regression method
ols<-TRUE  #TRUE=run ols, FALSE=run lasso

#Response variable:  User-defined function to compute finite differences
diff_x<-diff(x,delta)  #run user-defined function for x
diff_y<-diff(y,delta)
diff_z<-diff(z,delta)
x<-diff_x[[1]];y<-diff_y[[1]];z<-diff_z[[1]]  #times series same length differences
x.diff<-diff_x[[2]];y.diff<-diff_y[[2]];z.diff<-diff_z[[2]] #finite differences

#Regressors: user-defined function to compute multivariate polynomial expansion
poly.exp<-poly(m,Mp)
#Data check: compare order of terms in 'mpoly' with column names
termList2<-poly.exp[[1]]  #monomial terms
poly1<-mpoly(termList2);term<-terms(poly1);term
col.names<-poly.exp[[2]];col.names
#Monomials populated with data
data1<-poly.exp[[3]]  #populated data matrix
data2<-data1[complete.cases(data1),]  #remove last row with na's
data<-data2[,-1] #remove 1st column of 1's (regression packages don't require))

#Regression
if(ols){#User-defined function to run OLS
results_ols_x<-ols_udf(x.diff,data,m,Mp)  #x.diff
xp.ols<-results_ols_x[[1]]
table_ols_x<-results_ols_x[[2]]

results_ols_y<-ols_udf(y.diff,data,m,Mp)  #y.diff
yp.ols<-results_ols_y[[1]]
table_ols_y<-results_ols_y[[2]]

results_ols_z<-ols_udf(z.diff,data,m,Mp)  #z.diff
zp.ols<-results_ols_z[[1]]
table_ols_z<-results_ols_z[[2]]

xp<-xp.ols;yp<-yp.ols;zp<-zp.ols  #OLS parameters

} else

{#User-defined function to run LASSO
results_lasso_x<-lasso_udf(data,x.diff,alpha,weight,nu)
xp.min<-as.vector(results_lasso_x[[1]][,1])
xp.1se<-as.vector(results_lasso_x[[1]][,2])
dev_x<-results_lasso_x[[2]]

results_lasso_y<-lasso_udf(data,y.diff,alpha,weight,nu)
yp.min<-as.vector(results_lasso_y[[1]][,1])
yp.1se<-as.vector(results_lasso_y[[1]][,2])
dev_y<-results_lasso_y[[2]]

results_lasso_z<-lasso_udf(data,z.diff,alpha,weight,nu)
zp.min<-as.vector(results_lasso_z[[1]][,1])
zp.1se<-as.vector(results_lasso_z[[1]][,2])
dev_z<-results_lasso_z[[2]]
```

```
xp<-xp.min;yp<-yp.min;zp<-zp.min #LASSO min coefficients
#xp<-xp.1se;yp<-yp.1se;zp<-zp.1se #LASSO 1se coefficients
}

results_gf<-goodfit_3(x.diff,y.diff,z.diff,x,y,z,xp,yp,zp,Mp,ci)
fit_table<-results_gf[[4]]

#User-defined function to solve phenomenological model
solve_3<-function(x,y,z,xp,yp,zp,Mp,delta,method) {
library(deSolve) #ODE solver package

if(Mp==2) {
  state<-c(x=x[1],y=y[1],z=z[1]) #initial conditions
  parameters<-c(x1=xp[1],x2=xp[2],x3=xp[3],x4=xp[4],x5=xp[5],x6=xp[6],x7=xp[7],x8=xp[8],
    y1=yp[1],y2=yp[2],y3=yp[3],y4=yp[4],y5=yp[5],y6=yp[6],y7=yp[7],y8=yp[8],z1=zp[1],
    z2=zp[2],z3=zp[3],z4=zp[4],z5=zp[5],z6=zp[6],z7=zp[7],z8=zp[8])
  model<-function(t,state,parameters){
    with(as.list(c(state,parameters)),{
      dx <- x1+(x2*z)+(x3*y)+(x4*y*z)+(x5*x)+(x6*x*z)+(x7*x*y)+(x8*x*y*z)
      dy <- y1+(y2*z)+(y3*y)+(y4*y*z)+(y5*x)+(y6*x*z)+(y7*x*y)+(y8*x*y*z)
      dz <- z1+(z2*z)+(z3*y)+(z4*y*z)+(z5*x)+(z6*x*z)+(z7*x*y)+(z8*x*y*z)
    list(c(dx,dy,dz))
    }) #end with (as.list...
  } #end model function
  times<-seq(0,length(x),by=delta)  #specify integration step
  out<-ode(y=state,times=times,func=model,parms=parameters,
    method=method,verbose=TRUE)
  print(head(out))
} #end if Mp=2 statement

if(Mp==3) {
  state<-c(x=x[1],y=y[1],z=z[1]) #initial conditions
  parameters<-c(x1=xp[1],x2=xp[2],x3=xp[3],x4=xp[4],x5=xp[5],x6=xp[6],x7=xp[7],x8=xp[8],
    x9=xp[9],x10=xp[10],x11=xp[11],x12=xp[12],x13=xp[13],x14=xp[14],x15=xp[15],x16=xp[16],
    x17=xp[17],x18=xp[18],x19=xp[19],x20=xp[20],x21=xp[21],x22=xp[22],x23=xp[23],x24=xp[24],
    x25=xp[25],x26=xp[26],x27=xp[27],y1=yp[1],y2=yp[2],y3=yp[3],y4=yp[4],y5=yp[5],y6=yp[6],
    y7=yp[7],y8=yp[8],y9=yp[9],y10=yp[10],y11=yp[11],y12=yp[12],y13=yp[13],y14=yp[14],y15=yp[15],
    y16=yp[16],y17=yp[17],y18=yp[18],y19=yp[19],y20=yp[20],y21=yp[21],y22=yp[22],y23=yp[23],
    y24=yp[24],y25=yp[25],y26=yp[26],y27=yp[27],z1=zp[1],z2=zp[2],z3=zp[3],z4=zp[4],z5=zp[5],
    z6=zp[6],z7=zp[7],z8=zp[8],z9=zp[9],z10=zp[10],z11=zp[11],z12=zp[12],z13=zp[13],z14=zp[14],
    z15=zp[15],z16=zp[16],z17=zp[17],z18=zp[18],z19=zp[19],z20=zp[20],z21=zp[21],z22=zp[22],
    z23=zp[23],z24=zp[24],z25=zp[25],z26=zp[26],z27=zp[27])
  #model equations
  model<-function(t,state,parameters){
    with(as.list(c(state,parameters)),{
      dx <- x1+(x2*z)+(x3*z^2)+(x4*y)+(x5*y*z)+(x6*y*z^2)+(x7*y^2)+(x8*y^2*z)+(x9*y^2*z^2)+
        (x10*x)+(x11*x*z)+(x12*x*z^2)+(x13*x*y)+(x14*x*y*z)+(x15*x*y*z^2)+(x16*x*y^2)+
        (x17*x*y^2*z)+(x18*x*y^2*z^2)+(x19*x^2)+(x20*x^2*z)+(x21*x^2*z^2)+(x22*x^2*y)+
        (x23*x^2*y*z)+(x24*x^2*y*z^2)+(x25*x^2*y^2)+(x26*x^2*y^2*z)+(x27*x^2*y^2*z^2)
      dy <- y1+(y2*z)+(y3*z^2)+(y4*y)+(y5*y*z)+(y6*y*z^2)+(y7*y^2)+(y8*y^2*z)+(y9*y^2*z^2)+
        (y10*x)+(y11*x*z)+(y12*x*z^2)+(y13*x*y)+(y14*x*y*z)+(y15*x*y*z^2)+(y16*x*y^2)+
        (y17*x*y^2*z)+(y18*x*y^2*z^2)+(y19*x^2)+(y20*x^2*z)+(y21*x^2*z^2)+(y22*x^2*y)+
        (y23*x^2*y*z)+(y24*x^2*y*z^2)+(y25*x^2*y^2)+(y26*x^2*y^2*z)+(y27*x^2*y^2*z^2)
      dz <- z1+(z2*z)+(z3*z^2)+(z4*y)+(z5*y*z)+(z6*y*z^2)+(z7*y^2)+(z8*y^2*z)+(z9*y^2*z^2)+
        (z10*x)+(z11*x*z)+(z12*x*z^2)+(z13*x*y)+(z14*x*y*z)+(z15*x*y*z^2)+(z16*x*y^2)+
        (z17*x*y^2*z)+(z18*x*y^2*z^2)+(z19*x^2)+(z20*x^2*z)+(z21*x^2*z^2)+(z22*x^2*y)+
        (z23*x^2*y*z)+(z24*x^2*y*z^2)+(z25*x^2*y^2)+(z26*x^2*y^2*z)+(z27*x^2*y^2*z^2)
    list(c(dx,dy,dz))
    }) #end with (as.list...
  } #end model function
```

```
times<-seq(0,length(x),by=delta)  #specify integration step
#rk4,euler,ode23,ode45
out<-ode(y=state,times=times,func=model,parms=parameters,method=method)
print(head(out))
} #end if Mp =3 statement

results<-list(out)
return(results)
}   #end user-defined function

results_solve<-solve_3(x,y,z,xp,yp,zp,Mp,delta,method) #run user-defined function
out<-results_solve[[1]]  #output
plot(out) #plot variables
x.sim<-out[,"x"];length(x.sim)<-length(x)  #simulated variables
y.sim<-out[,"y"];length(y.sim)<-length(y)
z.sim<-out[,"z"];length(z.sim)<-length(z)
library(scatterplot3d)
scatterplot3d(x.sim,y.sim,z.sim,type='l',lwd=1)  #Simulated Lorenz attractor
```

The output of Code 9.6 gives the solutions of the phenomenological Lorenz model from the given initial conditions: *x.soln*, *y.soln* and *z.soln*. Table 9.6 shows the first few values.

Table 9.6 Code 9.6 output: solutions of phenomenological Lorenz model.

Time	x.soln	y.soln	z.soln
0.0	1.048823	1.523999	0.9731148
0.1	1.106071	1.785090	0.9437875
0.2	1.182875	2.062304	0.9191836
0.3	1.279149	2.360443	0.9001578
0.4	1.395262	2.684316	0.8878168
0.5	1.531994	3.038833	0.8835695

9.9 Phenomenological Model Extracted from Three Observed Variables

Code 9.7 combines Codes 9.1–9.6 into one program extracting a phenomenological model from three observed time series variables . In addition, we employ the user-defined function `spectral` (Code 6.8) to compute cycle lengths of the time series solutions from the model. Code 9.7 exports output tables summarizing OLS, LASSO or Adaptive LASSO estimated coefficients; goodness-of-fit statistics; and Lorenz solutions simulated by the phenomenological representation. It plots the simulated solutions along with the simulated Lorenz 'butterfly' attractor.

In running Code 9.7, our strategy is to tune parameters on a *trial-and-error* basis within ranges providing numerical stability and successful reproduction of empirically detected dynamics. We can set the number of equations in the phenomenological models with m and the polynomial range with M_p to fix the number of polynomial terms in the ODEs. We can select whether to use OLS, elastic-net or Adaptive LASSO regression coefficients to parameterize the model. If we select elastic-net regression, the mixing parameter α allows us to run ridge regression for $\alpha = 0$, LASSO for $\alpha = 1$ or a combination of the two regressions for intermediate values of α (Friedman *et al.*, 2015). If we select Adaptive LASSO, we can manipulate the tuning parameter nu (Zou, 2006). Finally, we can lower the integration step δ to decrease the error in computing finite differences (Gouesbet and Maquet, 1992) and to fine tune the numerical solution of the phenomenological model. Consistent with the Lorenz equations providing our data, we set the number of equations to $m = 3$, the number of polynomial terms to $M_p^m = 8$, and the integration step to $\delta = 0.01$. We found that OLS regression worked well to reproduce the interaction terms and associated coefficients in the original model.

```
#Code 9.7 Phenomenological model extracted from three observed variables
rm(list=ls(all=TRUE))  #remove objects between runs
options(scipen=999)  #disable scientific notation

#Directories
dir.udf<-as.character("C:/User Defined Functions")
dir.results<-as.character("C:/Results")

#Load user-defined functions
setwd(dir.udf)
dump("lorenz_udf", file="lorenz_udf.R")
source("lorenz_udf.R")
dump("finite_differences_udf", file="finite_differences_udf.R")
source("finite_differences_udf.R")
dump("polynomial_udf", file="polynomial_udf.R")
source("polynomial_udf.R")
dump("ols_udf", file="ols_udf.R")
source("ols_udf.R")
dump("lasso_udf", file="lasso_udf.R")
source("lasso_udf.R")
dump("goodfit_3_udf", file="goodfit_3_udf.R")
source("goodfit_3_udf.R")
dump("solve_ode_3_udf", file="solve_ode_3_udf.R")
source("solve_ode_3_udf.R")
dump("spectral_udf", file="spectral_udf.R")
source("spectral_udf.R")

#Generate Lorenz data with user-defined function 'lorenz_udf'
lorenz_results<-lorenz_udf(x0=1,y0=1,z0=1,sigma=10,beta=8/3,rho=28,
 t.end=50,delta=0.01)
x<-lorenz_results[[1]];y<-lorenz_results[[2]];z<-lorenz_results[[3]]

#Set parameters
delta<-0.01  #integration step
m<-3  #number of equations
Mp<-2  #polynomial range
alpha<-1  #glmnet elastic-net mixing parameter
weight<-FALSE  #run adaptive LASSO
nu<-1  #Adaptive LASSO non-negative tuning parameter
ci<-FALSE  #95% confidence intervals for NSE
method<-"lsoda"
method_spec<-TRUE

#Select regression method
ols<-FALSE  #TRUE=run ols, FALSE=run lasso
```

```
#Response variable: user-defined function to compute finite differences
diff_x<-diff(x,delta)  #run user-defined function for x
diff_y<-diff(y,delta)
diff_z<-diff(z,delta)
#times series same length as differences
x<-diff_x[[1]];y<-diff_y[[1]];z<-diff_z[[1]]
#finite differences
x.diff<-diff_x[[2]];y.diff<-diff_y[[2]];z.diff<-diff_z[[2]]

#Regressors: user-defined function to compute multivariate polynomial expansion
poly.exp<-poly(m,Mp)
#Data Check: compare order of terms in 'mpoly' with column names
termList2<-poly.exp[[1]]  #monomial terms
poly1<-mpoly(termList2);term<-terms(poly1);term
col.names<-poly.exp[[2]];col.names
#Monomials populated with data
data1<-poly.exp[[3]]  #populated data matrix
data2<-data1[complete.cases(data1),]  #remove last row with na's
data<-data2[,-1] #remove 1st column of 1's (regression packages don't require))

#Regression
if(ols){#User-defined function to run OLS
results_ols_x<-ols_udf(x.diff,data,m,Mp)  #x.diff
xp.ols<-results_ols_x[[1]]
table_ols_x<-results_ols_x[[2]]

results_ols_y<-ols_udf(y.diff,data,m,Mp)  #y.diff
yp.ols<-results_ols_y[[1]]
table_ols_y<-results_ols_y[[2]]

results_ols_z<-ols_udf(z.diff,data,m,Mp)  #z.diff
zp.ols<-results_ols_z[[1]]
table_ols_z<-results_ols_z[[2]]

xp<-xp.ols;yp<-yp.ols;zp<-zp.ols  #OLS parameters

} else

{#User-defined function to run LASSO
results_lasso_x<-lasso_udf(data,x.diff,alpha,weight,nu)
xp.min<-as.vector(results_lasso_x[[1]][,1])
xp.1se<-as.vector(results_lasso_x[[1]][,2])
dev_x<-results_lasso_x[[2]]

results_lasso_y<-lasso_udf(data,y.diff,alpha,weight,nu)
yp.min<-as.vector(results_lasso_y[[1]][,1])
yp.1se<-as.vector(results_lasso_y[[1]][,2])
dev_y<-results_lasso_y[[2]]

results_lasso_z<-lasso_udf(data,z.diff,alpha,weight,nu)
zp.min<-as.vector(results_lasso_z[[1]][,1])
zp.1se<-as.vector(results_lasso_z[[1]][,2])
dev_z<-results_lasso_z[[2]]

xp<-xp.min;yp<-yp.min;zp<-zp.min #LASSO min coefficients
#xp<-xp.1se;yp<-yp.1se;zp<-zp.1se #LASSO 1se coefficients
}

#User-defined function to assess goodness of fit
results_gf<-goodfit_3(x.diff,y.diff,z.diff,x,y,z,xp,yp,zp,Mp,ci)
```

```
dx<-results_gf[[1]];dy<-results_gf[[2]];dz<-results_gf[[3]]
fit_table<-results_gf[[4]]

#User-defined function to solve phenomenological model
results_solve<-solve_3(x,y,z,xp,yp,zp,Mp,delta,method)

#Solutions
out<-results_solve[[1]]
plot(out) #plot variables
x.sim<-out[,"x"];length(x.sim)<-length(x)
y.sim<-out[,"y"];length(y.sim)<-length(y)
z.sim<-out[,"z"];length(z.sim)<-length(z)

#Spectral analysis of ODE model solution for x (x.sim)
#methods="pgram", "ar"
x.spec<-spectral(x.sim,method_spec=TRUE)
cycles.x<-x.spec[[1]]
y.spec<-spectral(y.sim,method_spec=TRUE)
cycles.y<-y.spec[[1]]
z.spec<-spectral(z.sim,method_spec=TRUE)
cycles.z<-z.spec[[1]]
spec.table<-cbind(cycles.x,cycles.y,cycles.z)

#Plots
#Computed and simulated differentials
yrange<-range(dx,x.diff)
plot(dx,type='l',lwd=4,cex.lab=1.5,cex.axis=1,font=2,font.lab=2,col="dark grey")
lines(x.diff,lty=1,lwd=2)
yrange<-range(dy,y.diff)
plot(dy,type='l',lwd=4,cex.lab=1.5,cex.axis=1,font=2,font.lab=2,col="dark grey")
lines(y.diff,lty=1,lwd=2)
yrange<-range(dz,z.diff)
plot(dz,type='l',lwd=4,cex.lab=1.5,cex.axis=1,font=2,font.lab=2,col="dark grey")
lines(z.diff,lty=1,lwd=2)

#Signal and simulated signal
yrange<-range(x,x.sim)
plot(x,type='l',lwd=4,cex.lab=1.5,cex.axis=1,font=2,font.lab=2,col="dark grey")
lines(x.sim,lty=1,lwd=2)
yrange<-range(y,y.sim)
plot(y,type='l',lwd=4,cex.lab=1.5,cex.axis=1,font=2,font.lab=2,col="dark grey")
lines(y.sim,lty=1,lwd=2)
yrange<-range(z,z.sim)
plot(z,type='l',lwd=4,cex.lab=1.5,cex.axis=1,font=2,font.lab=2,col="dark grey")
lines(z.sim,lty=1,lwd=2)

#Reconstructed attractor
library(scatterplot3d)
scatterplot3d(x,y,z,type='l',lwd=1,cex.lab=1.5,cex.axis=1,
  lab=c(2,2),lab.z=2,font=2,font.lab=2)

#Simulated attractor
scatterplot3d(x.sim,y.sim,z.sim,type="l",lwd=1,cex.lab=1.5,cex.axis=1,
  lab=c(2,2),lab.z=2,font=2,font.lab=2)

#Export output
setwd(dir.results)

#Table with coefficients from all regressions
if(ols){
```

```
  coeffs_ols<-cbind(col.names,xp.ols,yp.ols,zp.ols)
  colnames(coeffs_ols)<-c("terms","xp.ols","yp.ols","zp.ols")
  write.table(coeffs_ols,sep=",","coeffs_ols.csv",row.names=FALSE)
} else{
  coeffs_lasso<-cbind(col.names,xp.min,yp.min,zp.min,xp.1se,yp.1se,zp.1se)
  colnames(coeffs_lasso)<-c("terms","xp.min","yp.min","zp.min",
                            "xp.1se","yp.1se","zp.1se")
  write.table(coeffs_lasso,sep=",","coeffs_lasso.csv",row.names=FALSE)
} #end if(ols)

#Goodness of fit
write.csv(fit_table,"fit_table.csv")

#Simulated values
sim.val<-cbind(x.sim,y.sim,z.sim)
write.table(sim.val,sep=",",col.names=c("x.sim","y.sim","z.sim"),
            "sim.val.csv",row.names=FALSE)

#Spectral results
write.table(spec.table,sep=",",col.names=c("cycles.x","cycles.y","cycles.z"),
  "cycle.length.csv",row.names=FALSE)
```

9.10 Phenomenological Model Extracted from a Single Observed Variable

If we are limited to observation of only a single time series variable x_t, we can extract a phenomenological model from x_t and its two delay-coordinate vectors x_{t+d} and x_{t+2d}, where d is the *embedding delay* (Brunton *et al.*, 2016). The unobserved variables are replaced with synthesized variables $y_t = x_{t+d}$ and $z_t = x_{t+2d}$. This is the same strategy we follow in time-delay embedding of a time series. As an illustration, Code 9.8 formulates and solves the phenomenological model extracted from a single solution of the Lorenz model integrated with time step $\delta = 0.001$, `lorenz_001`.

```
#Code 9.8 Phenomenological model extracted from single observed variable
rm(list=ls(all=TRUE))  #remove objects between runs
options(scipen=999)  #disable scientific notation

#Directories
dir.udf<-as.character("C:/User Defined Functions")
dir.results<-as.character("C:/Results")

#Load user-defined functions
setwd(dir.udf)
dump("lorenz_udf", file="lorenz_udf.R")
source("lorenz_udf.R")
dump("embed_udf", file="embed_udf.R")
source("embed_udf.R")
dump("finite_differences_udf", file="finite_differences_udf.R")
source("finite_differences_udf.R")
dump("polynomial_udf", file="polynomial_udf.R")
source("polynomial_udf.R")
dump("ols_udf", file="ols_udf.R")
source("ols_udf.R")
dump("lasso_udf", file="lasso_udf.R")
source("lasso_udf.R")
dump("goodfit_3_udf", file="goodfit_3_udf.R")
```

```
source("goodfit_3_udf.R")
dump("solve_ode_3_udf", file="solve_ode_3_udf.R")
source("solve_ode_3_udf.R")
dump("spectral_udf", file="spectral_udf.R")
source("spectral_udf.R")

#Generate Lorenz data with user-defined function 'lorenz_udf'
lorenz_results<-lorenz_udf(x0=1,y0=1,z0=1,sigma=10,beta=8/3,rho=28,
 t.end=50,delta=0.001)
x.ts<-lorenz_results[[1]]

#Set parameters
delta<-0.01  #integration step
m<-3  #number of equations
Mp<-3  #polynomial range
alpha<-0.5  #glmnet elastic-net mixing parameter
weight<-FALSE  #run adaptive LASSO
nu<-1  #Adaptive LASSO non-negative tuning parameter
ci<-FALSE  #95% confidence intervals for NSE
method<-"lsoda"
method_spec<-TRUE

#Select regression method
ols<-FALSE  #TRUE=run ols, FALSE=run lasso

#Synthesize dynamic variables by embedding observed time series
results.embed<-embed_udf(x.ts)  #use to compute embedding dimension
d<-results.embed[[1]]
library(tseriesChaos)
Mx<-embedd(x.ts,m=3,d) #embedded data matrix whose columns are x,y,z
x.emb<-Mx[,1];y.emb<-Mx[,2];z.emb<-Mx[,3] #Note that x,y,z are shorter than x.ts
library(scatterplot3d)
scatterplot3d(x.emb,y.emb,z.emb,type='l',lwd=1,cex.axis=1.1,
  cex.lab=1.5,font.axis=2,font.lab=2)

#Response variable:  user-defined function to compute finite differences
diff_x<-diff(x.emb,delta)  #run user-defined function for x
diff_y<-diff(y.emb,delta)
diff_z<-diff(z.emb,delta)
x<-diff_x[[1]];y<-diff_y[[1]];z<-diff_z[[1]]  #times series same length as differences
x.diff<-diff_x[[2]];y.diff<-diff_y[[2]];z.diff<-diff_z[[2]] #finite differences

#Regressors: user-defined function to compute multivariate polynomial expansion
poly.exp<-poly(m,Mp)
#Data check: compare order of terms in 'mpoly' with column names
termList2<-poly.exp[[1]]  #monomial terms
poly1<-mpoly(termList2);term<-terms(poly1);term
col.names<-poly.exp[[2]];col.names
#Monomials populated with data
data1<-poly.exp[[3]]  #populated data matrix
data2<-data1[complete.cases(data1),]  #remove last row with na's
data<-data2[,-1] #remove 1st column of 1's (regression packages don't require))

#Regression
if(ols){#User-defined function to run ols
results_ols_x<-ols_udf(x.diff,data,m,Mp)  #x.diff
xp.ols<-results_ols_x[[1]]
table_ols_x<-results_ols_x[[2]]

results_ols_y<-ols_udf(y.diff,data,m,Mp)  #y.diff
```

```
yp.ols<-results_ols_y[[1]]
table_ols_y<-results_ols_y[[2]]

results_ols_z<-ols_udf(z.diff,data,m,Mp)  #z.diff
zp.ols<-results_ols_z[[1]]
table_ols_z<-results_ols_z[[2]]

xp<-xp.ols;yp<-yp.ols;zp<-zp.ols  #OLS parameters

} else

{#User-defined function to run LASSO
results_lasso_x<-lasso_udf(data,x.diff,alpha,weight,nu)
xp.min<-as.vector(results_lasso_x[[1]][,1])
xp.1se<-as.vector(results_lasso_x[[1]][,2])
dev_x<-results_lasso_x[[2]]

results_lasso_y<-lasso_udf(data,y.diff,alpha,weight,nu)
yp.min<-as.vector(results_lasso_y[[1]][,1])
yp.1se<-as.vector(results_lasso_y[[1]][,2])
dev_y<-results_lasso_y[[2]]

results_lasso_z<-lasso_udf(data,z.diff,alpha,weight,nu)
zp.min<-as.vector(results_lasso_z[[1]][,1])
zp.1se<-as.vector(results_lasso_z[[1]][,2])
dev_z<-results_lasso_z[[2]]

xp<-xp.min;yp<-yp.min;zp<-zp.min #LASSO min coefficients
#xp<-xp.1se;yp<-yp.1se;zp<-zp.1se #LASSO 1se coefficients
}

#User-defined function to assess goodness of fit
results_gf<-goodfit_3(x.diff,y.diff,z.diff,x,y,z,xp,yp,zp,Mp,ci)
dx<-results_gf[[1]];dy<-results_gf[[2]];dz<-results_gf[[3]]
fit_table<-results_gf[[4]]

#User-defined function to solve phenomenological model
results_solve<-solve_3(x,y,z,xp,yp,zp,Mp,delta,method)

#Solutions
out<-results_solve[[1]]
plot(out) #plot variables
x.sim<-out[,"x"];length(x.sim)<-length(x)
y.sim<-out[,"y"];length(y.sim)<-length(y)
z.sim<-out[,"z"];length(z.sim)<-length(z)

#Spectral analysis of ODE model solution for x (x.sim)
#methods="pgram", "ar"
x.spec<-spectral(x.sim,method_spec=TRUE)
cycles.x<-x.spec[[1]]
y.spec<-spectral(y.sim,method_spec=TRUE)
cycles.y<-y.spec[[1]]
z.spec<-spectral(z.sim,method_spec=TRUE)
cycles.z<-z.spec[[1]]
spec.table<-cbind(cycles.x,cycles.y,cycles.z)

#Plots
#Computed and simulated differentials
par(mfrow=c(4,1))
yrange<-range(dx,x.diff)
```

```
plot(dx,type='l',lwd=4,cex.lab=1.5,cex.axis=1,font=2,font.lab=2,col="dark grey")
lines(x.diff,lty=1,lwd=2)
yrange<-range(dy,y.diff)
plot(dy,type='l',lwd=4,cex.lab=1.5,cex.axis=1,font=2,font.lab=2,col="dark grey")
lines(y.diff,lty=1,lwd=2)
yrange<-range(dz,z.diff)
plot(dz,type='l',lwd=4,cex.lab=1.5,cex.axis=1,font=2,font.lab=2,col="dark grey")
lines(z.diff,lty=1,lwd=2)

#Signal and simulated signal
par(mfrow=c(4,1))
yrange<-range(x,x.sim)
plot(x,type='l',lwd=4,cex.lab=1.5,cex.axis=1,font=2,font.lab=2,col="dark grey")
lines(x.sim,lty=1,lwd=2)
yrange<-range(y,y.sim)
plot(y,type='l',lwd=4,cex.lab=1.5,cex.axis=1,font=2,font.lab=2,col="dark grey")
lines(y.sim,lty=1,lwd=2)
yrange<-range(z,z.sim)
plot(z,type='l',lwd=4,cex.lab=1.5,cex.axis=1,font=2,font.lab=2,col="dark grey")
lines(z.sim,lty=1,lwd=2)

#Reconstructed attractor
library(scatterplot3d)
scatterplot3d(x,y,z,type='l',lwd=1,cex.lab=1.5,cex.axis=1,
  lab=c(2,2),lab.z=2,font=2,font.lab=2)

#Simulated attractor
scatterplot3d(x.sim,y.sim,z.sim,type="l",lwd=1,cex.lab=1.5,cex.axis=1,
  lab=c(2,2),lab.z=2,font=2,font.lab=2)

#Export output
setwd(dir.results)

#Table with coefficients from all regressions
if(ols){
  coeffs_ols<-cbind(col.names,xp.ols,yp.ols,zp.ols)
  colnames(coeffs_ols)<-c("terms","xp.ols","yp.ols","zp.ols")
  write.table(coeffs_ols,sep=",","coeffs_ols.csv",row.names=FALSE)
} else{
  coeffs_lasso<-cbind(col.names,xp.min,yp.min,zp.min,xp.1se,yp.1se,zp.1se)
  colnames(coeffs_lasso)<-c("terms","xp.min","yp.min","zp.min",
                            "xp.1se","yp.1se","zp.1se")
  write.table(coeffs_lasso,sep=",","coeffs_lasso.csv",row.names=FALSE)
} #end if(ols)

#Goodness of fit
write.csv(fit_table,"fit_table.csv")

#Simulated values
sim.val<-cbind(x.sim,y.sim,z.sim)
write.table(sim.val,sep=",",col.names=c("x.sim","y.sim","z.sim"),
            "sim.val.csv",row.names=FALSE)

#Spectral results
write.table(spec.table,sep=",",col.names=c("cycles.x","cycles.y","cycles.z"),
  "cycle.length.csv",row.names=FALSE)
```

Code 9.8 differs from Code 9.7 in the following two ways. First, Code 9.8 synthesizes the unobserved variables from the observed time series x_t with the user-defined function **embed_udf** (to compute the embedding delay d) and the routine **embedd** from the

R package `tseriesChaos` (to compute delayed-coordinate vectors) (Di Narzo, 2015). Second, Code 9.8 resets parameter values so that the phase space attractor constructed from the phenomenological model corresponds to that reconstructed from the observed solution x_t and its two delay-coordinate copies. In particular, we set the polynomial range parameter at $M_p = 3$, and obtain numerical stability of the solution by using estimated coefficients from a mixed ridge and unweighted LASSO regression ($\alpha = 0.5$), and fixing the integration step in `deSolve` at $\delta = 0.01$. We solve the model fit with the LASSO parameter vectors associated with $\lambda.min$: $xp.min$, $yp.min$ and $zp.min$.

Figure 9.2 shows a clear correspondence between the attractor reconstructed from the observed solution x_t (a) and that produced from the phenomenological model (b). Table 9.7 shows that, compared with OLS, the mixed ridge and LASSO regressions succeed in identifying parsimonious models requiring only a small fraction of the range of monomial terms. Table 9.8 shows that the estimated model provides an excellent fit to the data (x_t and its two delay-coordinate copies).

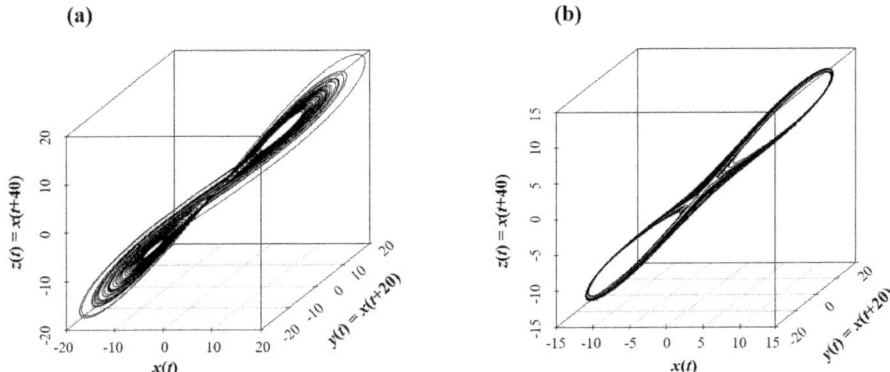

Fig. 9.2 Phenomenological Lorenz model extracted from a single solution: (a) attractor reconstructed from the observed solution $x(t)$ and delay-coordinate copies; (b) attractor reproduced from the phenomenological model.

9.11 Summary

Phenomenological modelling opens the door to the possibility that one or more observed time series variables can be distilled into a system of ODEs that analytically reproduces empirically reconstructed dynamics. Phenomenological models provide several benefits:

- They allow the data 'to speak' regarding their underlying dynamic structure rather than imposing this structure a priori with incompletely tested presumptions. In particular, they can provide a parsimonious representation of complex behavioural patterns composed of aperiodic (nonrepeating) oscillations. Armed with this information, we can characterize dynamic behaviour outside the range of observed data.

Table 9.7 Estimated coefficients for phenomenological models extracted from a single observed variable. The OLS coefficients are *xp.ols*, *yp.ols* and *zp.ols*. The mixed ridge and LASSO coefficients are *xp.min*, *yp.min* and *zp.min* (associated with $\lambda.min$) and *xp.1se*, *yp.1se* and *zp.1se* (associated with $\lambda.1se$).

Term	xp.ols	yp.ols	zp.ols	xp.min	yp.min	zp.min	xp.1se	yp.1se	zp.1se
Constant	0.00	0.00	0.00	0.01	0.00	0.01	0.00	0.00	0.01
z	−2.60	2.54	7.45	2.57	2.35	2.29	2.57	2.35	2.29
z^2	0.28	−0.13	0.23	0.00	0.00	0.00	0.00	0.00	0.00
y	10.25	−0.11	−9.83	0.00	0.00	0.00	0.00	0.00	0.00
yz	−1.11	0.50	−0.90	0.00	0.00	0.00	0.00	0.00	0.00
yz^2	0.06	−0.03	0.05	0.00	0.00	0.00	0.00	0.00	0.00
y^2	1.10	−0.49	0.89	0.00	0.00	0.00	0.00	0.00	0.00
y^2z	−0.10	0.05	−0.08	0.00	0.00	0.00	0.00	0.00	0.00
y^2z^2	0.00	0.00	0.00	0.00	0.00	0.00	0.00	0.00	0.00
x	−7.65	−2.43	2.38	−2.73	−2.35	−2.14	−2.72	−2.35	−2.14
xz	0.55	−0.25	0.45	0.00	0.00	0.00	0.00	0.00	0.00
xz^2	−0.14	0.06	−0.12	0.00	0.00	0.00	0.00	0.00	0.00
xy	−1.08	0.48	−0.88	0.00	0.00	0.00	0.00	0.00	0.00
xyz	0.39	−0.18	0.32	0.00	0.00	0.00	0.00	0.00	0.00
xyz^2	0.00	0.00	0.00	0.00	0.00	0.00	0.00	0.00	0.00
xy^2	−0.16	0.07	−0.13	0.00	0.00	0.00	0.00	0.00	0.00
xy^2z	0.00	0.00	0.00	0.00	0.00	0.00	0.00	0.00	0.00
xy^2z^2	0.00	0.00	0.00	0.00	0.00	0.00	0.00	0.00	0.00
x^2	0.26	−0.12	0.21	0.00	0.00	0.00	0.00	0.00	0.00
x^2z	−0.12	0.05	−0.10	0.00	0.00	0.00	0.00	0.00	0.00
x^2z^2	0.00	0.00	0.00	0.00	0.00	0.00	0.00	0.00	0.00
x^2y	0.08	−0.03	0.06	0.00	0.00	0.00	0.00	0.00	0.00
x^2yz	0.00	0.00	0.00	0.00	0.00	0.00	0.00	0.00	0.00
x^2yz^2	0.00	0.00	0.00	0.00	0.00	0.00	0.00	0.00	0.00
x^2y^2	0.00	0.00	0.00	0.00	0.00	0.00	0.00	0.00	0.00
x^2y^2z	0.00	0.00	0.00	0.00	0.00	0.00	0.00	0.00	0.00
$x^2y^2z^2$	0.00	0.00	0.00	0.00	0.00	0.00	0.00	0.00	0.00

Table 9.8 Goodness of fit for phenomenological model extracted from a single observed variable.

	x.ode	*y.ode*	*z.ode*
nse	0.98	0.80	0.98
p-value	0	0	0
Prob($nse \leq 0.65$)	0	0	0
Prob($0 < nse < 0.099$)	0	0	0
Prob($0.1 < nse < 0.199$)	0	0	0
Prob($0.2 < nse < 0.299$)	0	0	0
Prob($0.3 < nse < 0.399$)	0	0	0
Prob($0.4 < nse < 0.499$)	0	0	0
Prob($0.5 < nse < 0.599$)	0	0	0
Prob($0.6 < nse < 0.699$)	0	0	0
Prob($0.7 < nse < 0.799$)	0	0.67	0
Prob($0.8 < nse < 0.899$)	0	0.12	0
Prob($0.9 < nse < 1$)	1	0	1

- They can be used to characterize the dynamics of variable interactions: for example, whether an incremental increase in one variable drives a marginal increase/decrease in the growth rate of another, and whether these dynamic interactions follow systematic patterns over time.
- They provide an analytical framework for *data-driven* science still searching for credible theoretical explanation.
- They set a descriptive standard for how the real world operates so that theory is not misdirected in explaining fanciful behaviour.

The success of phenomenological modelling depends critically on selection of governing parameters. Model dimensionality and the time delays used to synthesize dynamic variables are guided by statistical tests run for phase space reconstruction. Other regression and numerical integration parameters can be set on a *trial-and-error* basis within ranges providing numerical stability and successful reproduction of empirically detected dynamics.

In this chapter, we have illustrated phenomenological modelling with solutions of the Lorenz model so that we would recognize the dynamics that needed to be reproduced. In Chapter 10, we extract a phenomenological model from real-world data.

10

Capstone: Application of NLTS to Real-World Data

```
'You know my methods.  Apply them.'
Arthur Conan Doyle, The Sign of the Four
```

In this *capstone* chapter, we illustrate how concepts in the book come together to allow us to diagnose real-world dynamics from observed time series data. In particular, we attempt to diagnose multi-strain infectious disease dynamics from weekly cases of scarlet fever, measles and pertussis in New York during the pre-vaccine period 1924–1948 (source: Project Tycho, http://www.tycho.pitt.edu/index.php). These data are found in the `disease_1924.csv` file in the `C:/Data` directory of the book's web materials. This file includes the observed time series data and the isolated signal and noise series.

Epidemiologists have relied on the susceptible–exposed–infective–removed (SEIR) model to study the dynamics of epidemic diseases. The classic SEIR model is a deterministic system of ordinary differential equations (ODEs) describing the rates at which a closed population moves through these epidemic stages (Kot *et al.*, 1988; Artalejo *et al.*, 2015). The dynamics of the classic model are limited to epidemics that reach a constant level or die out. Consequently, the classic model has been generalized through the years to square better with the persistently fluctuating behaviour that we observe, for example, in the New York data (Figure 10.1). The conventional approach in the literature is to attribute fluctuations to exogenous shocks best modelled with stochastic SEIR models (Artalejo *et al.*, 2015). These include *time series susceptible–infected–recovered (TSIR)* models that modify the 'deterministic skeleton' of an SEIR model with what are deemed to be inherently stochastic factors, including disease transmission rates and demographic variables (Bjørnstad *et al.*, 2002; Grenfell *et al.*, 2002). Alternatively, pioneering applications of nonlinear time series analysis by Schaffer and Kot (1985) and Kot *et al.* (1988) demonstrated that observed volatility in time series collected for several infectious diseases in major US and European cities is driven by low-dimensional and nonlinear dynamics, which they modelled by including a time-dependent infection rate term in the classic SEIR model.

The diagnostic question for our application is whether observed complexity in the New York data is most likely driven by linear stochastic dynamics best modelled with stochastic epidemic models or by low-dimensional nonlinear dynamics most informa-

Nonlinear Time Series Analysis with R, Ray Huffaker, Marco Bittelli and Rodolfo Rosa, Oxford University Press (2017).
© Ray Huffaker, Marco Bittelli and Rodolfo Rosa. DOI: 10.1093/oso/9780198782933.001.0001

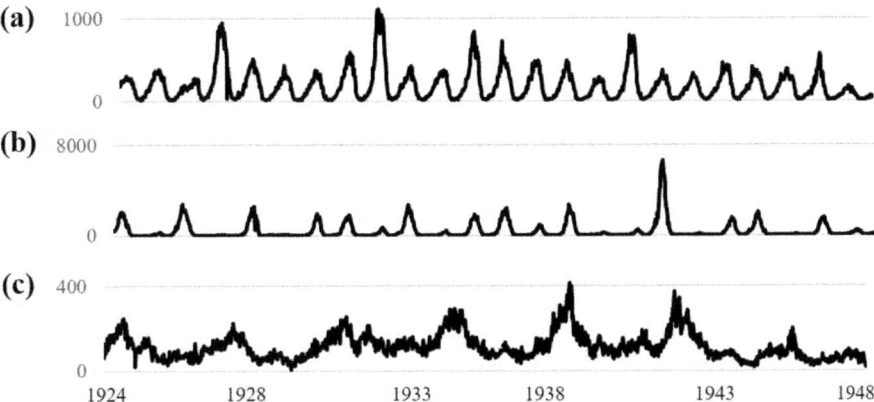

Fig. 10.1 Weekly cases of (a) scarlet fever, (b) measles and (c) pertussis in New York during the pre-vaccine period 1924–1948 (source: Project Tycho,http://www.tycho.pitt.edu/index.php).

tively modelled by increasing the structure of the deterministic models to increase nonlinear interactions among epidemic stages. Can observed complex epidemic dynamics be captured with deterministic and autonomous (i.e. non-time-dependent) nonlinear interactions?

10.1 Data Preprocessing

Our first consideration is whether a sufficiently strong signal can be isolated in each observed time series to merit further search for deterministic structure. We applied the Toeplitz method of *singular spectrum analysis (SSA)* to each mean-adjusted time series (Codes 6.9 and 6.11). Each time series had sporadic missing observations filled in with centred averages.

We isolated a strong signal in the scarlet fever time series accounting for 74% of total variation from the mean. We ran SSA with window length $L = 500$ representing 10 replications of the peak cycle length (50 weeks) uncovered in spectral analysis. The SSA diagnostic plots show that eigenvector pairs $(1, 2)$, $(3, 4)$ and $(5, 6)$ oscillate in phase quadrature at cycle lengths of 52 weeks (63%), 153 weeks (6%) and 26 weeks (5%), respectively, where the percentages in parentheses measure the portion of total variation explained in the observed time series (Figure 10.2a). Each of these cyclical components is insignificantly correlated with other paired groups and thus exhibits the necessary statistical independence (Figure 10.2b). We plot the observed time series and isolated signal in Figure 10.3a, the cyclical components in Figure 10.3b and the isolated noise (i.e. unstructured variation in the data) in Figure 10.3c.

There is a moderately strong signal in the measles time series accounting for 66% of total variation from the mean. We ran SSA with window length $L = 550$ representing 10 replications of the peak cycle length (55 weeks) uncovered in spectral analysis . Eigenvector pairs $(1, 2)$, $(3, 4)$, $(5, 6)$, $(7, 8)$ and $(9, 10)$ oscillate in phase quadrature

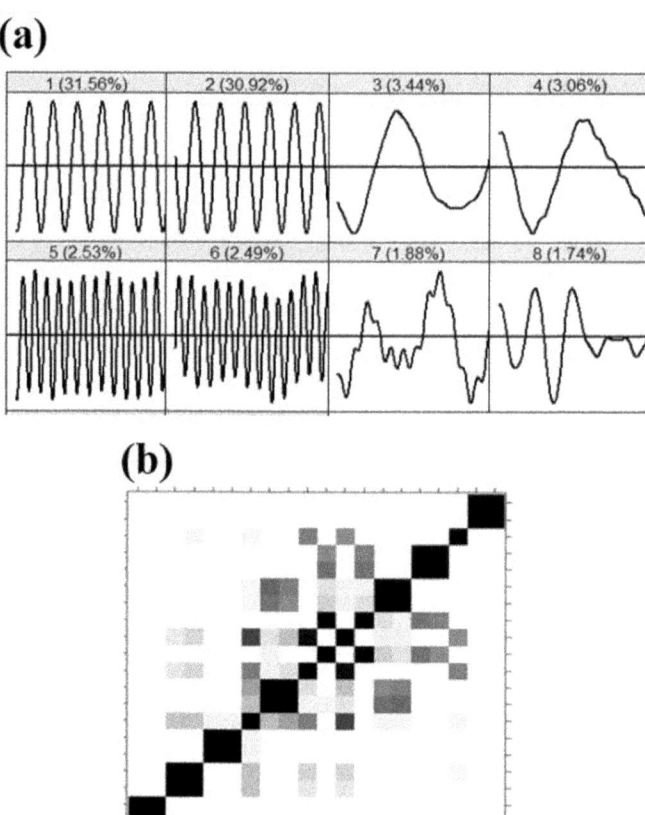

Fig. 10.2 SSA diagnostics for scarlet fever with window length $L = 500$. (a) Eigenvector plots. The first three pairs of eigenvectors oscillate in phase quadrature at 52-, 153- and 26-week cycle lengths. The signal is strong since these cycles together account for 74% of total variation in the time series from the mean. The 52-week cycle is dominant, accounting for 63% of total variation. (b) Weighted correlation matrix. Each of these cyclical components is insignificantly correlated with other paired groups and thus exhibits the necessary statistical independence.

at cycle lengths of 52 weeks (29%), 126 weeks (16%), 38 weeks (9%), 83 weeks (8%) and 27 weeks (4%), respectively (Figure 10.4a). Each paired group is insignificantly correlated with the others and thus can be grouped independently (Figure 10.4b). We plot the observed time series and isolated signal in Figure 10.5a, the three most dominant cyclical components in Figure 10.5b and the isolated noise in Figure 10.5c.

The pertussis time series contains a moderately strong signal accounting for 71% of total variation from the mean. We ran SSA with a window length equal to half of the time series length ($L = 626$), since spectral analysis did not detect a dominant frequency. Eigenvector pairs $(1, 2)$, $(3, 4)$ and $(9, 10)$ oscillate in phase quadrature at cycle lengths of 183 weeks (45%), 92 weeks (12%) and 61 weeks (3%), respectively.

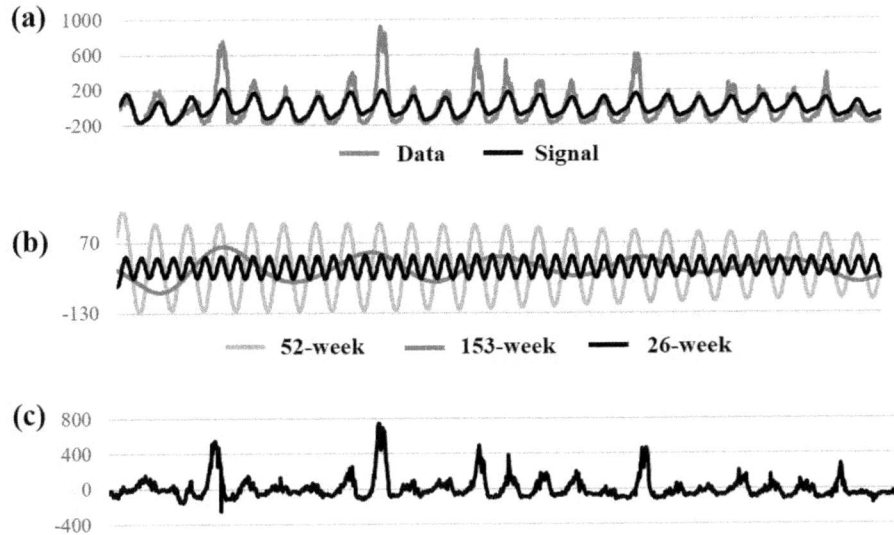

Fig. 10.3 Signal processing of scarlet fever. (a) Plots of observed mean-adjusted time series (grey curve) and reconstructed signal (black curve). (b) Plots of 52-, 153- and 26-week cycles comprising the signal. (c) Noise (unstructured variation) removed from the time series.

Eigenvector pair $(5, 6)$ is a low-frequency trend accounting for 11% of total variation (Figure 10.6a). The cyclical pairs are uncorrelated and thus register the necessary statistical independence (Figure 10.6b). We plot the observed time series and isolated signal in Figure 10.7a, the three most dominant cyclical components in Figure 10.7b and the isolated noise in Figure 10.7c.

We applied nonlinear cross prediction (Code 6.15) to find that each disease signal is stationary for the purposes of nonlinear analysis. In each signal, segments are highly skilled at cross-predicting the others (Figure 10.8).

10.2 Phase Space Reconstruction

Having isolated strong stationary signals in each disease time series, we can justifiably attempt to reconstruct epidemic system dynamics from each signal. Takens (1981) proved that the *shadow* phase space reconstructed from lagged copies of a single variable preserves essential dynamic properties of the original phase space if the shadow phase space has an embedding dimension sufficiently large to accommodate the fractal dimension of the original attractor. We reconstruct a shadow attractor from each signal, observe whether attractors exhibit apparent structure, and test the null hypothesis that observed structure is generated fortuitously by linear stochastic dynamics with surrogate data.

We ran Code 7.7 to reconstruct shadow attractors from each disease signal. The scarlet fever (Figure 10.9a), measles (Figure 10.10a), and pertussis (Figure 10.11a) shadow attractors exhibit apparent structure. Animating the attractors with the user-

(a)

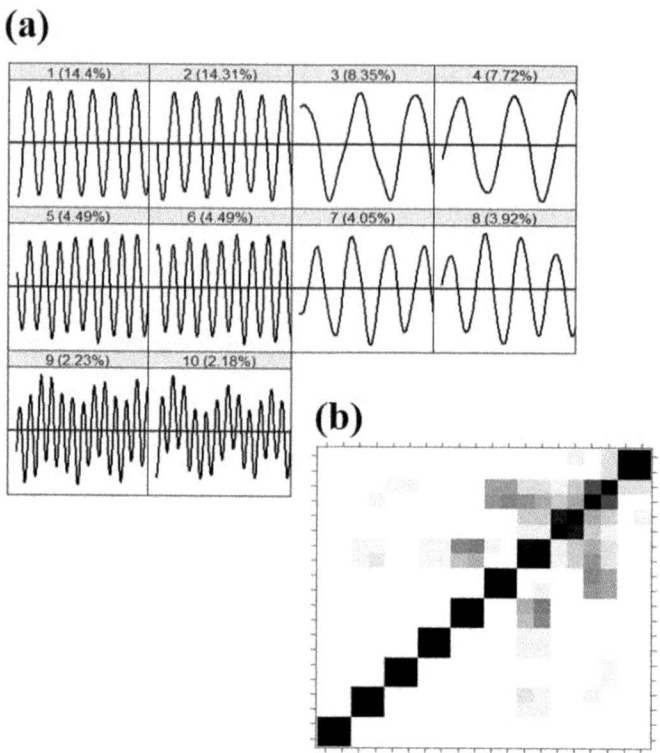

(b)

Fig. 10.4 SSA diagnostics for measles with window length $L = 550$. (a) Eigenvector plots. The first five pairs of eigenvectors oscillate in phase quadrature at 52-week (29%), 126-week (16%), 38-week (9%), 83-week (8%) and 27-week (4%) cycle lengths, where the percentages in parentheses measure the variation in the time series from the mean accounted for by each cycle. The signal is moderately strong since the cycles together account for 66% of total variation. (b) Weighted correlation matrix. Each of these cyclical components is insignificantly correlated with other paired groups and thus exhibits the necessary statistical independence.

defined function `animat3d_udf` (Code 6.3) demonstrates that each is composed of irregular oscillations that work their way forwards and backwards along a northeast axis. This rules out three-dimensional linear spiral centres that proceed in only one direction, but could still be consistent with noisy three-dimensional linear behaviour that we test with surrogate data. The correlation integrals plotted for the scarlet fever (Figure 10.9b), measles (Figure 10.10b) and pertussis (Figure 10.11b) signals exhibit possible scaling regions, but correlation dimensions computed with the Takens estimator (Takens, 1985) fail to saturate at increasing embedding dimensions (Figures 10.9c, 10.10c and 10.11c). Consequently, we do not use correlation dimension as a discriminating statistic in surrogate data testing of any signal. The semilogarithmic average separation plots calculated from the scarlet fever (Figure 10.9d) and measles (Figure 10.10d) signals do not exhibit a linear scaling region as required, so we do not use

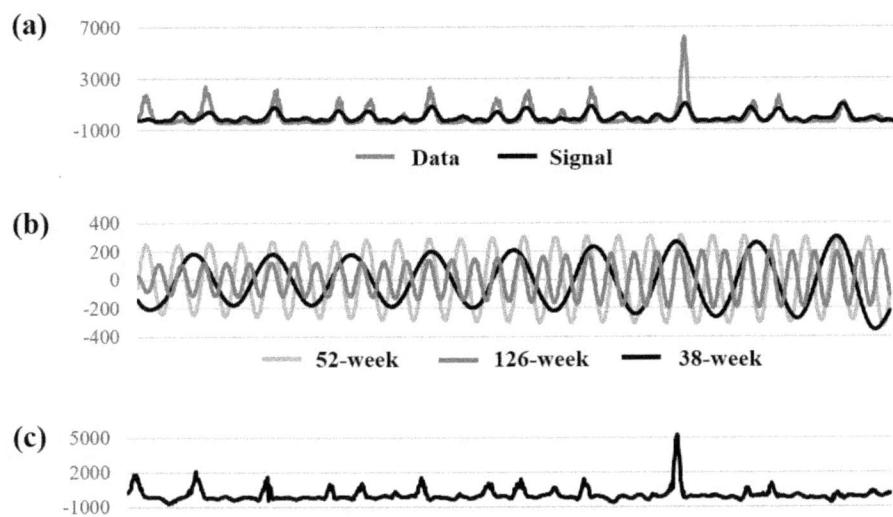

Fig. 10.5 Signal processing of measles. (a) Plots of observed mean-adjusted time series (grey curve) and reconstructed signal (black curve). (b) Plots of the three most dominant 52-, 126- and 38-week cycles comprising the signal. (c) Noise (unstructured variation) removed from the time series.

the maximum Lyapunov exponent as a discriminating statistic in surrogate data testing of these signals. Conversely, the average separation plot from the pertussis signal does exhibit a linear scaling region (Figure 10.11d, dashed curve), whose slope gives an estimated maximum Lyapunov exponent of 0.35. Accordingly, we use the maximum Lyapunov exponent as a discriminating statistic in surrogate data testing of the pertussis signal. Finally, the shadow attractors for each signal performed nearly perfectly in one-step nonlinear prediction, with Nash–Sutcliffe efficiencies ranging from 0.98 to 1. We therefore used nonlinear prediction skill and permutation entropy as discriminating statistics in surrogate data testing of each signal.

10.3 Surrogate Data Testing

We tested the null hypotheses that the disease signals used to reconstruct attractors are most likely generated by stochastic dynamics. We first tested for linear stochastic dynamics with AAFT surrogates, and then for randomly shifting periodic orbits with PPS surrogates. We set the probability of false rejection at $\alpha = 0.05$, and the rank-order test parameter at $k = 15$. This generated $(k/\alpha) - 1 = 299$ surrogate data vectors for the single-tailed tests and $S = (2k/\alpha) - 1 = 599$ for the two-tailed tests. We selected nonlinear prediction skill as a discriminating measure for all diseases, and we specified upper-tailed tests so that the null hypotheses were rejected only when empirical attractors (i.e. those reconstructed from a disease signal) predicted with significantly more skill than surrogate attractors. We also used permutation entropy as a discriminating statistic for all diseases, and we specified lower-tailed tests to

(a)

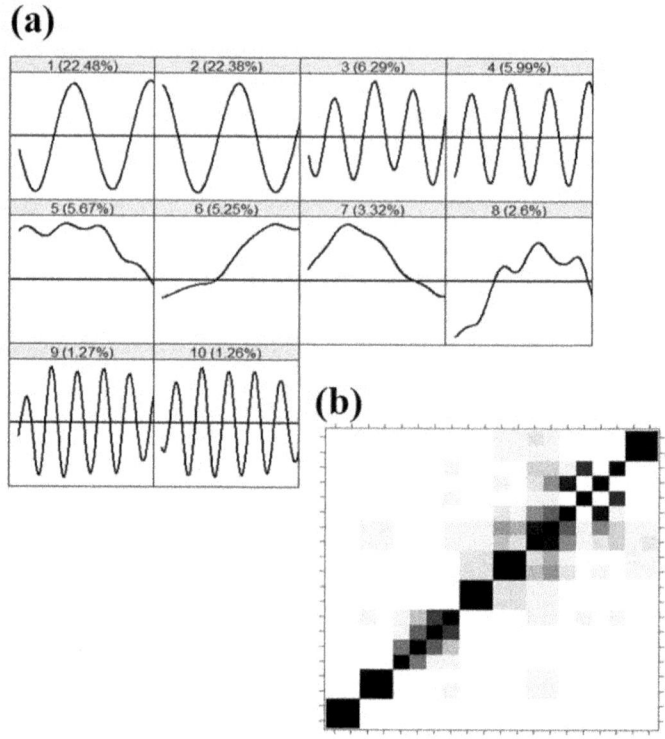

Fig. 10.6 SSA diagnostics for pertussis with window length $L = 626$. (a) Eigenvector plots. Eigenvector pairs $(1, 2)$, $(3, 4)$ and $(9, 10)$ oscillate in phase quadrature at 183-week (45%), 92-week (12%), and 61-week (3%) cycle lengths, respectively. The percentages in parentheses measure the variation in the time series from the mean accounted for by each cycle. Eigenvector pair $(5, 6)$ comprises a slow-moving trend (298 weeks) in the data. The signal is moderately strong since the cycles and trend together account for 71% of total variation. (b) Weighted correlation matrix. Each of these paired components is insignificantly correlated with other paired groups and thus exhibits the necessary statistical independence.

reject the null hypothesis when the measure for a signal was significantly lower than the surrogate measures. We used the maximum Lyapunov exponent as an additional discriminating statistic for the pertussis signal and ran a two-tailed test.

We report the results of the surrogate data tests in Table 10.1. When using non-linear prediction as the discriminating statistic, we weakly reject the null hypothesis of stochastic dynamics for both AAFT and PPS surrogates for all signals. The Nash–Sutcliffe efficiencies for scarlet fever (0.99), measles (0.99) and pertussis (1) lie barely above the respective lower bounds on the upper $k = 15$ surrogate efficiencies (reported in the AAFT and PPS columns in the table) and thus barely fall within the rejection interval. When using permutation entropy, we strongly reject the null hypothesis for measles and pertussis, since the measures taken from each signal (0.5 and 0.37, respectively) lie well below the upper bounds on the lower $k = 15$ measures for AAFT

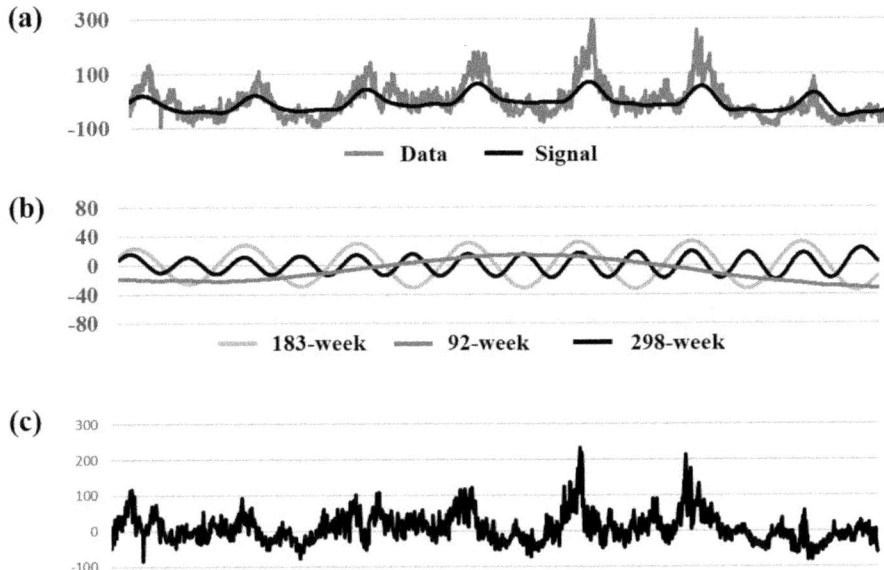

Fig. 10.7 Signal processing of pertussis. (a) Plots of observed mean-adjusted time series (grey curve) and reconstructed signal (black curve). (b) Plots of the three most dominant 183-, 92- and 298-week cycles comprising the signal. (c) Noise (unstructured variation) removed from the time series.

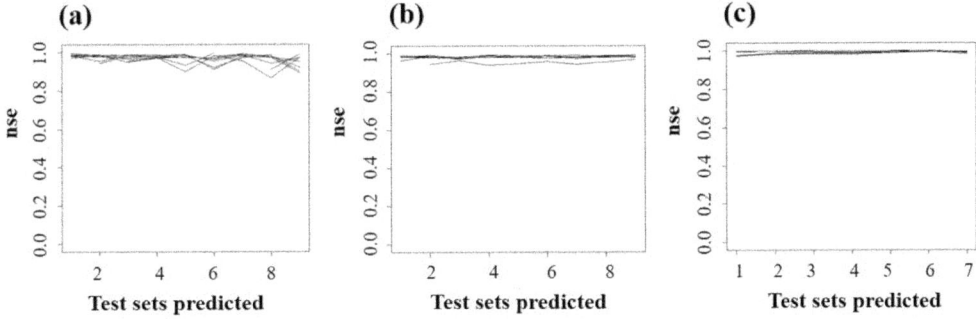

Fig. 10.8 Testing for nonlinear stationarity. Nonlinear cross-prediction plots show that the scarlet fever (a), measles (b) and pertussis (c) signals are stationary for the purposes of NLTS.

surrogates (0.55 and 0.67, respectively) and PPS surrogates (0.98 and 0.98, respectively). Following the same logic, we weakly reject the null hypothesis for scarlet fever when AAFT surrogates are used and strongly reject it for PPS surrogates. The results are mixed when we use the maximum Lyapunov exponent as a discriminating statistic in surrogate testing of the pertussis signal. We accept the null hypothesis for AAFT surrogates because the computed value for the signal is within the interval of

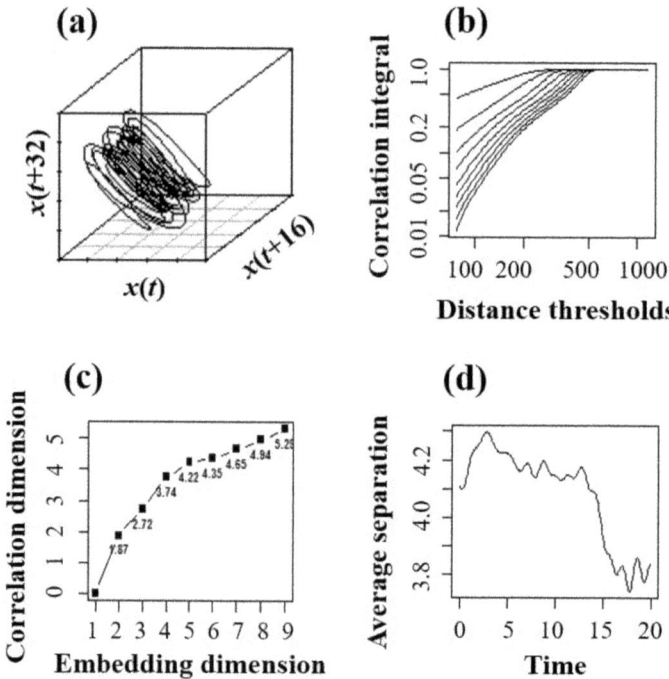

Fig. 10.9 Diagnostic plots for NLTS analysis of the scarlet fever signal. The reconstructed attractor shows visual regularity to be tested for deterministic structure with surrogate data (a). Plotted correlation integrals show potential scaling regions (b), but the Takens estimator of the correlation dimension does not reach a plateau (c), so we do not use it as a discriminating statistic in surrogate data testing. The semilogarithmic average separation plot does not exhibit an upward linear scaling region (d), so we do not use the maximum Lyapunov exponent as a discriminating statistic in surrogate data testing.

values reported in the AAFT column of the table. The lower value of the interval is the upper limit of the smallest k values, and the upper value is the lower limit of the largest k values. However, we reject the null hypothesis for PPS surrogates because the computed value for the signal rests above the upper limit of the acceptance interval of values.

We can conservatively claim that surrogate testing leaves the door open to nonlinear deterministic dynamics. Consequently, we do not rule out the possibility that the irregular oscillatory behaviour observed in the reconstructed attractors is systematic.

10.4 Convergent Cross Mapping

A key question in the epidemiological literature is whether infectious disease epidemics interact. Using linear correlation and regression techniques, Rohani *et al.* (2003) found evidence of interactive dynamics between measles and whooping cough in five Euro-

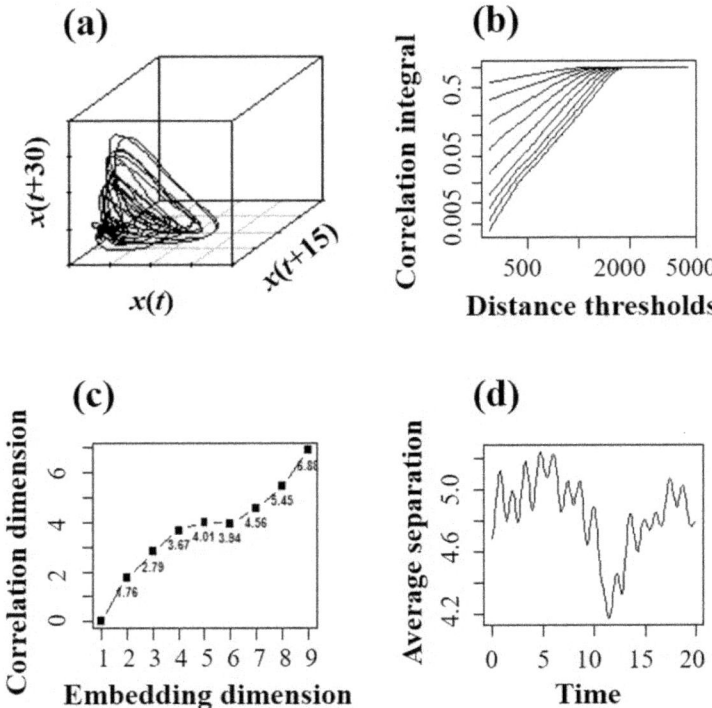

Fig. 10.10 Diagnostic plots for NLTS analysis of the measles signal. The reconstructed attractor exhibits visual regularity to be tested against surrogate data (a). Plotted correlation integrals show a potential scaling region (b), but the Takens estimator of the correlation dimension increases monotonically with the embedding dimension (c), so we do not use it as a discriminating statistic in surrogate data testing. The semilogarithmic average separation plot does not exhibit an upward linear scaling region (d), so we do not use the maximum Lyapunov exponent as a discriminating statistic in surrogate data testing.

pean cities. Given that our empirical diagnosis does not foreclose the possibility that the scarlet fever, measles and pertussis signals are driven by low-dimensional nonlinear deterministic dynamics, we apply convergent cross mapping (CCM) methods to test whether these multiple disease strains interact in the same real-world dynamic system.. We run CCM and delayed CCM with Code 8.4.

The CCM results indicate that measles is strongly forced by scarlet fever, since correlation coefficients for the cross mapping *measles* xmap *scarlet fever* reach an asymptote well exceeding our selected threshold of rho=0.65 (Figure 10.12a, black curve). Measles is not forced by pertussis, because the cross mapping *measles* xmap *pertussis* does not meet this threshold (Figure 10.12a, grey curve). Following this same reasoning, we see that scarlet fever is weakly forced by both measles and pertussis (Figure 10.12b) and that pertussis is strongly forced by measles and scarlet fever (Fig-

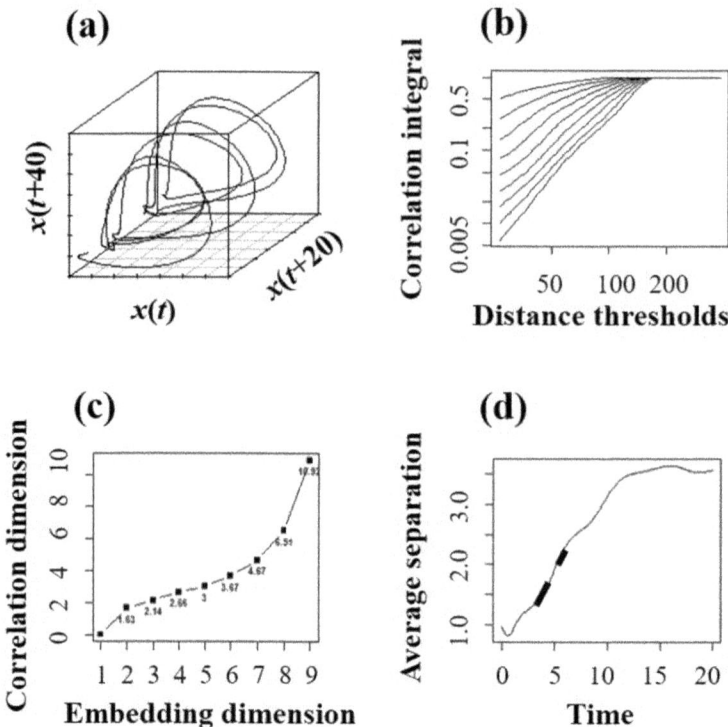

Fig. 10.11 Diagnostic plots for NLTS analysis of the pertussis signal. The reconstructed attractor exhibits visual regularity to be tested against surrogate data (a). Plotted correlation integrals show potential scaling regions (b), but the Takens estimator of the correlation dimension increases monotonically with the embedding dimension (c), so we do not use it as a discriminating statistic in surrogate data testing. The semilogarithmic average separation plot exhibits an upward linear scaling region (dashed line) whose slope estimates the maximum Lyapunov exponent (≈ 0.35) (d). Consequently, we use this as a discriminating statistic in surrogate data testing.

ure 10.12c). Delayed CCM does not identify any of these detected interactions as false positives, since the delayed cross-mapping curves when (d) measles, (e) scarlet fever or (f) pertussis is taken as the response variable are flat and thus do not achieve a maximum average correlation coefficient at a positive delay. We used Code 8.3 with the `tkplot` option to generate the corresponding network plot (Figure 10.12g).

10.5 Phenomenological Model

We extracted a phenomenological model from a time-delay embedding of the scarlet fever signal with an embedding delay of 16 weeks and an embedding dimension of three (Figure 10.9a). We selected the scarlet fever signal to simulate epidemic system

Table 10.1 Surrogate testing results. We test the null hypotheses that the disease signals are generated by stochastic dynamics with AAFT and PPS surrogates. We set the probability of false rejection at $\alpha = 0.05$ and the rank-order test parameter at $k = 15$, which generates 599 (299) for the two- and single-tailed tests, respectively. We specify an upper-tailed test when nonlinear prediction skill is used as the discriminating statistic, so that we reject the null hypothesis when the Nash–Sutcliffe efficiency measure for a signal rests above the lower limit of the upper $k = 15$ values reported in the AAFT and PPS columns. On this basis, we weakly reject the null hypothesis that these signals are generated by stochastic dynamics. We specify a lower-tailed test when permutation entropy is chosen as the discriminating statistic, and reject the null hypothesis when the measure for a signal is lower than the upper limit of the lower $k = 15$ values in the AAFT and PPS columns. We strongly reject the null hypothesis for all diseases, except for weak rejection for scarlet fever when AAFT surrogates are used. Finally, we specify a two-tailed test when the maximum Lyapunov exponent is used as a discriminating statistic, and reject the null hypothesis when the computed value for a signal results within the lowest or highest k values. The intervals reported in the AAFT and PPS columns give the upper (lower) limits on the lowest (largest) k values. Consequently, we accept the null hypothesis that the pertussis signal is generated by linear stochastic dynamics for AAFT surrogates (italicized values), but reject it for PPS surrogates. A conservative reading of the evidence is to weakly reject the null hypothesis.

Discriminating measure	Signal	AAFT	PPS
Nonlinear prediction:			
Scarlet fever	0.99	0.98	0.96
Measles	0.99	0.97	0.96
Pertussis	1	0.99	0.95
Permutation entropy:			
Scarlet fever	0.44	0.45	0.97
Measles	0.50	0.55	0.98
Pertussis	0.37	0.67	0.98
Maximum Lyapunov exponent:			
Pertussis	0.35	*(−0.14, 0.58)*	(−0.2, 0.19)

dynamics for the following reasons. First, CCM offered evidence that measles and pertussis dynamics are embedded in the attractor reconstructed from the scarlet fever signal. Second, we were able to tune available parameters to produce a numerically stable phenomenological model whose dynamic behaviour best matched the complexity of empirically detected multistrain dynamics. In particular, we set the number of equations to $m = 3$ and the polynomial range parameter to $M_p = 4$, resulting in $M_p^m = 64$ monomial terms in each equation. We ran ridge regression on each equation ($\alpha = 0$). Finally, we set the integration step at $\delta = 0.01$ and selected the *lsoda* integration method.

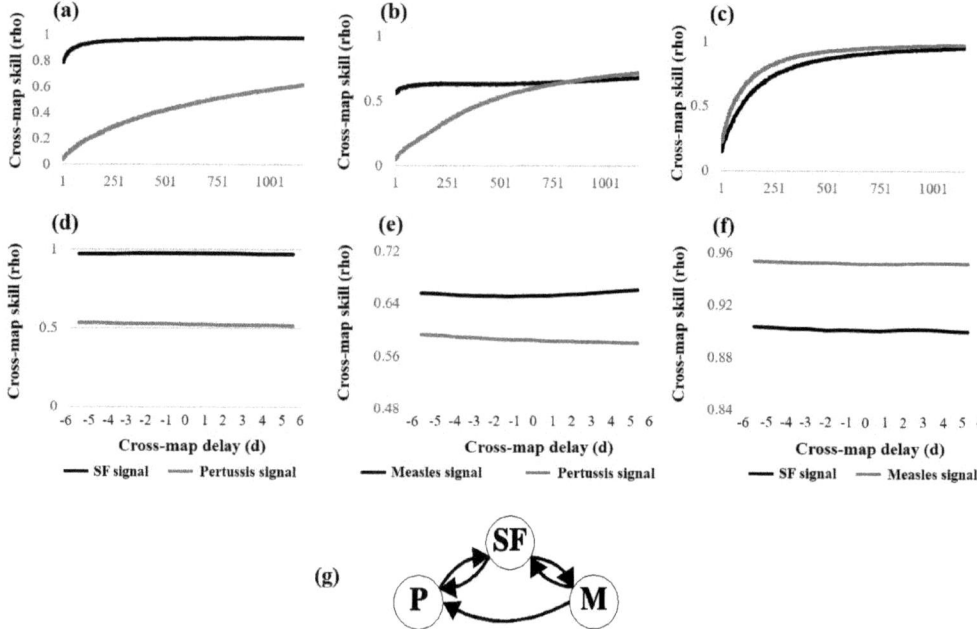

Fig. 10.12 Convergent cross mapping of disease signals. (a) Measles as the response variable. Measles is strongly forced by scarlet fever (SF), since the correlation coefficients for the cross mapping *measles* xmap *scarlet fever* reach an asymptote well exceeding our selected threshold for significance: `rho=0.65`. Measles is not forced by pertussis, because the cross mapping *measles* xmap *pertussis* does not exceed this threshold. (b) Scarlet fever as the response variable. Scarlet fever is weakly forced by both measles and pertussis. (c) Pertussis as the response variable. Pertussis is strongly forced by both measles and scarlet fever. (d) Delayed CCM with measles as the response variable. The result that measles is forced by scarlet fever is not a false positive, since the flat delayed cross-mapping curve (black) does not reach a maximum average correlation coefficient at a positive delay. (e) Delayed CCM with scarlet fever as the response variable. Neither measles nor pertussis is identified as a false driver of scarlet fever, since the delayed cross-mapping curves are flat. (f) Delayed CCM with pertussis as the response variable. Neither measles nor scarlet fever is identified as a false driver of pertussis, since the delayed cross-mapping curves are flat. (g) Network plot showing causal interactions detected by CCM.

This resulted in a parsimonious model composed of second-order polynomial ODEs containing only a fraction of the 64 possible monomial terms:

$$\dot{x} = -21.72 - 3.25z + 0.0003z^2 + 7.72y + 0.001yz - 0.002y^2 + 0.012x$$
$$- 0.01xz + 0.003xy + 0.001x^2$$

$$\dot{y} = -2.38 + 5.5z - 0.002z^2 - 0.12y + 0.001yz + 0.001y^2 - 5.69x$$
$$+ 0.003xz - 0.014xy \tag{10.1}$$

$$\dot{z} = -6.61 - 0.07z + 0.0005z^2 - 7.47y - 0.01yz + 0.0002y^2 + 3.26x$$
$$+ 0.001xz + 0.008xy - 0.002x^2$$

where $x = scarlet\,fever(t)$, $y = scarlet\,fever(t + 16)$ and $z = scarlet\,fever(t + 32)$.

The model produced a simulated signal (Figure 10.13a, black curve) successfully simulating the dominant 52-week cycle in the observed signal (Figure 10.13a, grey curve). Moreover, nonlinear interaction terms permitted the attractor reconstructed from lagged copies of the simulated signal (Figure 10.13b) to exhibit the forwards and backwards cycling characterizing the empirical attractor reconstructed from the scarlet fever signal (Figure 10.9a). The trajectory on the simulated attractor initially undergoes higher-amplitude cycling in a northeast direction and then reverses direction with lower-amplitude cycling. This behaviour distinguishes it from the unidirectional cycling of three-dimensional linear spiral sinks and centres, which was not observed in the empirically reconstructed attractors.

10.6 Summary

The nonlinear data diagnostics that we applied to the New York disease data provide a rigorous empirical benchmark of real-world epidemic behaviour that can be used, along with other expert information, to guide the specification and testing of subsequent epidemiological modelling. Signal processing uncovered dominant cycles in the data and phase space reconstruction provided geometric pictures of long-term epidemic dynamics, which a mechanistic model should reproduce to match the complexity of real-world behaviour. The three-dimensional embedding space of the empirical epidemic attractors indicates the minimum dimensionality required for a mechanistic model to reproduce diagnosed dynamics. The phenomenological model extracted from the observed scarlet fever signal shows that diagnosed dynamics can be captured by an ODE model composed of second-order polynomial interaction terms. This can guide the selection of appropriate functional responses in a mechanistic model.

In Chapter 11, we shall continue the disease application by investigating how extreme value statistics can be used to probabilistically model the noise isolated from observed data.

(a)

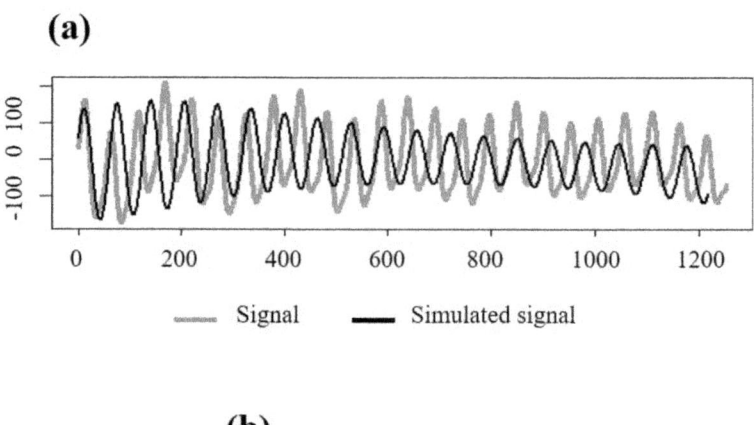

(b)

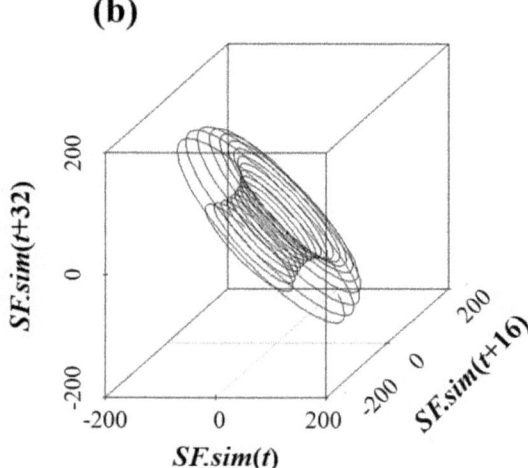

Fig. 10.13 System dynamics reconstructed from the phenomenological model extracted from a time-delay embedding of the scarlet fever signal with an embedding delay of 16 weeks and an embedding dimension of three. (a) The simulated signal (black curve) successfully reproduces the dominant 52-week cycle in the signal separated from the observed time series (grey curve). (b) The attractor reconstructed from lagged copies of the simulated signal successfully reproduces the forwards and backwards cycling characterizing the empirical attractor reconstructed from the scarlet fever signal (Figure 10.9a). The trajectory initially undergoes higher-amplitude cycling in a northeast direction and then reverses direction with lower-amplitude cycling. This behaviour distinguishes it from the unidirectional cycling of three-dimensional linear spiral sinks and centres, which was not observed in the empirically reconstructed attractors.

11
Extreme Value Statistics

'Mr. Holmes, they were the footprints of a gigantic hound!'
    Arthur Conan Doyle, The Hound of the Baskervilles

11.1   Introduction

In Chapter 6, we used singular spectrum analysis to separate signal (capturing structured variation) from noise (capturing residual unstructured variation) in an observed time series. Positive (negative) noise measures the amount by which the time series exceeds (falls below) the signal. These discrepancies may include irregular behaviour falling outside detected behavioural patterns, behaviour following detected oscillatory patterns but at unexpectedly high or low amplitudes, or low-frequency oscillatory behaviour that cannot be resolved with limited available data. A core objective of public policy is to mitigate extreme events such as floods, droughts or epidemics, whether they occur as signal or noise. We investigate the use of extreme value statistics (EVS) to probabilistically model extreme events separated as unstructured noise. EVS models extreme events that are statistically independent (Katz, 2010). This condition is promoted by applying EVS to unstructured residual noise separated from observed time series data in signal processing. We apply a version of EVS that computes the likelihood of extreme discrepancies exceeding a selected threshold value within a given time interval. In theory, exceedances follow a generalized Pareto (GP) distribution, and we run diagnostics to determine how well the data actually fit this distribution. If we find a reasonable fit, we can invert the GP distribution to solve for quantiles providing a useful noise diagnostic: *return-level plots*. Return-level plots show the *return periods* expected before particular extreme noise levels (*return levels*) are realized.

11.2   The Generalized Pareto Distribution

The GP cumulative distribution gives the conditional probability that a random variable X exceeds an observed value x given that X exceeds a threshold value u (Katz et al., 2005):

$$P\{X > x \mid X > u\} = \begin{cases} 1 - \left(1 + \dfrac{\xi(x - u)}{\sigma}\right)^{-1/\xi} & \text{for } \xi \neq 0 \\[2mm] 1 - e^{-(x-u)/\sigma} & \text{as } \xi \to 0 \end{cases} \qquad (11.1)$$

Nonlinear Time Series Analysis with R. Ray Huffaker, Marco Bittelli and Rodolfo Rosa, Oxford University Press (2017).
© Ray Huffaker, Marco Bittelli and Rodolfo Rosa. DOI: 10.1093/oso/9780198782933.001.0001

In eqn (11.1), $\sigma > 0$ is the *scale* parameter, and $-\infty < \xi < \infty$ is the *shape* parameter determining whether the GP conforms to a light-tailed exponential distribution ($\xi = 0$), a heavy-tailed Pareto distribution ($\xi > 0$) or a bounded beta distribution ($\xi < 0$). Applying the definition of conditional probability

$$P\{X > x\} = P\{X > x \mid X > u\}P\{X > u\}$$

we can convert eqn (11.1) to the more convenient unconditional distribution

$$P\{X > x\} = \begin{cases} \left[1 - \left(1 + \dfrac{\xi(x-u)}{\sigma}\right)^{-1/\xi}\right]\dfrac{k}{n} & \text{for } \xi \neq 0 \\[3mm] \left(1 - e^{-(x-u)/\sigma}\right)\dfrac{k}{n} & \text{as } \xi \to 0 \end{cases} \tag{11.2}$$

In this equation, we estimate $P\{X > u\}$ as the sample proportion of observations over the threshold u: k/n, where k is the number of exceedances and n is sample size.

We compute quantiles by inverting the unconditional GP distribution (11.2) to solve for x in terms of $P\{X > x\}$. In particular, we solve for the value of x, x_m, that exceeds the threshold value u once every m periods. Thus, we set $P\{X > x\} = 1/m$ and solve for x_m:

$$P\{X > x\} = 1/m = \begin{cases} \left[1 - \left(1 + \dfrac{\xi(x_m-u)}{\sigma}\right)^{-1/\xi}\right]\dfrac{k}{n} & \text{for } \xi \neq 0 \\[3mm] \left(1 - e^{-(x_m-u)/\sigma}\right)\dfrac{k}{n} & \text{as } \xi \to 0 \end{cases} \tag{11.3}$$

The resulting quantiles are (Gilleland and Katz, 2013)

$$x_m = \begin{cases} u + \dfrac{\sigma}{\xi}\left[\left(\dfrac{mk}{n}\right)^{\xi} - 1\right] & \text{for } \xi \neq 0 \\[3mm] u + \sigma \log\left(\dfrac{mk}{n}\right) & \text{as } \xi \to 0 \end{cases} \tag{11.4}$$

The *return level* x_m is the noise level that we expect to be exceeded every m observations or time periods. Put another way, the *return time* m is the expected waiting time before the noise level x_m is exceeded. The *return-level plot* graphs return levels against increasing return times.

11.3 Extreme Value Statistics with R

Code 11.1 computes return-level plots with a 95% confidence band for positive and negative noise discrepancies. Negative noise is simply positive noise multiplied by -1. We first model the positive noise separated from the pre-vaccine New York scarlet fever cases in Chapter 10. Initially, we plot the noise to visualize an initial estimate of the threshold parameter (Figure 11.1a). The threshold should allow for sufficient discrepancies to provide a good fit of the GP distribution, but not so many that discrepancies are not *extreme* values. In Step 1, we build loop i to compute *n.per* points on the return-level plot. We rely on the R package **extRemes** (Gilleland and Katz, 2013) to estimate the parameters of the GP distribution (Step 1a). We then compute the quantiles giving the return levels in the plot for a given return time using the sample proportion of observations over the threshold provided by **extRemes** (Step 1b).

```
#Code 11.1  Extreme value statistics
rm(list=ls(all=TRUE))

library(extRemes)  #gpd mle regression in boot function
library(boot)  #goodness of fit


#Directories
dir.data<-as.character("C:/Data")
dir.results<-as.character("C:/Results")


#Load data
setwd(dir.data)
ts<-read.csv("disease_1924.csv")


x<-ts$SF_noise  #positive discrepancies
th<-200  #threshold for positive discrepancies
R<-2000 #bootstrap replications for positive discrepancies

#x<--ts$SF_noise  #negative discrepancies
#th<-60  #threshold for negative discrepancies
#R<-1500 #bootstrap replications for negative discrepancies


#Plot noise
plot(x,type='l',lwd=2,cex.axis=1.2,cex.lab=1.2,font=2)

#Step 1:  Loop i computes 'n.per' points on the return-level plot
n.per<-100  #has to be greater than rt0 computed below (first return time
  #for which return level is above threshold)

xm.hold<-matrix(0,n.per,1)  #store return levels

for(i in 1:n.per){
  #Step 1a: Fit noise to GP distribution
  fit<-fevd(x,threshold=th,type=c("GP"),method=c("MLE"))
  epsilon<-as.numeric(fit$results$par[2]) #GP shape parameter
  omega<-as.numeric(fit$results$par[1])   #GP scale parameter
  proportion<-fit$rate  #sample proportion of observations over threshold

  #Step 1b: Compute Quantiles (xm exceeded on average once every m observations)
  if (epsilon==0) {xm<-th+(omega*log(i*proportion))} else
    {xm<-th+((omega/epsilon)*(((i*proportion)^epsilon)-1))}
  xm.hold[i,]<-xm  #store return levels
}

  #Step 1c: Diagnostics plot for GP fit
plot(fit,lwd=2,cex.axis=1.2,cex.label=1.2, font=2)

  #Step 1d: Start return plot at return level immediately above threshold
xm<-xm.hold  #return levels
```

```
return.time<-which(xm>=th)  #return times when xm at or above threshold
rt0<-return.time[1]  #first return time time when xm above threshold
return.level<-xm[return.time]  #associated return level

#Step 2: Bootstrap confidence intervals around each point of return-level plot
th.per<-n.per-(rt0-1)  #number of points above threshold
lower<-matrix(0,th.per,1)  #store lower confidence bound
upper<-matrix(0,th.per,1)  #store upper confidence bound

for(j in 1:th.per){
  #Step 2a: Define bootstrap user-defined function. Begin bootstrap at
    #rt0 (first return time when xm above threshold)
  boot.xm<-function(data,indices,threshold=th,m=(j+(rt0-1))){
    y<-data[indices]
    fit<-fevd(y,threshold=th,type=c("GP"),method=c("MLE"))
    epsilon<-as.numeric(fit$results$par[2]) #GP shape parameter
    omega<-as.numeric(fit$results$par[1])   #GP scale parameter
    proportion<-fit$rate  #sample proportion of observations over threshold

    #Compute Quantiles
    if (epsilon==0) {xm<-th+(omega*log(m*proportion))} else
      {xm<-th+((omega/epsilon)*(((m*proportion)^epsilon)-1))}

return(xm)
} #end user-defined function boot.xm

#Step 3: Run bootstrap user-defined function
b<-boot(data=x,statistic=boot.xm,R=R)
#b$t  #bootstrapped return levels
ci<-boot.ci(b,conf=0.95,type='bca')  #bootstrapped confidence intervals
low<-ci$bca[4];print(low)
up<-ci$bca[5];print(up)
lower[j,]<-low  #lower 95% confidence bound
upper[j,]<-up   #upper 95% confidence bound
}

#Step 4: Return-level plot with 95% confidence bands

#Return-level plot from noise data
plot(return.time,return.level,type="l",xlab= "return time",ylab="return level",
  ylim=range(return.level,lower,upper),cex.lab=1.5,las=1,cex.axis=1.5,font=2,
  lwd=2)

#Add confidence band and overlay return-level plot
time<-rt0:n.per
#rev(time) is time in reverse order
#c(time,rev(time)) is time in forward order followed by rev(time)
polygon(c(time,rev(time)),c(lower,rev(upper)),
  col="dark gray",border=NA)
lines(time,return.level,lty="solid",col="black",lwd=6) #plot regression line

#Export return levels and confidence band limits to external *.csv file
setwd(dir.results)
data1<-cbind(lower,return.level,upper)
names<-c("lower","level","upper")
write.table(data1,sep=",","return.csv",col.names=names,row.names=FALSE)

#Errors
#"estimated adjustment 'a' is NA" generated by R boot package
#Correction:  Increase bootstrap replications [R in boot()]
```

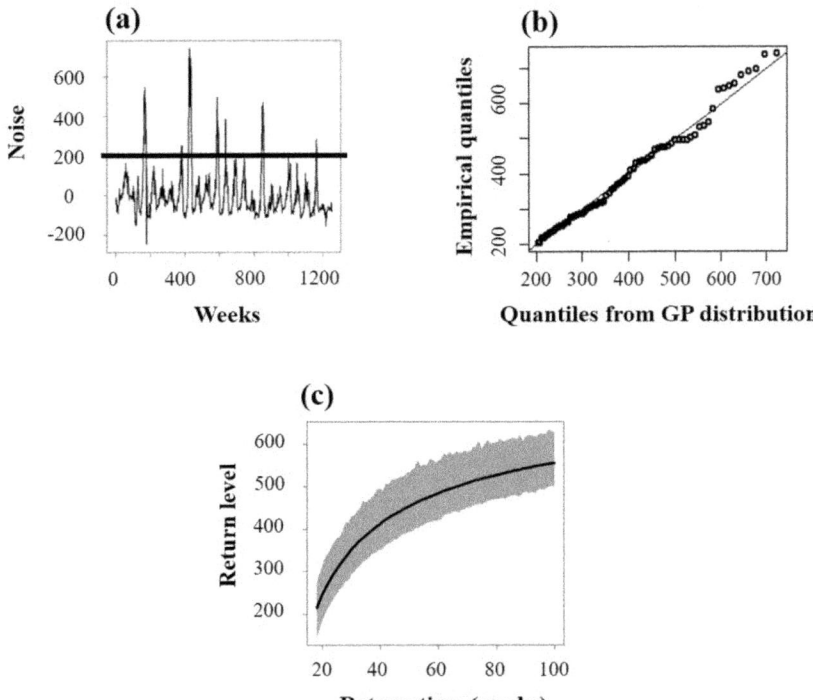

Fig. 11.1 Analyzing positive noise separated from scarlet fever cases. (a) Time series of noise, with the horizontal line at the selected threshold of 200 cases. (b) Q–Q plot of empirical quantiles against those computed from the GP distribution. The data fit the GP distribution well, as indicated by points lying mostly on the 45° line. (c) Return-level plot showing the noise magnitudes expected on average for given return times (weeks). The shaded band is the bootstrapped 95% confidence interval.

In Step 1c, we call the Q–Q diagnostic plot generated by extRemes to evaluate how well the noise data fit the GP distribution. A Q–Q *plot* plots quantiles computed from the fitted GP distribution against those estimated from the data using an empirical estimator (Hyndman and Fan, 1996). If the data provide a good fit to the GP distribution, the points in the Q–Q plot rest approximately on the 45° line. The Q–Q plot for positive scarlet fever noise (conditioned on a threshold equalling 200 scarlet fever cases) exhibits an excellent fit up to a return time of about 500 weeks, and then continues to disperse tightly around the 45° line (Figure 11.1b). In Step 1d, we adjust the return levels and return times so that the return-level plot begins at the first return time ($rt0$) for which the return level is above the threshold. The return-level plot will be undefined if the number of points generated in the plot ($n.per$) do not exceed this initial return time $rt0$.

In Step 2, we bootstrap a point-by-point 95% confidence band around the return-level plot. We first formulate a user-defined function boot.xm that assigns the return

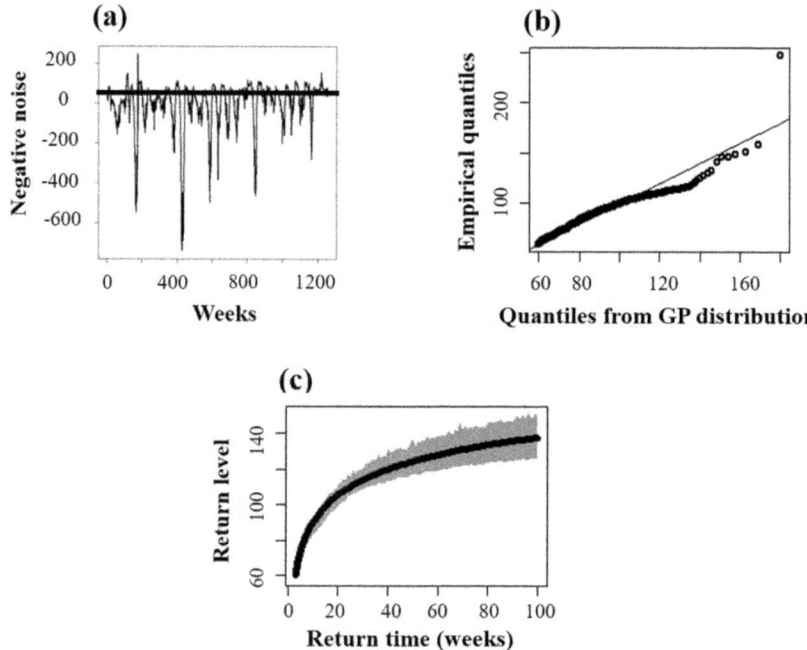

Fig. 11.2 Analyzing negative noise separated from scarlet fever cases: (a) Time series of noise, with the horizontal line at the selected threshold of 60 cases. (b) Q–Q plot of empirical quantiles against those computed from the GP distribution. The data fit the GP distribution well, as indicated by points lying mostly on (or close to) the 45° line. (c) Return-level plot showing noise magnitudes expected on average for given return times (weeks). The shaded band is the bootstrapped 95% confidence interval.

level xm as the bootstrapped statistic to be run in the subsequent `boot` command (Step 3). The user-defined function follows the same procedure used in Step 1 to fit the bootstrapped noise to the GP distribution and compute quantiles. We run the bootstrap over each point on the return-level plot in loop j, being careful to run the loop only over points exceeding the threshold and from the initial return time that the threshold is exceeded ($rt0$). Finally, we plot the return-level plot and subsequently overlay the confidence band (Step 4). The return-level plot for positive scarlet fever noise and the shaded confidence band are shown in Figure 11.1c. We expect noise of threshold magnitude (200 cases) on average every 18 weeks. We expect noise of double the threshold magnitude on average every 40 weeks.

We now turn to the negative scarlet fever noise plotted in Figure 11.2a. The Q–Q plot (conditioned on a threshold of 60 cases) demonstrates an excellent fit for about 120 weeks, and then begins to deviate close to the 45° line (Figure 11.2b). The return-level plot shows that we can expect noise at threshold magnitude on average every 3 weeks, and noise at 100 cases about every 20 weeks (Figure 11.2c).

Appendix A

A.1 Probability Density for the Logistic Map

To derive the probability density $Beta(0.5, 0.5)$ for the logistic map (2.25), we will follow Alligood *et al.* (1997). We introduce the *tent map* $t(x)$, which is given by the equation

$$x_{t+1} = r \left(1 - 2\left|x_t - \tfrac{1}{2}\right|\right) = \begin{cases} 2rx_t & \text{if } 0 \leqslant x_t < \tfrac{1}{2} \\ 2r(1 - x_t) & \text{if } \tfrac{1}{2} \leqslant x_t \leqslant 1 \end{cases} \tag{A.1}$$

As can be seen from Figure A.1, the plot of this equation calls to mind a tent as drawn by a child – hence the name of the map. Let us fix $r = 1$ and take x_t to lie in the interval $[0, 1]$. We seek the probability $\rho(x_t)$ that x_t is in the interval $[a, b]$ at time t. Figure A.1 shows that for values of x_t lying in $[a, b]$, there correspond two intervals of x_{t-1}, namely $\left[\tfrac{1}{2}a, \tfrac{1}{2}b\right]$ and $\left[1 - \tfrac{1}{2}b, 1 - \tfrac{1}{2}a\right]$.

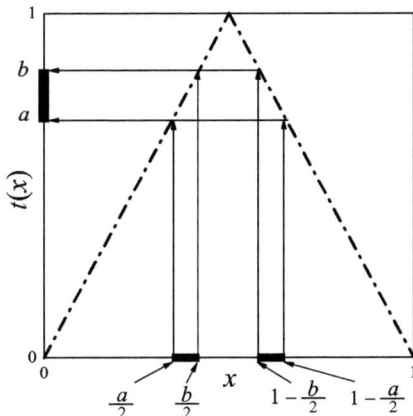

Fig. A.1 Scheme to compute the density $\rho_t(x)$ for the tent map $t(x)$ (A.1) with $r = 1$.

The probabilities that $x_t \in [a, b]$ and that $x_{t-1} \in \left[\tfrac{1}{2}a, \tfrac{1}{2}b\right] \cup \left[1 - \tfrac{1}{2}b, 1 - \tfrac{1}{2}a\right]$ are thus the same:

$$\int_a^b \rho_t(x)\, dx = \int_{\frac{1}{2}a}^{\frac{1}{2}b} \rho_{t-1}(x)\, dx + \int_{1-\frac{1}{2}b}^{1-\frac{1}{2}a} \rho_{t-1}(x)\, dx$$

On the right-hand side, let $x = \tfrac{1}{2}y$ in the first integral and $x = 1 - \tfrac{1}{2}y$ in the second:

$$\int_a^b \rho_t(x)\, dx = \int_a^b \left[\tfrac{1}{2}\rho_{t-1}\left(\tfrac{1}{2}y\right) + \tfrac{1}{2}\rho_{t-1}\left(1 - \tfrac{1}{2}y\right) \right] dy$$

Then

$$\rho_t(x) = \tfrac{1}{2}\left[\rho_{t-1}\left(\tfrac{1}{2}x\right)\right] + \tfrac{1}{2}\left[\rho_{t-1}\left(1 - \tfrac{1}{2}x\right)\right]$$

the solution of which is (removing subscripts)

$$\rho(x) = 1, \quad 0 \leqslant x \leqslant 1 \tag{A.2}$$

The density $\rho(x)$ is uniform, which means that the probability that an iterate is between x and $x + dx$ is constant in the whole interval $[0, 1]$.

The tent map is conjugate to the logistic map with $r = 4$, which means that, roughly speaking, the two maps have the same dynamical behaviour (see Alligood *et al.* (1997), pp. 114–124). So, if we write

$$x = \sin^2 \tfrac{1}{2}\pi y$$

we find that the two intervals $[y, y + dy]$ and $[x, x + dx]$ are visited with the same frequency by the tent map and the logistic map, respectively. We can then write

$$\rho^{\text{lo4}}(x)\, dx = \rho^{\text{tent}}(y)\, dy$$

or

$$\rho^{\text{lo4}}(x) = \frac{dy}{dx}\, \rho^{\text{tent}}(y)$$

Since $\rho^{\text{tent}}(y) = 1$ and

$$\frac{dx}{dy} = \frac{d}{dy}\sin^2\tfrac{1}{2}\pi y = \pi \underbrace{\sin\tfrac{1}{2}\pi y}_{=\sqrt{x}}\ \underbrace{\cos\tfrac{1}{2}\pi y}_{=\sqrt{1-x}} = \pi\sqrt{x(1-x)}$$

we obtain (2.25) as depicted by the dashed line in Figure 2.26b. From (2.25), we can see that the density assumes an infinite value for $x = 0$ and $x = 1$, which are the iterates of the point $x_{\max} = 0.5$ corresponding to the maximum value of the map:

$$f(0.5) = 1 \qquad \text{and} \qquad f(1) = 0$$

A.2 Elements of Ergodic Theory

We give here a brief account of ergodic theory, following the book by Ott (2002). Eckmann and Ruelle (1985) write (their italics):

Ergodic theory says that a time average equals a space average. The weight with which the space average has to be taken is an *invariant measure*.

The word 'space' in 'space average' means the phase space, and 'measure' means a real function with certain properties. We will speak of 'measure' rather than 'density';

indeed, the first notion is better defined from a mathematical point of view. Very briefly, the measure is a function that assigns a 'dimension' to a subset of the phase space. For instance, consider the Lebesgue measure on the interval $[0, 1]$; the measure of the interval $[a, b] \subset [0, 1]$ is its length $b - a$. In particular, we will use the *probability measure*. There are many excellent books on these topics, among which see, for instance, Billingsley (1995) and Ash and Doléans-Dade (2000). We shall give here a brief description of some basic concepts, referring to the literature for more extensive discussions and mathematical insights.

Let Ω be an arbitrary set. The notion of an *algebra* (or field) is introduced as a collection $\mathcal{F}$ of subsets of Ω such that the following hold:

(1) $\Omega \in \mathcal{F}$.

(2) If $S \in \mathcal{F}$, then $S^c \in \mathcal{F}$, where S^c denotes the complement of S.

(3) If $S_1, S_2, \ldots, S_n \in \mathcal{F}$, then $\bigcup_{i=1}^{n} S_i \in \mathcal{F}$.

In other words, $\mathcal{F}$ is *closed under complementation and finite union*. If the union in (3) is replaced by a *countable* union, i.e.

(3′) If $S_1, S_2, \cdots \in \mathcal{F}$, then $\bigcup_{i=1}^{\infty} S_i \in \mathcal{F}$

(in other words, $\mathcal{F}$ is *closed under complementation and countable union*), then $\mathcal{F}$ is called a *σ-algebra* (or σ-field). For instance, the smallest σ-algebra of subsets of Ω is $\mathcal{F} = \{\varnothing, \Omega\}$, where $\varnothing$ is the empty set; while the largest σ-algebra is the collection of all subsets of Ω, including $\varnothing$ and Ω itself.

We define a measure of the elements (sets) of $\mathcal{F}$; in probability theory, these elements are called *events* and a probability (measure) is assigned. Consider the function

$$\mu : \mathcal{F} \longrightarrow [0, +\infty]$$

The function μ is a *measure* on a σ-algebra $\mathcal{F}$ of subsets of Ω if the following hold:

(1) $\mu(S) \geqslant 0 \ \forall S \in \mathcal{F}$ (*non-negativity axiom*).

(2) If $S_i \in \mathcal{F} \ \forall i \in I \subseteq \mathbb{N}$, such that $S_i \bigcap S_j = \varnothing \ \forall i, j \in I, i \neq j$, then

$$\mu\left(\bigcup_{i \in I} S_i\right) = \sum_{i \in I} \mu(S_i) \qquad (\sigma\text{-}additivity \ axiom)$$

The σ-additivity axiom says that the measure of the union of the disjoint subsets S_i is equal to the sum of the measures of the subsets themselves. Note that $S_1, S_2, \ldots$ may form a finite or countably infinite collection of sets in $\mathcal{F}$. The triple $(\Omega, \mathcal{F}, \mu)$ is called *measure space*.

We define the *Dirac measure* $\delta_x \ (x \in \Omega)$ by

$$\delta_x(A) = \begin{cases} 1 & \text{if } x \in A \\ 0 & \text{if } x \notin A \end{cases} \tag{A.3}$$

This is a measure centred on some fixed point x in the measurable space $(\Omega, \mathcal{F})$.

We give a couple of definitions:

If a measure μ is such that $\mu(\Omega) < \infty$, then it is called a *finite measure*.

If a measure μ is such that $\mu(\Omega) = 1$, then it is called a *probability measure* or a *probability*.

If μ is a probability measure, then the triple $(\Omega, \mathcal{F}, \mu)$ forms a *probability space*. The probability μ is then defined as a particular measure associated with the measurable space $(\Omega, \mathcal{F})$.

Before going further into the formal aspects of the subject, let us recall that the word 'ergodic' derives from the Greek $\epsilon\rho\gamma o\nu$ (*ergon*: 'work', 'energy') and $o\delta\delta\sigma$ (*odon*: 'route', 'path'). The term was introduced by Boltzmann in statistical mechanics to describe the 'work trajectory' covered in phase space by the representative point of the system. The system is ergodic if its work trajectory visits, sooner or later, all the points of phase space allowed by energy constraints. This hypothesis was later called the *ergodic hypothesis*; see also Brush (1986) and von Plato (1994).

Let us give some more definitions.

A measure μ is *invariant* with respect to the map f if

$$\mu(S) = \mu\big(f^{-1}(S)\big) \tag{A.4}$$

where $S \subseteq \mathbb{R}$ and $\mu\big(f^{-1}(S)\big)$ is the preimage (inverse image) of S under f.

In analogy to the notion of 'natural density', we can define the *natural measure* of the set S. Consider the measure $\mu(x_0, S)$: if it is the same for every initial condition x_0 (except for a a set of measure zero), then $\mu(x_0, S) = \mu(S)$ is a natural measure of S. In fact, there is a close connection between the invariant density and the invariant measure: an invariant density ρ induces an invariant probability measure μ on Ω. Indeed, it can be proved that for every subset $S \subseteq \mathbb{R}$,

$$\mu(S) = \int_S \rho(x)\, dx$$

and that $\mu(S)$ is an invariant probability measure. In addition to an invariant measure with respect to a map, we can define a map preserving a measure: the map f preserves the measure μ if eqn (A.4) holds.

More generally, instead of 'map', one speaks of 'transformation' and μ is a probability measure on $\mathbb{R}^n$. Then the transformation f is *measure-preserving* if

$$\mu(\mathbf{S}) = \mu\big(f^{-1}(\mathbf{S})\big) \quad \forall \mathbf{S} \subset \mathbb{R}^n$$

For instance, if $S = [0, 1]$ and μ is the invariant probability measure corresponding to the probability density function $Beta(0.5, 0.5)$ in eqn (2.25), then the logistic map, with $r = 4$, preserves the measure μ.

On the basis of these definitions, we introduce the notion of *ergodic measure*. An probability measure μ that is invariant with respect to a map f is called ergodic (also metrically decomposable or metrically transitive) if the following equality *does not* hold:

$$\mu = p\mu_1 + (1 - p)\mu_2, \quad 0 < p < 1 \tag{A.5}$$

where μ_1 and μ_2 ($\mu_1 \neq \mu_2$) are two invariant probability measures.

We can also define an *invariant set*: a set S is invariant with respect to a transformation (map) f if $f^t(S) = S \; \forall t > 0$. This means that a trajectory starting inside an invariant set will remain forever inside that set. From this and the earlier definitions, we can then say that a probability measure μ is invariant with respect to a map f if $\mu(S) = 0$ or $\mu(S) = 1$ for every set S that is invariant with respect to f. This means that a trajectory described by the map f cannot remain trapped in any subspace of phase space.

The well-known *ergodic theorem* due to G. Birkhoff states that if f is measure-preserving and μ is an ergodic probability measure, then, for ϕ a 'sufficiently regular' (integrable) function, it holds that

$$\lim_{n \to \infty} \frac{1}{n} \sum_{i=0}^{n-1} \phi\big(f^{[i]}(x_0)\big) = \int \phi(x) \, d\mu(x) \tag{A.6}$$

where $d\mu(x) = \rho(x) \, dx$, if the density $\rho(x)$ exists. This holds for all initial conditions x_0, except for a a set of measure zero. The theorem means that the *time average* of the function $\phi(x)$ computed along a trajectory starting from an initial condition x_0 is equal to the *space average*, i.e. the average in phase space, weighted according its natural measure μ. We say that a system (a transformation, a map, a process) is ergodic, with respect to a given measure μ, if eqn (A.6) holds.

As mentioned before, the notion of ergodicity originates in Boltzmann's statistical mechanics. For instance, the average velocity of the molecules of a gas may be estimated, in principle, by computing the velocity of a single molecule at a certain instant and dividing the result by the number of molecules. Alternatively, the time average of a single molecule can be computed, if it is hypothesized that the molecule assumes over long periods of time all possible velocities allowed by the energy constraints. In Maxwell's words: '(it) is that the system, if left to itself in its actual state of motion, will, sooner or later, pass through every phase which is consistent with the equation of energy.' Here 'phase' means 'dynamical state'. In fact, in its strong version, the ergodic hypothesis has been proved false, so one speaks of the *quasi-ergodic hypothesis*, meaning that a typical trajectory, in the long run, will eventually pass arbitrarily close to every point of phase space. The ergodic theorem may be viewed as a kind of *law of large numbers* for dependent data. In their well-known book *Limit Distributions for Sums of Independent Random Variables*, Gnedenko and Kolmogorov (1954) write, as a statement of the underlying philosophical concept of *statistical stability*, that

In fact, all epistemological value of the theory of probability is based on this: that large-scale random phenomena in their collective action create strict, nonrandom regularity.

We know from the *law of large numbers* that if $\{X_n\}$ is a sequence of independent and identically distributed (i.i.d.) random variables, with expected values $\mathrm{E}[X_n] < \infty$, then, for the sequence of the variables' sample means $\overline{X}_n = \sum_{i=1}^{n} X_i/n$, we have

$$\overline{X}_n \xrightarrow[n \to \infty]{} \mathrm{E}[X_n]$$

The law of large numbers may be weak or strong, depending on the type of convergence – in probability or almost sure.

The ergodic theorem gives a similar convergence result for a sequence of random variables that are *not i.i.d.* but dependent. In the case of i.i.d. random variables, the convergence means that a sufficiently large sample is representative of the entire population, while in the case of dependent random variables, the ergodic theorem ensures that if over a long period of time the process assumes a large number of possible values, then a 'time sample' of these values is equivalent to a sample extracted from all the values accessible to the system. In other words, a single trajectory is representative of all the possible trajectories allowed to the system.

We have seen in Section 2.7.1 how a small distance δx_0 close to the point x_0 expands (contracts) after one iteration, which, for the sake of convenience, we rewrite here as

$$\delta x_1 = \left| \frac{df(x_t)}{dx_t} \right|_{x_t=x_0} \delta x_0$$

We define the *amplification rate* of the map $f : [0,1] \to [0,1]$ at the point x as the quantity

$$\log \left| \frac{df(x)}{dx} \right|$$

and we introduce the *characteristic exponent* for the measure μ as the average rate of growth over the whole interval $[0,1]$:

$$\lambda = \int_0^1 \log \left| \frac{df(x)}{dx} \right| d\mu \tag{A.7}$$

We have mentioned that for an invariant measure μ, we can write $d\mu(x) = \rho(x)\,dx$, if the density $\rho(x)$ exists. For the logistic map, the density does indeed exist and is given by eqn (2.25). Therefore, for the logistic map with $r = 4$, eqn (A.7) becomes

$$\lambda = \frac{1}{\pi} \int_0^1 \frac{\log|4(1-2x)|}{\sqrt{x(1-x)}} \, dx = \log 2$$

which is the Lyapunov exponent derived in Section 2.7.2 with a different method.

A.3   Dirac Delta Function

We describe here some properties of the Dirac $\delta(x)$ function. This is not a function in the strictest sense; rather it is a *distribution*. It is defined by

$$\delta(x) = 0, \quad x \neq 0$$

and

$$\int_{-\infty}^{+\infty} \delta(x)\,dx = 1$$

It holds that, for a function $h(x)$,

$$\int_{-\infty}^{+\infty} h(x)\delta(x-a)\,dx = h(a)$$

We can think of $\delta(x)$ as the limit of a sequence of functions $\delta(x)$ that are different from zero only inside an interval around $x = 0$, with their peaks becoming more and more spiky with increasing n but with their integrals remaining equal to 1.

One approach to deriving $\delta(x-x_0)$ is sketched in Figure A.2. Consider the function

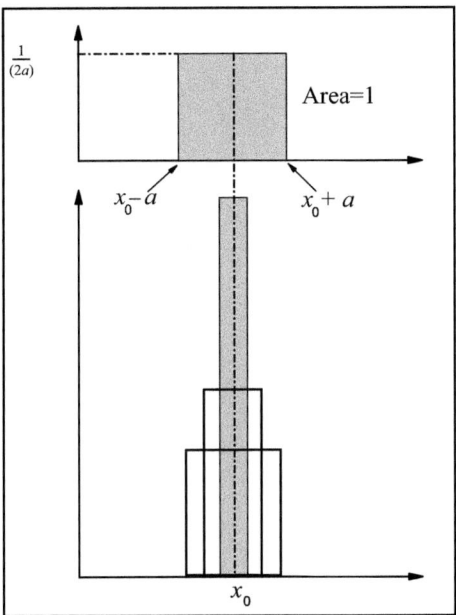

Fig. A.2 Sketch of a derivation of $\delta(x - x_0)$.

that has the value $1/(2a)$ (where a is positive but otherwise arbitrary) in the interval $[x - a, x + a]$ and is 0 elsewhere, as depicted by the rectangle at the top (note that the figure is not to scale), whose area is equal to 1. Let us now imagine narrowing the base of this rectangle, i.e. the interval where the function is nonzero, but keeping the area equal to 1. As $a \to 0$, we arrive at a 'rectangle' with base $= 0$, height $= \infty$ and area $= 1$, which is precisely $\delta(x - x_0)$.

The Dirac delta function has the following properties:

(1) $\delta(x) = \delta(-x)$

(2) $\delta(ax) = \dfrac{1}{a}\delta(x), \quad a > 0$

(3) $h(x)\delta(x - a) = h(a)\delta(x - a)$

(4) $\displaystyle\int \delta(x-y)\delta(y-a)\,dy = \delta(x-a)$

(5) $\displaystyle h(x) = \int \delta(x-y)h(y)\,dy$

If the integral in (5) is definite, then we have

(6) $\displaystyle\int_a^b \delta(x-y)h(y)\,dy = \begin{cases} h(x) & \text{if } a < x < b \\ 0 & \text{if } x \notin (a,b) \end{cases}$

Appendix B

B.1 Introduction to the Bootstrap

Given a sample

$$\mathbf{x} = (5.24, 5.55, 4.69, 4.39, 6.87, 5.15, 4.61, 5.20, 5.49, 4.81,$$
$$2.74, 3.50, 5.19, 5.40, 3.81, 6.49, 6.34, 4.45, 5.10, 3.17) \tag{B.1}$$

with mean $\bar{x} = 4.91$, let us compute the standard error $\widehat{se}(\bar{x})$ of $\bar{x}$:

$$\widehat{se}(\bar{x}) = \left[\frac{1}{n(n-1)} \sum_{i=1}^{n} (x_i - \bar{x})^2 \right]^{1/2} \tag{B.2}$$

where $n = 20$. We find $\widehat{se}(\bar{x}) = 0.24$. The question arises as to whether there exists another method to compute the standard error of the mean. In fact, there are a number of ways of doing this. Let us look at one of these methods. We make $n = 20$ random draws with replacement from the original data set (B.1), obtaining, for instance, the vector

$$\mathbf{x}_1^* = (4.39, 6.34, 5.20, 4.61, 5.19, 5.19, 4.69, 5.15, 3.50, 5.19,$$
$$2.74, 2.74, 2.74, 3.50, 4.45, 6.34, 4.69, 3.81, 4.45, 5.15) \tag{B.3}$$

Notice that some of the numbers are selected once, some twice, etc., and some never. The mean of $\mathbf{x}_1^*$ turns out to be $\bar{x}_1^* = 4.50$. We repeat the same procedure for, say, $B = 15$ times, obtaining the vector $\widehat{\mathbf{x}}^*$ formed by the 15 means:

$$\widehat{\mathbf{x}}^* = (\bar{x}_1^*, \bar{x}_2^*, \ldots, \bar{x}_{15}^*)$$

For our example,

$$\widehat{\mathbf{x}}^* = (4.50, 5.27, 5.27, 5.12, 5.23, 5.23, 4.89, 5.03, 4.55, 4.78,$$
$$4.81, 4.66, 4.91, 5.03, 4.99)$$

We compute the standard deviation of $\widehat{\mathbf{x}}^*$ with the equation

$$\widehat{se}_B(\widehat{\mathbf{x}}^*) = \left[\sum_{b=1}^{B} \frac{(\bar{x}_b^* - \bar{x}^*)^2}{B - 1} \right]^{1/2}$$

where

$$\bar{x}^* = \frac{1}{B} \sum_{b=1}^{B} \bar{x}_b^*$$

In our case, with $B = 15$, $x^* = 4.95$ and the final result is $\widehat{se}_B(\hat{\mathbf{x}}^*) = 0.25$. This result is used as an estimate of the standard error of the mean $\bar{x}$, instead of $\widehat{se}(\bar{x})$.

What we have just described is the simplest form of the bootstrap method, introduced by Efron (1979) and described more fully in Efron and Tibshirani (1993) and Davison and Hinkley (1997). We shall now present a more formal sketch of this method, adopting the same nomenclature as Efron and Tibshirani (1993). Let

$$\mathbf{x} = (x_1, x_2, \ldots, x_n)$$

be a sample of n independent observations. Regard this sample as a realization of the random variables $(X_1, X_2, \ldots, X_n)$ from an unknown underlying probability distribution F. Note that the individual data x_i can be numbers, or vectors, or indeed anything else. We shall suppose that the variables $(X_1, X_2, \ldots, X_n)$ have the same F distribution and are independent (*i.i.d.*), but, in general, this it is not guaranteed. Let

$$\theta = t(F) \tag{B.4}$$

be an unknown parameter of interest (e.g. the mean, median or correlation coefficient). For example, the expected value $\mathrm{E}_F[X]$ of a real-valued X, with density $p(x)$, can be regarded as the parameter θ (if X is a continuous variable):

$$\theta = t(F) = \mathrm{E}_F[X] = \int_{-\infty}^{+\infty} x p(x)\, dx$$

The subscript F here means that the expected value is computed with respect to the distribution F. If F is the normal distribution, with mean μ and variance σ^2, i.e. $X \sim \mathcal{N}(\mu, \sigma^2)$, then $t(F) = \mu$. In eqn (B.4), 't' represents the procedure to be applied to F to obtain θ; t may comprise simple formulae or computer-based methods. For instance, if $F = \mathcal{N}(\mu, \sigma^2)$ and $\theta = \mathrm{E}_F[X]$, then 't' represents the integral

$$\int_{-\infty}^{+\infty} x \frac{1}{\sqrt{2\pi\sigma^2}} \exp\left[-\frac{1}{2}\left(\frac{x-\mu}{\sigma}\right)^2\right] dx$$

Clearly, $t(F)$ can be the variance, median, correlation coefficient, etc. But F is unknown, and so we must exploit what we have at hand, namely the sample $\mathbf{x}$, to make an inference on, i.e. to estimate, θ. One possible choice is to replace F by $\widehat{F}$, the *empirical distribution function* for the data, which places equal probabilities $1/n$ at each sample value x_i. We estimate the function $\theta = t(F)$ of the probability distribution F with the *same* function of the empirical distribution function $\widehat{F}$. This is the *plug-in principle*: '[...] the oldest idea in the statistical book: substitute the empirical distribution of the data for the (unknown) true distribution in anything you wish to estimate' (Efron, 1988). For example, the plug-in estimate of the expected value $\mathrm{E}_F[X]$ of X is the sample mean $\hat{\theta} = n^{-1}\sum_{i=1}^{n} x_i$, usually denoted by $\bar{x}$. The bootstrap is exploited for *studying the distribution of the plug-in estimate* $\hat{\theta} = t(\widehat{F})$, and then for the estimates of standard error and distortion, but *not to make the estimate.*

B.2 Bootstrap Standard Error

To see how the bootstrap works, we will use it for estimating the standard error of the sample mean. Let X be a random variable with probability distribution function F. Let $\mu_F = \mathrm{E}_F[X]$ be the expected value of X and σ_F^2 the variance:

$$\mathrm{Var}_F[X] = \sigma_F^2 = \mathrm{E}_F\big[(X - \mu_F)^2\big] \tag{B.5}$$

The *standard deviation* σ_F of X is defined as

$$\sigma_F = \sqrt{\mathrm{Var}_F[X]}$$

Let $\overline{X}$ be the random variable *sample mean*. The standard error $se_F(\overline{X})$ is given by

$$se_F(\overline{X}) = \sqrt{\mathrm{Var}_\mathrm{F}\big[\,\overline{X}\,\big]} = \sigma_F/\sqrt{n}$$

The standard error is a frequent way to give a measure of statistical accuracy on the basis of the central limit theorem, which (under quite general conditions on the F) states that, for large sample sizes, the sample mean $\overline{X}$ is approximately distributed normally with mean μ_F and variance σ_F^2/n:

$$\overline{X} \approx \mathcal{N}(\mu_F, \sigma_F^2/n)$$

Note that 'approximately' here refers to normality, not to μ_F and σ_F^2/n, which are exact. From the properties of the normal distribution, it follows that

$$\mathrm{P}\big\{|\bar{x} - \mu_F| < \sigma_F/\sqrt{n}\big\} = 0.683 \quad \text{and} \quad \mathrm{P}\big\{|\bar{x} - \mu_F| < 2\sigma_F/\sqrt{n}\big\} = 0.954$$

Let $\bar{x}$ be a realization of $\overline{X}$, i.e. the mean of the sample $\mathbf{x} = (x_1, x_2, \ldots, x_n)$. In other words, we can expect $\bar{x}$ to be within one standard error of the mean μ_F about 68% of the time and within two standard errors about 95% of the time.

In the case of the sample mean, the estimate $\hat{\sigma}_F^2$ of σ_F^2 is given by

$$\hat{\sigma}_F^2 = \frac{1}{n-1} \sum_{i=1}^{n} (x_i - \bar{x})^2 \tag{B.6}$$

Then the estimate $\widehat{se}(\bar{x})$ of $se(\overline{X})$ is given by

$$\widehat{se}(\bar{x}) = \left[\frac{1}{n(n-1)} \sum_{i=1}^{n} (x_i - \bar{x})^2 \right]^{1/2} \tag{B.7}$$

Applying the plug-in principle, we replace $\mu_F = \mathrm{E}_F[X]$ with $\mu_{\widehat{F}} = \mathrm{E}_{\widehat{F}}[X]$, and we write $\mu_{\widehat{F}} = \bar{x}$. Then eqn (B.5) becomes

$$\sigma_{\widehat{F}}^2 = \mathrm{E}_{\widehat{F}}\left[(X - \mu_{\widehat{F}})^2\right] = \frac{1}{n}\sum_{i=1}^{n}(x_i - \bar{x})^2 \tag{B.8}$$

The final result is the same as eqn (B.7), but with n^2 in place of $n(n-1)$. Obviously, if n is not too small, one estimate is like the other.

We now estimate the standard error with the bootstrap. Let $\mathbf{x} = (x_1, x_2, \ldots, x_n)$ be the observed sample, derived from an unknown probability distribution F. We have to estimate a certain parameter $\theta = t(F)$ on the basis of $\mathbf{x}$. For this purpose, we compute an estimate $\hat{\theta}$ of θ:

$$\hat{\theta} = s(\mathbf{x}) = s(x_1, x_2, \ldots, x_n) \tag{B.9}$$

The function $s(\mathbf{x})$ may be the plug-in estimate $t(\widehat{F})$, but not necessarily. In fact, $s(\mathbf{x})$ may be a very complex function of the data, but it may be also a numerical procedure, a computer code, with input the sample $\mathbf{x}$ and output the estimate $\hat{\theta}$. As a measure of statistical accuracy, we choose the bootstrap estimate of the standard error. We will see that this estimate:

(1) is completely automatic;

(2) does not require theoretical calculations;

(3) can be obtained for any $s(\mathbf{x})$ (function or code).

The bootstrap is based on two notions: *bootstrap sample* and *bootstrap replication*. A bootstrap sample $\mathbf{x}^*$ is a random sample of dimension n drawn from $\widehat{F}$:

$$\mathbf{x}^* = (x_1^*, x_2^*, \ldots, x_n^*)$$

We can regard $\mathbf{x} = (x_1, x_2, \ldots, x_n)$ as the *population* of n objects $(x_1, x_2, \ldots, x_n)$ and $(x_1^*, x_2^*, \ldots, x_n^*)$ as a random sample of dimension n drawn from the population. An example of a bootstrap sample is eqn (B.3).

With the bootstrap sample $\mathbf{x}^*$, we compute the bootstrap replication $\hat{\theta}^*$:

$$\hat{\theta}^* = s(\mathbf{x}^*) = s(x_1^*, x_2^*, \ldots, x_n^*)$$

It is immediately evident that we have applied to $\mathbf{x}^*$ the same function (or code) we applied to $\mathbf{x}$. For instance, if $s(\mathbf{x})$ is the mean calculated on the sample $\mathbf{x}$ (in our case $\bar{x} = 4.91$), then $\mathbf{x}^*$ is the mean of the bootstrap sample $\mathbf{x}^*$, i.e.

$$\bar{x}^* = \frac{1}{n}\sum_{i=1}^{n} x_i^*$$

In our case, $\bar{x}^* = 4.50$. We derive $se_{\widehat{F}}(\hat{\theta}^*)$ as the bootstrap estimate of $se_F(\hat{\theta})$. Symbolically,

$$se_{\widehat{F}}(\hat{\theta}^*) \xleftarrow{\text{bootstrap estimate}} se_F(\hat{\theta})$$
$$\text{plug-in} \nearrow \qquad\qquad\qquad \nwarrow \text{unknown}$$

However, to compute the empirical distribution function $\widehat{F}$, we need *all* the bootstrap samples and we compute the replications $\hat{\theta}^*$ for all the samples. The number of bootstrap samples is given by the number of k-combinations with repetition:

$$C_{n,k}^R = \binom{n+k-1}{k}$$

In our case $k = n$, and then, putting $C_{n,n}^R = q$, the number of distinct bootstrap samples is

$$q = \binom{2n-1}{n} = \frac{(2n-1)!}{n!(n-1)!}$$

If $n = 20$, as in our example, then q is the rather large number $68\,923\,264\,410$. Therefore, the computation of $se_{\widehat{F}}(\hat{\theta}^*)$ turns out to be impracticable, so it is rightly called the *ideal bootstrap estimate*. One way to overcome this difficulty is to generate by Monte Carlo methods only a number B of bootstrap samples, obtaining the so-called *Monte Carlo bootstrap estimate* $se_B(\hat{\theta}^*)$, or, more simply, se_B. The *bootstrap algorithm* can be outlined as follows:

1. Let $\mathbf{x} = (x_1, x_2, \ldots, x_n)$ be the observed sample.

2. Through the empirical distribution function $\widehat{F}$, which puts mass $1/n$ on each observation x_i, generate B bootstrap samples $\mathbf{x}_b^*$, $b = 1, \ldots, B$, with dimension n and independent of each other:

$$\mathbf{x}^* = (\mathbf{x}_1^*, \mathbf{x}_2^*, \ldots, x_B^*)$$

3. For each bootstrap sample, calculate the corresponding bootstrap replication, obtaining B replications:

$$\hat{\theta}_1^* = s(\mathbf{x}_1^*), \quad \hat{\theta}_2^* = s(\mathbf{x}_2^*), \quad \ldots, \quad \hat{\theta}_B^* = s(\mathbf{x}_B^*)$$

Estimate $se_{\widehat{F}}(\hat{\theta}^*)$ with the estimator standard deviation of the B replications $(\hat{\theta}_1^*, \hat{\theta}_2^*, \ldots, \hat{\theta}_B^*)$:

$$se_B(\hat{\theta}^*) = \left[\sum_{b=1}^{B} \frac{(\hat{\theta}_b^* - \hat{\theta}^*)^2}{B-1} \right]^{1/2} \tag{B.10}$$

where

$$\hat{\theta}^* = \frac{1}{B} \sum_{b=1}^{B} \hat{\theta}_b^*$$

There are no fixed rules for choosing the value of B, but the choice does depend on the problem under consideration. Sensible values of B are in the range 100–1000. The whole logical path can be very briefly sketched as follows:

$\underbrace{se_F(\hat{\theta})}$	$\xleftarrow{\text{estimated by}}$	$\underbrace{se_{\widehat{F}}(\hat{\theta}^*)}$	$\xleftarrow{\text{estimated by}}$	$\underbrace{se_B(\hat{\theta}^*)}$
se of the statistic $\hat{\theta}$		std. dev. of *all* the replics.		std. dev. of B replics.

B.3 Bootstrapping in R

There are two main packages for bootstrapping in R: **bootstrap** and **boot**. As stated in the first of these, **bootstrap** is of particular help for the book by Efron and Tibshirani (1993), and it gives software (bootstrap, cross-validation, and jackknife) and data discussed there. The package **boot** offers a number of functions to deal with a variety of situations. For a detailed description of the main components of the **boot** package and their use, the article by Canty (2002) is recommended. For the purposes of illustration, we show here the use of the function **boot** as applied to the sample (B.1). The following is the Code B.1 **boot_std dev.R** for obtaining bootstrap estimates of the standard deviation of the distribution of the mean:

```
#Code B.1 Bootstrap estimates of the standard deviation
# of the distribution of the mean

#install.packages("boot", dep = TRUE)     # if necessary

library(boot)
x<- c(5.24, 5.55, 4.69, 4.39, 6.87, 5.15 ,4.61, 5.20, 5.49, 4.81,
      2.74, 3.50, 5.19, 5.40, 3.81, 6.49, 6.34, 4.45, 5.10, 3.17)   # observed sample
set.seed(3)

samplemean <- function(x, d) {           # estimation function
   return(mean(x[d]))         }

b = boot(x, samplemean, R=1000)          # 1000 replications
print(b)
plot(b)
```

This code calls the estimation function R times, and, each time, a new bootstrap sample is generated, on which the statistic is calculated. Note that the number of replications, denoted by R in the code, is the number denoted by B in our earlier discussion. The result consists of a table and a figure. The table reports the evaluated statistic of the original observed sample:

```
Bootstrap Statistics :
    original      bias      std. error
t1*   4.9095  -0.0099205     0.228821
```

Recall that we found $\widehat{se}(\bar{x}) = 0.24$ with the analytical formula (B.2). The code also produces a graphical plot of the bootstrap replications **t***, as shown in Figure B.1.

It is possible to select other estimation functions: for example, if one desires to obtain the median, the instruction is

```
samplemedian <- function(x, d) {
return(median(x[d]))         }
```

The replication plot can also be made by adding to Code B.1 the lines

```
quant<-quantile(b$t[,1], probs = c(0.025, 0.16, 0.5, 0.84, 0.975))
quant
i68.half<- (quant[4]-quant[2])/2
i68.half
windows()
```

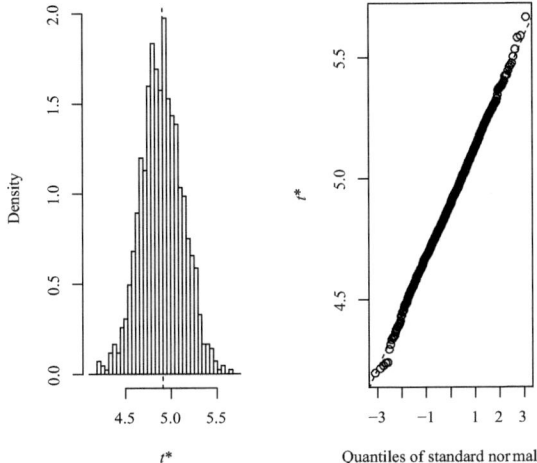

Fig. B.1 Plots of the bootstrap output: (a) bootstrap replications; (b) normal Q–Q plot of bootstrap estimate **t***.

```
hist(b$t[,1],lty=1,main="",xlim=range(thb),cex.lab=1.2,xlab="replications",ylab="frequency")
abline(v=quant[2],lty=2,col="black",lwd=2)
abline(v=quant[4],lty=2,col="black",lwd=2)
abline(v=quant[3],lty=3,col="black",lwd=4)
```

The quantiles are printed in the output:

```
    2.5%       16%       50%       84%     97.5%
4.451925  4.680420  4.892750  5.126320  5.364087
```

Figure B.2 shows the histogram of the replications, as in Figure B.1a, but with vertical lines corresponding to the 16%, 50% and 84% quantiles.

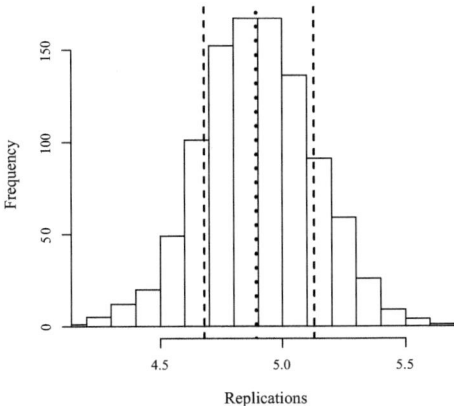

Fig. B.2 Bootstrap replications as in Figure B.1a, but with vertical lines corresponding to the 16%, 50% and 84% quantiles.

Half of the width of the interval $[16\%, 84\%]$, equal to 0.224, may be considered as the 68% bootstrap confidence interval, in this case almost equal to the standard error $\widehat{se}(\bar{x}) = 0.24$ (eqn B.2).

Confidence intervals can be computed with the function **boot.ci**. This function takes the output b of the instruction `boot(x, samplemean, R=1000)` and returns up to five different types of bootstrap confidence interval. See DiCiccio and Efron (1996) for bootstrap confidence intervals. For example, consider computation of the bootstrap normal interval and the bootstrap confidence interval. The bootstrap normal interval assumes an asymptotic normal distribution and uses the bootstrap estimation of the standard deviation. Choosing the confidence levels as 68%, 90% and 95%, the following line is added to Code B.1:

```
boot.ci(b,conf=c(0.68,0.90,0.95),type=c("norm","perc"))
```

The function **boot.ci** returns a matrix with three columns ,with the first being the level and the other two the boundaries of the intervals:

```
Intervals :
Level Normal Percentile
68% ( 4.692, 5.147 ) ( 4.680, 5.128 )
90% ( 4.543, 5.296 ) ( 4.530, 5.278 )
95% ( 4.471, 5.368 ) ( 4.449, 5.367 )
```

The small differences with the values reported earlier are due to different sequences of random numbers.

B.4 Comments

There is a version of the bootstrap method called *parametric bootstrap*, and, in contrast, the bootstrap we have been discussing is also known as *nonparametric bootstrap*. In the parametric version, the distribution function F is no longer estimated by the empirical distribution function $\widehat{F}$, but by the distribution $\widehat{F}_{\text{par}}$ derived from a parametric model for the data. So, instead of sampling with replacement from the data, we generate B samples of size n from $\widehat{F}_{\text{par}}$. After that, we then proceed exactly as in the nonparametric version. To use the parametric bootstrap with Code B.1, the statement `sim = "parametric"` has to be added to the `boot()` call. The default is `sim = "ordinary"` for the nonparametric bootstrap. The parametric bootstrap may be more useful to physicists than to statisticians when it is applied to problems of error propagation. Suppose that the probability distribution of each initial experimental datum is known; that is, each measurement leads to results with normally distributed errors, with zero mean and known standard deviation, which may be a meter sensitivity or another quantity determined once for all through appropriate experimental operations. We have measured n quantities, obtaining the results $x_i \pm \sigma_i, i = 1, \ldots, n$, where σ_i are the standard deviations, assumed to be known. These measures are the input of a complex computer code to obtain the value of the derived quantity ξ as output. The problem is that of the evaluation of the error σ_ξ of the derived quantity in terms of the combined effect of the errors of the direct measurements. The guiding idea is to mimic the process of performing a sequence of measurements each of which is a repetition of the only one actually made. We sample from a normal distribution of mean ξ and standard deviation σ_i, B values in order to construct B 'measured' values of each quantity, after which we continue, drawing the bootstrap samples as we have seen before.

There are situations, for instance time series analysis, in which the observations can no longer be considered the realization of mutually independent random variables with the same distribution function F. In these cases, the bootstrap (nonparametric or parametric) is not applicable in the form that we have outlined, because the dependence structure of the data is disregarded. So the bootstrap for independent and identically distributed ($i.i.d$) variables has to be replaced by techniques based on *block resampling*. This improvement amounts to 'robustifying' the original bootstrap against serial dependence. Several variants of block-resampling bootstrap have been proposed. We outline here the so-called *moving block bootstrap (MBB)*

We define blocks of l observations each and consider all possible contiguous blocks of length l. Since the blocks are taken as overlapping, we obtain m possible blocks, with $m = n - l + 1$. The original data can be wrapped around in a circle before blocking them. From these m blocks, we draw at random with replacement k blocks, with $k \times l = n$. In analogy with the $i.i.d.$ scheme, a suitable number B of MBB replications is formed, from each of which the statistic of interest is computed and the MBB estimate of the standard error is derived.

The underlying idea is that if the block length l is large enough, observations belonging to different blocks are nearly independent of one another, so the blocks are actually $i.i.d.$ random variables under the MBB scheme. At the same time, inside each block, the correlation is retained. For an application of the MBB to statistical mechanics, see Mignani and Rosa (2001).

In the package `boot`, the function **tsboot** allows one to deal with different resampling methods for time series. For reviews of such methods, see the article by Politis (2003) and the book by Lahiri (2003). It must be noted, however, that, even if this is not mentioned in the statistical literature, the idea of the bootstrap for dependent data appeared in a paper on high-energy physics by Gottlieb *et al.* (1986). These authors, referring explicitly to Efron (1979), explained clearly the advantages of bootstrapping blocks when handling dependent data.

The name 'bootstrap' comes from the saying 'to pull oneself up by one's bootstraps', as Baron von Münchhausen did in the book written in 1785 by Rudolf Erich Raspe. This name refers clearly to the fact that in the original observed sample, after suitable handling, all the properties of the whole population are readable. 'It is as if statisticians had discovered the statistical analogue of the hologram' (Diaconis and Efron, 1983), referring to the fact that each piece of the surface of a hologram retains the entire original scene, even though the pieces are broken off. But, as these authors warn:

There are always a few samples for which the bootstrap does not work, and one cannot know in advance which they are. The limitation is not so much a failure of the bootstrap procedure as it is a restatement of the conditions of uncertainty under which all statistical analyses must proceed.

Appendix C

C.1 Properties of Square Matrices

We consider the properties of a 2×2 square matrix A:

$$A = \begin{bmatrix} a_{11} & a_{12} \\ a_{21} & a_{22} \end{bmatrix} \tag{C.1}$$

The *trace* and *determinant* of A are given by

$$\text{Tr}(A) = a_{11} + a_{22} \quad \text{and} \quad |A| = a_{11}a_{22} - a_{12}a_{21} \tag{C.2}$$

respectively. The eigenvalues of A satisfy

$$|A - \lambda I| = \begin{vmatrix} a_{11} - \lambda & a_{12} \\ a_{21} & a_{22} - \lambda \end{vmatrix} = 0 \tag{C.3}$$

where I is the 2×2 identity matrix. Expanding the middle determinant in eqn (C.3) and using the expressions for $\text{Tr}(A)$ and $|A|$ from eqns (C.2) gives the *characteristic equation*:

$$\lambda^2 - \text{Tr}(A)\,\lambda + |A| = 0 \tag{C.4}$$

The eigenvalues are thus the roots of the characteristic equation:

$$\begin{bmatrix} \lambda_1 \\ \lambda_2 \end{bmatrix} = \begin{bmatrix} \dfrac{\text{Tr}(A) + \sqrt{[\text{Tr}(A)]^2 - 4|A|}}{2} \\ \dfrac{\text{Tr}(A) - \sqrt{[\text{Tr}(A)]^2 - 4|A|}}{2} \end{bmatrix} \tag{C.5}$$

When the discriminant (the quantity under the square root) is negative, that is, $[\text{Tr}(A)]^2 - 4|A| < 0$, the eigenvalues of A are a *complex conjugate pair* with real part α and imaginary part $\pm\beta$: $\lambda_1 = \alpha + \beta i$ and $\lambda_2 = \alpha - \beta i$.

The roots λ_1 and λ_2 can be paired in 11 possible configurations depending on whether they are real, complex, distinct or nonzero (Figure C.1).

Each eigenvalue has a corresponding 2×1 eigenvector. The eigenvector can be computed for real eigenvalues λ_i ($i = 1, 2$) in either of the following two ways:

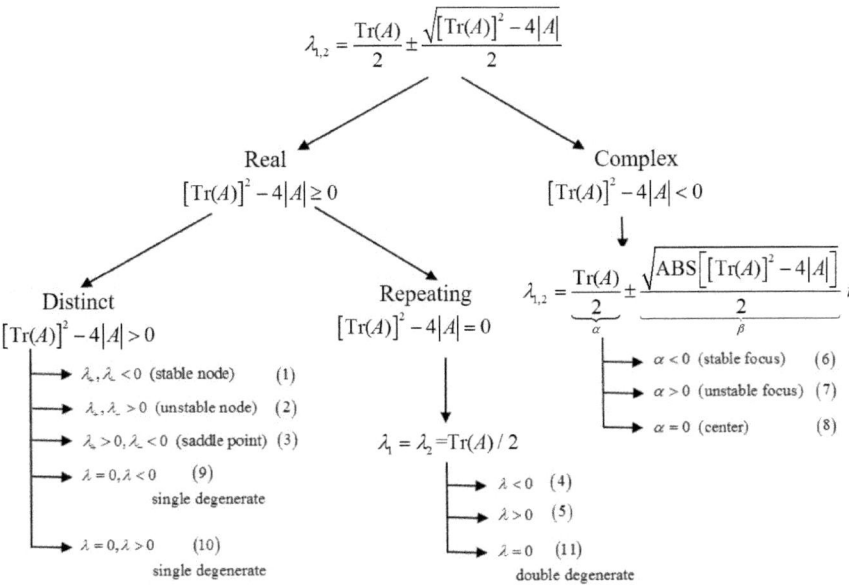

Fig. C.1 Stability configurations of a linear two-dimensional ODE system.

$$\begin{bmatrix} v_1^i \\ v_2^i \end{bmatrix} = \begin{bmatrix} v_1^i \\ -\dfrac{a_{11} - \lambda_i}{a_{12}} v_1^i \end{bmatrix} \quad \text{or} \quad \begin{bmatrix} v_1^i \\ -\dfrac{a_{21}}{a_{22} - \lambda_i} v_1^i \end{bmatrix} \tag{C.6}$$

where v_1^i is free to take any arbitrary value. The slope of an eigenvector in the v_1–v_2 vector space is the ratio of the eigenvector elements:

$$\frac{v_2^i}{v_1^i} = -\frac{a_{11} - \lambda_i}{a_{12}} \quad \text{or} \quad -\frac{a_{21}}{a_{22} - \lambda_i} \tag{C.7}$$

Consequently, the eigenvector can be plotted in v_1–v_2 space as

$$v_2^i = -\frac{a_{11} - \lambda_i}{a_{12}} v_1^i \tag{C.8}$$

To compute the eigenvectors associated with complex eigenvalues, we proceed similarly, for each of the complex conjugate eigenvalues $\lambda_1 = \alpha + \beta i$ and $\lambda_2 = \alpha - \beta i$:

$$\begin{bmatrix} v_1^1 \\ v_2^1 \end{bmatrix} = \begin{bmatrix} v_1^1 \\ -\dfrac{a_{11} - (\alpha + \beta i)}{a_{12}} v_1^1 \end{bmatrix} \tag{C.9}$$

$$\begin{bmatrix} v_1^2 \\ v_2^2 \end{bmatrix} = \begin{bmatrix} v_1^2 \\ -\dfrac{a_{11} - (\alpha - \beta i)}{a_{12}} v_1^2 \end{bmatrix} \tag{C.10}$$

Again v_1^1 and v_1^2 can take arbitrary values, and by taking them both equal to $-a_{12}$, we obtain

$$\begin{bmatrix} v_1^1 \\ v_2^1 \end{bmatrix} = \begin{bmatrix} -a_{12} \\ a_{11} - \alpha - \beta i \end{bmatrix} = \begin{bmatrix} -a_{12} \\ a_{11} - \alpha \end{bmatrix} + \begin{bmatrix} 0 \\ -\beta \end{bmatrix} i \tag{C.11}$$

$$\begin{bmatrix} v_1^2 \\ v_2^2 \end{bmatrix} = \begin{bmatrix} -a_{12} \\ a_{11} - \alpha + \beta i \end{bmatrix} = \begin{bmatrix} -a_{12} \\ a_{11} - \alpha \end{bmatrix} + \begin{bmatrix} 0 \\ \beta \end{bmatrix} i \tag{C.12}$$

Example: Real eigenvalues.   We compute the eigenvalues and eigenvectors associated with

$$A = \begin{bmatrix} 1 & 1 \\ -2 & 4 \end{bmatrix} \tag{C.13}$$

The trace of A is $\mathrm{Tr}(A) = 1 + 4 = 5$, the determinant is $|A| = (1)(4) - (-2)(1) = 6$ and the characteristic equation is

$$\lambda^2 - 5\lambda + 6 = 0 \tag{C.14}$$

with eigenvalues $\lambda_1 = 3$ and $\lambda_2 = 2$.

The eigenvectors are (taking the arbitrary v_1^1 and v_1^2 both equal to 1)

$$v^1 = \begin{bmatrix} v_1^1 \\ v_2^1 \end{bmatrix} = \begin{bmatrix} v_1^1 \\ -\frac{1-3}{1} v_1^1 \end{bmatrix} = \begin{bmatrix} v_1^1 \\ 2v_1^1 \end{bmatrix} = \begin{bmatrix} 1 \\ 2 \end{bmatrix} \tag{C.15}$$

$$v^2 = \begin{bmatrix} v_1^2 \\ v_2^2 \end{bmatrix} = \begin{bmatrix} v_1^2 \\ -\frac{1-2}{1} v_1^2 \end{bmatrix} = \begin{bmatrix} v_1^2 \\ v_1^2 \end{bmatrix} = \begin{bmatrix} 1 \\ 1 \end{bmatrix} \tag{C.16}$$

Example: Complex eigenvalues.   We compute the eigenvalues and eigenvectors associated with the matrix

$$A = \begin{bmatrix} 2 & -36 \\ 1 & 2 \end{bmatrix} \tag{C.17}$$

The trace of A is $\text{Tr}(A) = 2 + 2 = 4$, the determinant is $|A| = (2)(2) - (2)(-36) = 40$ and the characteristic equation is

$$\lambda^2 - 4\lambda + 40 = 0 \tag{C.18}$$

with eigenvalues $\lambda_{1,2} = 2 \pm 6i$ and complex eigenvectors

$$v^1 = \begin{bmatrix} -\underbrace{(-36)}_{a_{12}} \\ \underbrace{2}_{a_{11}} - \underbrace{2}_{\alpha} \end{bmatrix} + \begin{bmatrix} 0 \\ -\underbrace{6}_{\beta} \end{bmatrix} i = \begin{bmatrix} 36 \\ 0 \end{bmatrix} + \begin{bmatrix} 0 \\ -6 \end{bmatrix} i \tag{C.19}$$

$$v^2 = \begin{bmatrix} -(-36) \\ 2-2 \end{bmatrix} + \begin{bmatrix} 0 \\ 6 \end{bmatrix} i = \begin{bmatrix} 36 \\ 0 \end{bmatrix} + \begin{bmatrix} 0 \\ 6 \end{bmatrix} i \tag{C.20}$$

C.2   Analytical Construction of a Phase Diagram

We illustrate the analytical construction of a phase diagram with the following system:

$$\begin{bmatrix} \dot{x}_t \\ \dot{y}_t \end{bmatrix} = \underbrace{\begin{bmatrix} -3 & 2 \\ 2 & -3 \end{bmatrix}}_{A} \begin{bmatrix} x_t \\ y_t \end{bmatrix} \tag{C.21}$$

The nullclines setting each ODE to zero are

$$\dot{x}_t = -3x_t + 2y_t = 0 \quad \Longrightarrow \quad y\big|_{\dot{x}=0} = \tfrac{3}{2}x(t) \tag{C.22}$$

$$\dot{y}_t = 2x_t - 3y_t = 0 \quad \Longrightarrow \quad y\big|_{\dot{x}=0} = \tfrac{2}{3}x(t) \tag{C.23}$$

The nullclines (black lines) are plotted in phase space in Figure C.2a.

The nullclines partition phase space into four *nullsectors*. The black arrows in the background of Figure C.2 show the directions that trajectories follow in each nullsector (the *vector field*). We can use the system (C.21) to qualitatively determine these directions by using the nullclines as reference points. In one experiment, we determine motion away from the $\dot{x} = 0$ nullcline. We know that x_t is constant along the nullcline, since growth in x_t is zero. In the left-hand plot in Figure C.2b, we incrementally increase y_t off the nullcline (the upward-pointing dashed arrow). Growth in x_t marginally increases, since $\partial \dot{x}/\partial y = a_{12} = 2 > 0$. This means that x_t increases above the nullcline, as indicated by the rightward-pointing arrow. The opposite occurs below the nullcline. We would get the same result if we experimented with an incremental increase in x_t off the nullcline and repeated the above reasoning with $\partial \dot{x}/\partial x = a_{11} = -3 < 0$. In the other experiment, we determine motion away from the $\dot{y} = 0$ nullcline. In the right-hand diagram in Figure C.2b, we incrementally increase x_t off the nullcline (the rightward-pointing dashed arrow). Growth in y_t marginally increases, since $\partial \dot{y}/\partial x = a_{21} = 2 > 0$. This means that y_t increases below the nullcline

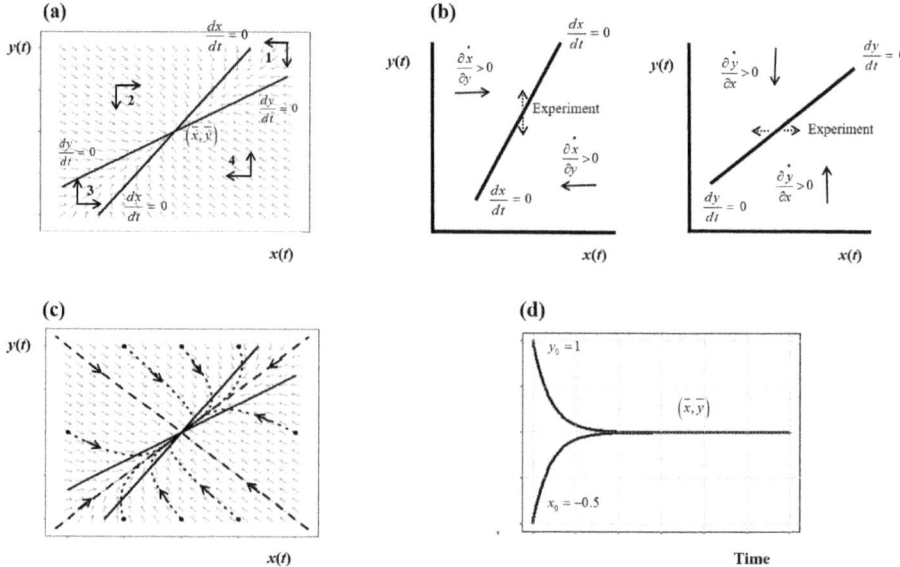

Fig. C.2 Phase diagram constructed for the case of stable node equilibrium: (a) nullclines and vector field; (b) experiments run to determine directions of motion; (c) phase diagram of a stable node equilibrium; (d) time series plot of variables along a trajectory.

(indicated by the upward-pointing arrow). The opposite happens above the nullcline. We would get the same result if we experimented with an incremental decrease in y_t off the nullcline and repeated the above reasoning with $\partial \dot{y}/\partial y = a_{22} = -3 < 0$.

We can see from the directions of motion that the equilibrium is *globally* stable, since it attracts all initial conditions. We can confirm this by running a stability check with the eigenvalues. The eigenvalues are $\lambda_1 = -1$, $\lambda_2 = -5$, confirming that each term in the system (C.21) decays to equilibrium. This is a *stable node* equilibrium. The corresponding eigenvectors are

$$\begin{bmatrix} v_1^1 \\ v_2^1 \end{bmatrix} = \begin{bmatrix} 2 \\ 2 \end{bmatrix} \qquad (\lambda_1 = -1) \tag{C.24}$$

$$\begin{bmatrix} v_1^2 \\ v_2^2 \end{bmatrix} = \begin{bmatrix} 2 \\ -2 \end{bmatrix} \qquad (\lambda_2 = -5) \tag{C.25}$$

These eigenvectors are plotted in Figure C.2c as lines going through the equilibrium (origin) with slope $= 1$ for $\lambda_1 = -1$ and slope $= -1$ for $\lambda_2 = -5$ (dashed lines). Eigenvectors are key trajectories in phase space that can be viewed as 'positioning' other trajectories. Other trajectories run parallel to an eigenvector towards equilibrium (if the corresponding eigenvalue is negative) or away from equilibrium (if the corresponding eigenvalue is positive). The figure also shows phase space trajectories (dotted curves) each beginning at initial values denoted by the black dots and each running parallel to an eigenvector. Note finally that the trajectories cross the $\dot{x} = 0$ nullcline perpendicular to the x axis (indicating that x_t is constant on the nullcline), and cross the $\dot{y} = 0$ nullcline perpendicular to the y axis (indicating that y_t is constant on the nullcline). Finally, Figure C.2d shows the time series plots corresponding to the trajectory with initial values $(x_0 = -0.5, y_0 = 1)$.

List of Symbols

Lower case Latin

d	time delay
$f^{[n]}(x)$	nth iterate of the map $f(x)$
m	embedding dimension
m	$\left(= \left.\dfrac{df(x_t)}{dx_t}\right|_{x_t=x^*}\right)$ fixed point multiplier
r	control parameter of the logistic map
r_∞^*	accumulation point, logistic map: ≈ 3.5699, birth of chaos
$s_1, s_2, \ldots$	measured time series
w	pendulum: angular frequency of the driving force
$\mathbf{y}_1, \mathbf{y}_2, \ldots$	reconstructed time series

Upper case Latin

$\mathcal{C}$	Cantor set
$C(\epsilon)$	correlation integral
$\widehat{C}(\epsilon)$	correlation integral estimate
$\mathbf{CR}_{i,j}$	cross recurrence plot matrix
$\mathbf{D}$	derivative matrix
D_0	Hausdorff–Besicovitch dimension
D_2	correlation dimension
$\widehat{D}_2$	correlation dimension estimate
F	pendulum: amplitude of the driving force
H	Shannon entropy
$I(X;Y)$	mutual information between random variables X and Y
$\mathbf{J}$	Jacobian matrix
$\mathbf{JR}_{i,j}$	joint recurrence plot matrix
$\mathbf{R}_{i,j}$	recurrence plot matrix
S_ρ	metric entropy measure

Lower case Greek

α	Feigenbaum alpha (≈ 2.5029)
δ	Feigenbaum delta (≈ 4.6692)
δ	pendulum: damping constant
δ_{ij}	distance between two points at times i and j
λ	maximum characteristic Lyapunov exponent
$\hat{\lambda}$	maximum characteristic Lyapunov exponent estimate
$\lambda_1, \lambda_2, \ldots$	Lyapunov spectrum
ν	pendulum: frequency
$\rho(x, x_0)$	density of points in $[x, x + dx]$, initial point x_0
θ	pendulum: angle with respect to the vertical
ω	pendulum: angular frequency

Upper case Greek

$\Theta(\cdot)$	Heaviside function

List of R Codes

References

Abarbanel, H. D. I. (1996). *Analysis of Observed Chaotic Data*. Springer, New York.

Alligood, K. T., Sauer, T., and Yorke, J. A. (1997). *Chaos: An Introduction to Dynamical Systems*. Springer, New York.

Andersen, H. and Hepburn, B. (2016). Scientific method. In *The Stanford Encyclopedia of Philosophy* (Summer 2016 edn) (ed. E. N. Zalta). Metaphysics Research Lab, Stanford University. https://plato.stanford.edu/archives/sum2016/entries/scientific-method/.

Artalejo, J. R., Economou, A., and Lopez-Herrero, M. J. (2015). The stochastic SEIR model before extinction: computational approaches. *Applied Mathematics and Computation*, **265**, 1026–1043.

Ash, R. and Doléans-Dade, C. (2000). *Probability and Measure Theory* (2nd edn). Academic Press, USA.

Baker, G. L. and Gollub, J. P. (1996). *Chaotic Dynamics: An Introduction* (2nd edn). Cambridge University Press.

Baker, G. L., Gollub, J. P., and Blackburn, J. (1996). Inverting chaos: extracting system parameters from experimental data. *Chaos*, **6**, 528–533.

Berra, T. (2009). *Charles Darwin: The Concise Story of an Extraordinary Man*. Johns Hopkins University Press, Baltimore.

Billingsley, P. (1995). *Probability and Measure* (3rd edn). Wiley, New York.

Bjørnstad, O., Finkenstädt, B., and Grenfell, B. (2002). Dynamics of measles epidemics: estimating scaling of transmission rates using a time series SIR model. *Ecological Monographs*, **72**, 169–184.

Blanchard, P., Devaney, R., and Hall, G. (2012). *Differential Equations* (4th edn). Brooks/Cole, Boston.

Borovkova, S., Rosa, R., and Sardonini, L. (2011). Resolving statistical uncertainty in correlation dimension estimation. *Chaos*, **21**, Article 023124.

Brading, K. and Castellani, E. (2013). Symmetry and symmetry breaking. In *The Stanford Encyclopedia of Philosophy* (Spring 2013 edn) (ed. E. N. Zalta). Metaphysics Research Lab, Stanford University. https://plato.stanford.edu/archives/spr2013/entries/symmetry-breaking/.

Brandmaier, A. (2015). Package `pdc`: Permutation distribution clustering.

Brandt, C. and Pompe, B. (2002). Permutation entropy: a natural complexity measure for time series. *Physical Review Letters*, **88**, 174102.

Breeden, J. and Hubler, A. (1990). Reconstructing equations of motion from experimental data with unobserved variables. *Physical Review A*, **42**, 5817–5826.

Brock, W., Dechert, W., and Scheinkman, J. (1986). A test for independence based on the correlation dimension. *Econometric Reviews*, **15**, 197–235.

Brown, R., Bryant, P., and Abarbanel, H. D. I. (1991). Computing the Lyapunov spectrum of a dynamical system from an observed time series. *Physical Review A*, **43**, 2787–2806.

Brown, T. (1996). Measuring chaos using the Lyapunov exponent. In *Chaos Theory in the Social Sciences: Foundations and Applications* (ed. E. Kiel and E. Elliott). University of Michigan Press, Ann Arbor.

Brunton, S., Proctor, J., and Kurtz, J. (2016). Discovering governing equations from data by sparse identification of nonlinear dynamical systems. *Proceedings of the National Academy of Sciences of the USA*, **113**, 3932–3937.

Brush, S. G. (1986). *The Kind of Motion We Call Heat: A History of the Kinetic Theory of Gases in the 19th Century*. North-Holland, Amsterdam.

Bühlmann, P. (1997). Sieve bootstrap for time series. *Bernoulli*, **3**, 123–148.

Canty, A. J. (2002). Resampling methods in R: the boot package. *R News*, **2/3**, 2–7.

Chan, K. and Tong, H. (2001). *Chaos: A Statistical Perspective*. Springer, New York.

Clark, A. (2015). Package `multispatialCCM`: Multispatial convergent cross mapping.

Coco, M. I. and Dale, R. (2014). Cross-recurrence quantification analysis of categorical and continuous time series: an R package. *Frontiers in Psychology*, **5**, Article 510.

Colby, S. and Baskerville, E. (2016). Limits to causal inference with state-space reconstruction for infectious disease. *PLoS ONE*, **11**(12), e0169050.

Constantine, W. and Percival, D. (2015). Package `fractal`: Fractal time series modeling and analysis.

Crawley, M. (2015). *Statistics: An Introduction Using R* (2nd edn). Wiley, Chichester.

Csardi, G. (2015). Package `igraph`: Network analysis and visualization.

Cvitanovic, P. (1989). *Universality in Chaos* (2nd edn). Taylor & Francis, New York.

da Vinci, Leonardo (1452–1519). *Notebooks*. Oxford University Press, Oxford.

Davison, A. C. and Hinkley, D. V. (1997). *Bootstrap Methods and their Application*. Cambridge University Press, Cambridge.

Denbigh, K. G. and Denbigh, J. S. (1985). *Entropy in Relation to Incomplete Knowledge*. Cambridge University Press, Cambridge.

Di Narzo, A. F. (2015). Package `tseriesChaos`: Analysis of nonlinear time series.

Diaconis, P. and Efron, B. (1983). Computer-intensive methods in statistics. *Scientific American*, **248**(May), 116–130.

DiCiccio, T. J. and Efron, B. (1996). Bootstrap confidence intervals. *Statistical Science*, **11**, 189–228.

Didonato, M. W., Martin, C. L., and England, D. (2014). Gendered interactions and their consequences in a dynamical perspective. In *Gender and Development* (ed. P. J. Leman and H. R. Tenenbaum), pp. 20–42. Psychology Press, New York.

Diks, C. (1996). Estimating invariants of noisy attractors. *Physical Review E*, **53**, R4263–R4266.

Diks, C. (1999). *Nonlinear Time Series Analysis: Methods and Application*. World Scientific, Singapore.

Ding, M., Grebogi, C., Ott, E., Sauer, T., and Yorke, J. A. (1993). Plateau onset for correlation dimension: When does it occur? *Physical Review Letters*, **70**, 3872–3873.

Dong, L. Wei, Ren, K., Cluzel, S. Meunier Guttin, and Gouesbet, G. (2015). Global vector-field reconstruction of nonlinear dynamical systems from a time series with

SVD method and validation with Lyapunov exponents. *Chinese Physics*, **12**, 1366–1373.

Eckmann, J.-P., Kamphorst, S. O., and Ruelle, D. (1987). Recurrence plots of dynamical systems. *Europhysics Letters*, **5**, 973–977.

Eckmann, J.-P., Kamphorst, S. O., Ruelle, D., and Ciliberto, S. (1986). Lyapunov exponents from a time series. *Physical Review A*, **34**, 4971–4979.

Eckmann, J.-P. and Ruelle, D. (1985). Ergodic theory of chaos and strange attractors. *Reviews of Modern Physics*, **57**, 617–656.

Economist, The (2016*a*). Big economic ideas: Breakthroughs and brickbats. *The Economist* (21 July).

Economist, The (2016*b*). Financial stability: Minsky's moment. *The Economist* (30 July).

Economist, The (2016*c*). If economists reformed themselves: A less dismal science. *The Economist* (16 May).

Efron, B. (1979). Bootstrap methods: another look at the jackknife. *Annals of Statistics*, **7**, 1–26.

Efron, B. (1988). Three examples of computer-intensive statistical inference. *Sankhyā: The Indian Journal of Statistics, Series A*, **50**, 338–362. Also presented at the 20th Interface Symposium on Computing Science and Statistics and at Compstat 88.

Efron, B. and Tibshirani, R. J. (1993). *An Introduction to the Bootstrap*. Chapman & Hall, New York.

Ekstrom, C. (2015). Package MESS: Miscellaneous esoteric statistical scripts.

Elsner, J. and Tsonsis, A. (2010). *Singular Spectrum Analysis*. Plenum Press, New York.

Falconer, K. (2014). *Fractal Geometry: Mathematical Foundations and Applications* (3rd edn). Wiley, Chichester.

Fama, E. (1970). Efficient capital markets: a review of theory and empirical work. *Journal of Finance*, **25**, 383–417.

Feder, G. (1979). Pesticides, information, and pest management under uncertainty. *American Journal of Agricultural Economics*, **61**, 97–103.

Feigenbaum, M. J. (1979). The universal metric properties of nonlinear transformations. *Journal of Statistical Physics*, **21**, 669–706.

Friedman, J., Hastie, T., Simon, N., and Tibshirani, R. (2015). Package glmnet: Lasso and elastic-net regularized generalized linear models.

Geman, D. and Geman, S. (2016). Science in the age of selfies. *Proceedings of the National Academy of Sciences of the USA*, **113**, 9384–9387.

Ghil, M., Allen, M., Dettinger, M., Ide, K., Kondrashov, D., Mann, M., Robertson, A., Saunders, A., Tian, Y., Varadi, F., and Yiou, P. (2001). Advanced spectral methods for climatic time series. *Reviews of Geophysics*, **40**, 1–41.

Giannerini, S. (2012). The quest for nonlinearity in time series. In *Handbook of Statistics: Time Series* (ed. T. Subba Rao, S. Subba Rao, and C. Rao), Volume 30, pp. 43–63. Elsevier, Oxford.

Giannerini, S. (2015). Package tseriesEntropy: Entropy based analysis and tests for time series.

Giannerini, S., Maasoumi, E., and Bee Dagum, E. (2007*a*). Entropy testing for nonlinearity in time series. *Bulletin of the International Statistical Institute* (56th session), 3620–3623.

Giannerini, S., Maasoumi, E., and Bee Dagum, E. (2015). Entropy testing for nonlinear serial dependence in time series. *Biometrika*, **102**, 661–675.

Giannerini, S. and Rosa, R. (2002). Caos, statistica e metodi di ricampionamento. *Statistica*, **62**, 359–378.

Giannerini, S., Rosa, R., and Gonzalez, D. L. (2007*b*). Testing chaotic dynamics in systems with two positive Lyapunov exponents: a bootstrap solution. *International Journal of Bifurcation and Chaos*, **17**, 169–182.

Gilleland, E. and Katz, R. W. (2013). in2extRemes: Into the R package extRemes – extreme value analysis for weather and climate applications. National Center for Atmospheric Research, Technical Note NCAR/TN-523+STR.

Gilmore, C. G. (1993). A new test for chaos. *Journal of Economic Behaviour and Organization*, **22**, 209–237.

Glendinning, P. (1994). *Stability, Instability and Chaos: An Introduction to the Theory of Nonlinear Differential Equations*. Cambridge University Press, Cambridge.

Gnedenko, B. V. and Kolmogorov, A. N. (1954). *Limit Distributions for Sums of Independent Random Variables*. Addison-Wesley, Cambridge, MA.

Golyandina, N., Nekrutkin, V., and Zhigljavsky, A. (2001). *Analysis of Time Series Structure*. Chapman & Hall/CRC, New York.

Gonzalez, D. L., Giannerini, S., and Rosa, R. (2006). Detecting structure in parity binary sequences: Error correction and detection in DNA. *IEEE Engineering in Medicine and Biology Magazine*, **25**(1), 69–81.

Gonzalez, D. L., Giannerini, S., and Rosa, R. (2008). Strong short-range correlations and dichotomic codon classes in coding DNA sequences. *Physical Review E*, **78**, 051918.

Gotoda, H. and Kobayashi, H. (2017). Chaotic dynamics of a swirling flame front instability generated by a change in gravitational orientation. *Physical Review E*, **95**, 022201.

Gottlieb, S., Mackenzie, P. B., Thacker, H. B., and Weingarten, D. (1986). Hadronic coupling constants in lattice gauge theory. *Nuclear Physics B*, **263**, 704–730.

Gouesbet, G. and Maquet, J. (1992). Construction of phenomenological models from numerical scalar time series. *Physica D*, **58**, 202–215.

Granger, C. (1969). Investigating causal relations by econometric models and cross spectral methods. *Econometrica*, **37**, 424–438.

Granger, C. W., Maasoumi, E., and Racine, J. (2004). A dependence metric for possibly nonlinear processes. *Journal of Time Series Analysis*, **25**, 649–669.

Grassberger, P. and Procaccia, I. (1983*a*). Characterization of strange attractors. *Physical Review Letters*, **50**, 346–349.

Grassberger, P. and Procaccia, I. (1983*b*). Measuring the strangeness of strange attractors. *Physica D*, **9**, 189–208.

Grayling, M. (2015). Package phaseR: Phase plane analysis of one- and two-dimensional autonomous ODE systems.

Grebogi, C., Hammel, S. M., Yorke, J. A., and Sauer, T. (1990). Shadowing of physical trajectories in chaotic dynamics: containment and refinement. *Physical*

Review Letters, **65**, 1527–1530.

Greco, G., Rosa, R., Beskin, G., Karpov, S., Romano, L., Guarnieri, A., Bartolini, C., and Bedogni, R. (2011). Evidence of deterministic components in the apparent randomness of GRBs: clues of a chaotic dynamic. *Scientific Reports (Nature)*, **1**, Article 91.

Grenfell, B., Bjørnstad, O., and Finkenstädt, O. (2002). Dynamics of measles epidemics: scaling noise, determinism, and predictability with the TSIR model. *Ecological Monographs*, **72**, 185–202.

Gross, D. J. (1996). The role of symmetry in fundamental physics. *Proceedings of the National Academy of Sciences of the USA*, **93**, 14256–14259.

Haack, S. (1999). Defending science – within reason. *Principia*, **3**(2), 187–211.

Hassani, H. (2007). Singular spectrum analysis: methodology and comparison. *Journal of Data Science*, **5**, 239–257.

Heathcote, A. and Elliott, D. (2011). Nonlinear dynamic analysis of noisy time series. In *Nonlinear Dynamical Systems Analysis for the Behavioral Sciences Using Real Data* (ed. S. Guastello and R. Gregson), pp. 103–134. CRC Press, Boca Raton, FL.

Hegger, R., Kantz, H., and Schreiber, T. (1999). Practical implementation of nonlinear time series methods: The TISEAN package. *Chaos*, **9**, 413–435.

Hentschel, H. G. E. and Procaccia, I. (1983). The infinite number of generalized dimensions of fractals and strange attractors. *Physica D*, **8**, 435–444.

Hilborn, R. and Mangel, M. (1997). *The Ecological Detective: Confronting Models with Data*. Princeton University Press, Princeton, NJ.

Hirsch, M., Smale, S., and Devaney, R. (2004). *Differential Equations, Dynamical Systems, and an Introduction to Chaos*. Academic Press, San Diego.

Hofstadter, D. R. (1981). Strange attractors: mathematical patterns delicately poised between order and chaos. *Scientific American*, **245**(November), 22–43.

Horan, B. (1994). The statistical character of evolutionary theory. *Philosophy of Science*, **61**, 76–95.

Hornberger, G. and Spear, R. (1981). An approach to the preliminary analysis of environmental systems. *Journal of Environmental Management*, **12**, 7–18.

Hotelling, H. (1933). Analysis of a complex of statistical variables into principal components. *Journal of Educational Psychology*, **24**, 417–441, 498–520.

Huffaker, R. (2015). Building economic models corresponding to the real world. *Applied Economic Perspectives and Policy*, **37**, 537–552.

Huffaker, R., Canavari, M., and Munoz-Carpena, R. (2016a). Distinguishing between endogenous and exogenous price volatility in food security assessment: An empirical nonlinear dynamics approach. *Agricultural Systems* (in press). doi:10.1016/j.agsy.2016.09.019.

Huffaker, R., Munoz-Carpena, R., Campo-Bescos, M., and Southworth, J. (2016b). Demonstrating correspondence between decision-support models and dynamics of real-world environmental systems. *Environmental Modeling and Software*, **83**, 74–87.

Hyndman, R. and Fan, Y. (1996). Sample quantiles in statistical packages. *American Statistician*, **50**, 361–365.

Ide, T. and Inoue, K. (2005). Knowledge discovery from heterogeneous dynamic systems using change point correlations. In *Proceedings of 2005 SIAM International Conference on Data Mining (SDM 05), Newport Beach, CA* (ed. H. Kargupta, J. Srivastava, C. Kamath, and A. Goodman), pp. 571–576. SIAM, Philadelphia.

Itoh, N. and Kurths, J. (2011). Change-point detection based on ssa in precipitation time series. In *Intelligent Automation and Systems Engineering* (ed. S. Ao, M. Amouzega, M. Rieger, and B. Burghard), Lecture Notes in Electrical Engineering, Volume 103, pp. 285–293. Springer, Berlin.

Itoh, N. and Marwan, N. (2013). An extended singular spectrum analysis transformation (SST) for the investigation of Kenyan precipitation data. *Nonlinear Processes in Geophysics*, **20**, 467–481.

Jolliffe, I. T. (2002). *Principal Component Analysis* (2nd edn). Springer, New York.

Kahle, D. (2015). Package `mpoly`: Symbolic computation and more with multivariate polynomials.

Kantz, H. (1994). A robust method to estimate the maximal Lyapunov exponent of a time series. *Physics Letters A*, **185**, 77–87.

Kantz, H. and Schreiber, T. (1997). *Nonlinear Time Series Analysis*. Cambridge University Press, Cambridge.

Kaplan, D. and Glass, L. (1995). *Understanding Nonlinear Dynamics*. Springer, New York.

Katz, R. (2010). Statistics of extremes in climate change. *Climate Change*, **100**, 71–76.

Katz, R., Brush, G., and Parlange, M. (2005). Statistics of extremes: modeling ecological disturbances. *Ecology*, **86**, 1124–1134.

Kennel, M., Brown, R., and Abarbanel, H. D. I. (1992). Determining embedding dimension for phase-space reconstruction using a geometrical construction. *Physical Review A*, **45**, 3403–3411.

Korobeynikov, A., Shlemov, A., Usevich, K., and Golyandina, N. (2015). Package `Rssa`: A collection of methods for singular spectrum analysis.

Kot, M., Schaffer, W., Truty, G., Graser, D., and Olsen, L. (1988). Changing criteria for imposing order. *Ecological Modeling*, **43**, 75–110.

Kowalski, A., Martin, M., Plastino, A., and Judge, G. (2012). On extracting probability distribution information from time series. *Entropy*, **14**, 1829–1841.

Kugiumtzis, D. (1999). Test your surrogate data before you test for nonlinearity. *Physical Review E*, **60**, 2808–2816.

Kugiumtzis, D. (2001). On the reliability of the surrogate data test for nonlinearity in the analysis of noisy time series. *International Journal of Bifurcation and Chaos*, **11**, 1881–1896.

Kugiumtzis, D. (2002). Surrogate data test on time series. In *Modelling and Forecasting Financial Data, Techniques of Nonlinear Dynamics* (ed. A. Soofi and L. Cao), pp. 267–282. Springer, New York.

Kuhn, T. S. (1962). *The Structure of Scientific Revolutions*. University of Chicago Press, Chicago.

Lahiri, S. N. (2003). *Resampling Methods for Dependent Data*. Springer, New York.

Lenton, T. and Livina, V. (2016). Detecting and anticipating climate tipping points. In *Extreme Events: Observations, Modeling, and Economics* (ed. M. Chavez,

M. Ghil, and J. Urrutia-Fucugauchi), AGU Geophysical Monograph Series, Volume 214, pp. 51–62. Wiley, Hoboken, NJ.

Li, T.-Y. and Yorke, J. A. (1975). Period three implies chaos. *American Mathematical Monthly*, **82**, 985–992.

Liebovitch, L. S. (1998). *Fractals and Chaos Simplified for the Life Sciences*. Oxford University Press, New York.

Livina, V., Kwasniok, F., Lohmann, G., Kantelhardt, J., and Lenton, T. (2011). Changing climate states and stability: from Pliocene to present. *Climate Dynamics*, **37**, 2437–2453.

Lorenz, E. N. (1963). Deterministic nonperiodic flow. *Journal of the Atmospheric Sciences*, **20**, 130–141.

Lorenz, E. N. (1995). *The Essence of Chaos* (2nd edn). University of Washington Press, Seattle.

Maasoumi, E. and Racine, J. (2002). Entropy and predictability of stock market returns. *Journal of Econometrics*, **107**, 291–312.

Mandelbrot, B. (1982). *The Fractal Geometry of Nature*. Freeman, San Francisco.

Mañé, R. (1981). On the dimension of the compact invariant set of certain non-linear maps. In *Dynamical Systems and Turbulence: Proceedings of a Symposium Held at the University of Warwick 1979/80* (ed. D. A. Rand and L.-S. Young), pp. 230–242. Springer, Berlin.

Marwan, N. (2008). A historical review of recurrence plots. *European Physical Journal, Special Topics*, **164**, 3–12.

Marwan, N. (2011). How to avoid potential pitfalls in recurrence plot based data analysis. *International Journal of Bifurcation and Chaos*, **21**, 1003–1017.

Marwan, N. and Kurths, J. (2002). Nonlinear analysis of bivariate data with cross recurrence plots. *Physics Letters A*, **302**, 299–307.

Marwan, N. and Kurths, J. (2004). Cross recurrence plots and their applications. In *Mathematical Physics Research at the Cutting Edge* (ed. C. V. Benton), pp. 101–139. Nova Science Publishers, Hauppauge, NY.

Marwan, N., Romano, M. C., Thiel, M., and Kurths, J. (2007). Recurrence plots for the analysis of complex systems. *Physics Reports*, **438**, 237–329.

Marwan, N. and Webber, C. L. Jr. (2015). Mathematical and computational foundations of recurrence quantifications. In *Understanding Complex Systems: Recurrence Quantification Analysis – Theory and Best Practices* (ed. C. L. Webber Jr. and N. Marwan), pp. 3–43. Springer, Cham, Switzerland.

Marwan, N., Wessel, N., Meyerfeldt, U., Schirdewan, A, and Kurths, J. (2002). Recurrence-plot-based measures of complexity and their application to heart-rate-variability data. *Physical Review E*, **66**, 026702.

McCaffrey, D. F., Ellner, S., Gallant, A. R., and Nychka, D. W. (1992). Estimating the Lyapunov exponent of a chaotic system with nonparametric regression. *Journal of the American Statistical Association*, **87**, 682–695.

McGuinness, M. (1983). The fractal dimension of the Lorenz attractor. *Physics Letters A*, **99**, 5–9.

McSharry, P. (2011). The danger of wishing for chaos. In *Nonlinear Dynamical Systems Analysis for the Behavioral Sciences Using Real Data* (ed. S. Guastello and

R. Gregson), pp. 539–558. CRC Press, Boca Raton, FL.

Medina, M. (2016). R code for animation of a 3D attractor. PhD Thesis, University of Florida.

Metropolis, N., Rosenbluth, A. W., Rosenbluth, M. N., Teller, A. H., and Teller, E. (1953). Equations of state calculations by fast computing machine. *Journal of Chemical Physics*, **21**, 1087–1091.

Mignani, S. and Rosa, R. (2001). Markov chain Monte Carlo in statistical mechanics: the problem of accuracy. *Technometrics*, **43**, 347–355.

Minsky, H. (1992). The financial instability hypothesis. Levy Economics Institute of Bard College, Working Paper No. 74.

Moskvina, V. and Zhigljavsky, A. (2003). An algorithm based on singular spectrum analysis for change-point detection. *Communications in Statistics*, **32**, 319–353.

Muir, J. (1911). *My First Summer in the Sierra*. Houghton Mifflin, Boston.

Nash, J. and Sutcliffe, J. (1970). River flow forecasting through conceptual models,: Part I – A discussion of principles. *Journal of Hydrology*, **70**, 90255–90256.

Nychka, D. W., Ellner, S., Gallant, A. R., and McCaffrey, D. F. (1992). Finding chaos in noisy systems. *Journal of the Royal Statistical Society, Series B*, **54**, 399–426.

Odum, E. (1953). *Fundamentals of Ecology*. W. B. Saunders, Philadelphia.

Oreskes, N., Shrader-Frechette, K., and Belitz, K. (1994). Verification, validation, and confirmation of numerical models in the earth sciences. *Science*, **263**, 641–646.

Osborne, A. R. and Provenzale, A. (1989). Finite correlation dimension for stochastic systems with power-law spectra. *Physica D*, **35**, 357–381.

Oseledec, V. L. (1968). A multiplicative ergodic theorem: Lyapunov characteristic numbers for dynamical systems. *Transactions of the Moscow Mathematical Society*, **19**, 197–231.

Ott, E. (2002). *Chaos in Dynamical Systems* (2nd edn). Cambridge University Press, Cambridge.

Packard, N. H., Crutchfield, J. P., Farmer, J. D., and Shaw, R. S. (1980). Geometry from time series. *Physical Review Letters*, **45**, 712–716.

Pearson, K. (1901). On lines and planes of closest fit to systems of points in space. *Philosophical Magazine*, **2**, 559–572.

Poincaré, J. H. (1890). Sur le problème des trois corps et les équations de la dynamique. *Acta Mathematica*, **13**, 1–271.

Poincaré, J. H. (1952). Analyse des travaux scientifiques faite par Henry Poincaré. In *Œuvres de Henry Poincaré,* tome 7. Gauthier-Villars, Paris.

Politis, D. N. (2003). The impact of bootstrap methods on time series analysis. *Statistical Science*, **11**, 219–230.

Provenzale, A., Smith, L. A., Vio, R., and Murante, G. (1992). Distinguishing between low-dimensional dynamics and randomness in measured time series. *Physica D*, **58**, 31–49.

PSMSL (2016). Permanent Service for Mean Sea Level. http://www.psmsl.org/data/obtaining/stations/316.php (accessed August 2016).

Ritter, A. and Munoz-Carpena, R. (2013). Performance evaluation of hydrologic models: statistical significance for reducing subjectivity in goodness-of-fit assessments. *Journal of Hydrology*, **480**, 33–45.

Rohani, P., Green, C., Manitlla-Beniers, N., and Grenfell, B. (2003). Ecological interference between fatal diseases. *Nature*, **422**, 885–888.

Romano, M., Thiel, M., Kurths, J., and von Bloh, W. (2004). Multivariate recurrence plots. *Physics Letters A*, **330**, 214–223.

Rosenstein, M. T., Collins, J. J., and DeLuca, C. J. (1993). A practical method for calculating largest Lyapunov exponents from small data sets. *Physica D*, **65**, 117–134.

Rössler, O. E. (1976). An equation for continuous chaos. *Physics Letters A*, **57**, 397–398.

Saha, P. and Strogatz, S. H. (1995). The birth of period three. *Mathematics Magazine*, **68**, 42–47.

Saltelli, A. and Funtowitz, S. (2014). When all models are wrong. *Issues in Science and Technology* (Winter 2014), 79–85.

Sano, M. and Sawada, Y. (1985). Measurement of the Lyapunov spectrum from a chaotic time series. *Physical Review Letters*, **55**, 1082–1085.

Sauer, T., Yorke, J. A., and Casdagli, M. (1991). Embedology. *Journal of Statistical Physics*, **65**, 579–616.

Schaffer, W. and Kot, M. (1985). Nearly one dimensional dynamics in an epidemic. *Journal of Theoretical Biology*, **112**, 403–427.

Schreiber, T. (1997). Detecting and analyzing nonstationarity in a time series with nonlinear cross predictions. *Physical Review Letters*, **78**, 843–846.

Schreiber, T. (1999). Interdisciplinary application of nonlinear time series methods. *Physics Reports*, **308**, 1–64.

Schreiber, T. and Schmitz, A. (2000). Surrogate time series. *Physica D*, **142**, 346–382.

Shannon, C. E. (1948). A mathematical theory of communication. *Bell System Technical Journal*, **27**, 379–423.

Small, M. and Tse, C. (2003). Detecting determinism in time series: The method of surrogate data. *IEEE Transactions on Circuits and Systems*, **50**, 663–672.

Soetaert, K., Petzoldt, T., and Setzer, R. (2016). Package `deSolve`: Solvers for initial value problems of differential equations.

Sprott, J. (2016). Lyapunov exponent and dimension of the lorenz attractor. http://sprott.physics.wisc.edu/chaos/lorenzle.htm (accessed December 2016).

Sugihara, G., May, R., Hao, Y., Chih-hao, H., Deyle, E., and Fogarty, M. (2012). Detecting causality in complex ecosystems. *Science*, **338**, 496–500.

Takens, F. (1981). Detecting strange attractors in turbulence. In *Dynamical Systems and Turbulence: Proceedings of a Symposium Held at the University of Warwick 1979/80* (ed. D. A. Rand and L.-S. Young), pp. 366–381. Springer, Berlin.

Takens, F. (1985). On the numerical determination of the dimension of an attractor. In *Dynamical Systems and Bifurcations* (ed. B. L. J. Braaksma, H. W. Broer, and F. Takens), Lecture Notes in Mathematics, Volume 1125, pp. 99–106. Springer, Berlin.

Theiler, J. (1986). Spurious dimension from correlation algorithms applied to limited time series data. *Physical Review A*, **34**, 2427–2432.

Theiler, J. (1990). Estimating the fractal dimension of chaotic time series. *Lincoln Laboratory Journal*, **3**, 63–86.

Theiler, J., Eubank, S., Longtin, A., Galdrikian, B., and Farmer, J. (1992). Testing for nonlinearity in time series: The method of surrogate data. *Physica D*, **58**, 77–94.

Tibshirani, R. (1996). Regression shrinkage and selection via the lasso. *Journal of the Royal Statistical Society, Series B*, **58**, 267–288.

Tribus, M. and McIrvine, E. C. (1971). Energy and information. *Scientific American*, **225**(September), 179–188.

Trulla, L. L., Giuliani, A., Zbilut, J. P., and Webber, C. L. Jr. (1996). Recurrence quantification analysis of the logistic equation with transients. *Physics Letters A*, **223**, 255–260.

US Congress Subcommittee on Science and Technology (2010). *Building a Science of Economics for the Real World: Hearing before the Subcommittee on Investigations and Oversight, Committee on Science and Technology, House of Representatives, One Hundred Eleventh Congress, second session, July 20, 2010*. US Government Printing Office, Washington, DC.

Uusitalo, L., Lehikoinen, A., Helle, I., and Myrberg, K. (2015). An overview of methods to evaluate uncertainty of deterministic models in decision support. *Environmental Modeling and Software*, **63**, 24–31.

Vautard, R. (1999). Patterns in time: Ssa and mssa in analysis of climate. In *Analysis of Climate Variability* (ed. H. von Storch and A. Navarra). Springer, New York.

von Plato, J. (1994). *Creating Modern Probability*. Cambridge University Press.

Warnes, G. R., Bolker, B., Bonebakker, L., Gentleman, R., Liaw, W. H. A., Lumley, T., Maechler, M., Magnusson, Arni., Moeller, S., Schwartz, M., and Venables, B. (2016). Package `gplots`: Various R programming tools for plotting data.

Webber, C. L. Jr. and Zbilut, J. P. (1994). Dynamical assessment of physiological systems and states using recurrence plot strategies. *Journal of Applied Physiology*, **76**, 965–973.

Williams, G. (1997). *Chaos Theory Tamed*. John Henry Press, Washington, DC.

Wolf, A., Swift, J. B., Swinney, H. L., and Vastano, J. A. (1985). Determining Lyapunov exponents from a time series. *Physica D*, **16**, 285–317.

Xie, Y. (2017). Package `animation`: A gallery of animations in statistics and utilities to create animations.

Ye, H., Deyle, E., Gilarranz, L., and Sugihara, G. (2015). Distinguishing time-delayed causal interactions using convergent cross mapping. *Scientific Reports (Nature)*, **5**, Article 14750.

Zbilut, J. P., Giuliani, A., and Webber, C. L. Jr. (1998). Detecting deterministic signals in exceptionally noisy environments using cross-recurrence quantification. *Physics Letters A*, **246**, 122–128.

Zbilut, J. P. and Webber, C. L. Jr. (1992). Embeddings and delays as derived from quantification of recurrence plots. *Physics Letters A*, **171**, 199–203.

Zou, H. (2006). The adaptive lasso and its oracle properties. *Journal of the American Statistical Association*, **101**, 1418–1429.

Index